THINKING PHILOSOPHICALLY

DEDICATED

To my wife,
Dr. Diane Wallick Creel, an administrator who loves faculty,

And to my friends,
Dr. Stephen Bickham, Mansfield University,
    and
Dr. James Keller, Wofford College,
for their criticism and encouragement of my work over the years, and for their many contributions to philosophy.

Thinking Philosophically

An Introduction to Critical Reflection and Rational Dialogue

RICHARD E. CREEL

Copyright © Richard E. Creel 2001

The right of Richard E. Creel to be identified as author of this work has been asserted in accordance with the Copyright, Designs and Patents Act 1988.

First published 2001

2 4 6 8 10 9 7 5 3 1

Blackwell Publishers Inc.
350 Main Street
Malden, Massachusetts 02148
USA

Blackwell Publishers Ltd
108 Cowley Road
Oxford OX4 1JF
UK

All rights reserved. Except for the quotation of short passages for the purposes of criticism and review, no part of this publication may be reproduced, stored in a retrieval system, or transmitted, in any form or by any means, electronic, mechanical, photocopying, recording or otherwise, without the prior permission of the publisher.

Except in the United States of America, this book is sold subject to the condition that it shall not, by way of trade or otherwise, be lent, resold, hired out, or otherwise circulated without the publisher's prior consent in any form of binding or cover other than that in which it is published and without a similar condition including this condition being imposed on the subsequent purchaser.

Library of Congress Cataloging-in-Publication Data

Creel, Richard E., 1940–
  Thinking philosophically : an introduction to critical reflection and rational dialogue / Richard E. Creel
     p.   cm.
  Includes bibliographical references and index.
  ISBN 0-631-21934-X (hardcover : alk. paper) — ISBN 0-631-21935-8 (pbk. : alk. paper)
  1. Philosophy—Introductions. I. Title.

BD21 .C74 2001
100—dc21

00-060795

British Library Cataloguing in Publication Data
A CIP catalogue record for this book is available from the British Library.

Typeset in 10 on 12.5 pt Ehrhardt
by Ace Filmsetting Ltd, Frome, Somerset
Printed in Great Britain by Biddles, Guildford, Surrey

This book is printed on acid-free paper.

A Preface to Teachers x
Acknowledgments xiii
Philosophers in *Thinking Philosophically* xiv

PART I METAPHILOSOPHY 1

1 Introduction 3
Three ways into philosophy 3
The nature of philosophy 4
The three most basic problems in philosophy 6
Developing a philosophy of your own 10

2 What is Philosophy? 11
Before philosophy 11
The historical beginnings of western philosophy 12
The literal meaning of "philosophy" 15
The basic problems and areas of philosophy 15
The interconnectedness of the issues of philosophy 18
A definition of philosophy 20
Clue words to areas in philosophy 21
Sample statements and questions in different areas 22

3 Why We Do Philosophy 24
The noetic motive 24
The cathartic motive 25
The mystical motive 26
The wisdom motive 27
The sport motive 28

4	**The Two Most Basic Causes of Philosophy**	30
	Ambiguity	30
	Curiosity	33
	Vagueness, ambivalence, and ambiguity	33
5	**Reason, Philosophy, and Other Disciplines**	35
	An expanded definition of philosophy	35
	Philosophy and Religion	35
	Philosophy and Science	37
	Philosophy and Mathematics	39
	Philosophy and History	40
6	**Methods for Doing Philosophy**	43
	The Socratic Method	43
	Running out the permutations	45
	Rational dialogue	47
7	**Things Philosophers Do**	53
	Exposit	54
	Analyze	56
	Synthesize	56
	Describe	57
	Speculate	60
	Prescribe	62
	Criticize	64
8	**A Healthy Philosophical Attitude**	69
	Caring rather than indifferent	69
	Courageous rather than timid	70
	Open rather than closed	71
	Grateful rather than resentful	72
	Assertive rather than passive	73
9	**Alternatives to Philosophy**	75
	Neglect	76
	Skepticism	77
	Dogmatism	81
	Solitude	84
	Philosophy is important and inescapable	86
	We are responsible for our beliefs	86
	Philosophy not adversarial	88

PART II EPISTEMOLOGY   91

10 What is Truth?   93
  Non-epistemic uses of "true"   94
  The kind of thing that is true or false   94
  The nature of truth   95
  What makes an assertion true or false   95
  Competing conceptions of truth   96
  Why truth is important   98
  Three laws of thought   99
  Six sources of truth   100

11 What is Knowledge?   104
  Hope   104
  Faith   105
  Belief based on evidence   106
  True belief based on evidence   107
  Justified belief   107
  Justified true belief   108
  The justification theory of knowledge   108
  The causal theory of knowledge   109

12 Logic – Understanding and Evaluating Arguments   112
  What "argument" means in logic   113
  The deductive argument   115
  The inductive argument   119

PART III THEORY OF VALUE   127

13 Axiology and Happiness   129
  Our innate craving for happiness   130
  Aristotle's definition of happiness   132
  Critiques of happiness   133
  Axiology: Its nature and purposes   137
  The Good of enjoyment vs. Enjoyment of the good   138
  Three conceptions of the good   139
  The interaction theory of experience   141
  The package theory of alternatives   141
  Ignorance as cause / Intelligence as cure   142
  Jeremy Bentham's Hedonic Calculus   145
  John Stuart Mill's criticism of Bentham   149
  Intrinsic values and Instrumental values   150
  A summary of concepts in value theory   153

14 Ethics and Morality	**158**
Good and Bad; Right and Wrong; Self-interest and Morality	159
Different meanings of "right" and "wrong"	161
Ethical Nihilism	162
Ethical Relativism	164
Individual Relativism	164
Social Relativism	166
Ethical Absolutism	171
Theocentric Theories of Ethics	**175**
Divine Command Ethics	175
Perfect Being Ethics	178
Anthropocentric Theories of Ethics	**181**
Rationalistic (Deontological) Ethics	181
Altruistic (Utilitarian) Ethics	183
Universal Eudaemonism and Moral Happiness	186
Why be moral?	190
The need for moral education	193

PART IV METAPHYSICS 203

15 Freedom and Determinism	**205**
Can we do what we ought to do?	206
Words, Concepts, Positions, Justifications, and Criticisms	207
Philosophical Anthropology	209
Objective freedom and Subjective freedom	209
Libertarianism	212
Universal Determinism	216
Theistic Determinism	217
Naturalistic Determinism	219
Soft Determinism (Compatibilism)	223
Hard Determinism (Incompatibilism)	226
Criticisms	229
16 The Mind/Body Problem	**237**
Dualistic Interactionism	239
Occasionalism	247
Parallelism	248
Epiphenomenalism	250
Physical Monism	255
Psychic Monism	265

	Neutral Monism	271
	Phenomenalism	275
17	Philosophical Theism	278
	Terminology: polytheism, deism, theism, pantheism, atheism, agnosticism	278
	Philosophical Worldviews	279
	Hard and soft beliefs	282
	Religious Theism and Philosophical Theism	283
	How can we tell what God is like?	284
	Four arguments for belief in the existence of God	287
18	Metaphysical Materialism	296
	Criticisms of arguments for belief in the existence of God	296
	Four arguments against belief in the existence of God	298
	Materialism as a worldview	302
19	Metaphysical Idealism	310
	Popular idealism and metaphysical idealism	311
	A general justification of metaphysical idealism	312
	Subjective Idealism (theistic): Berkeley and Hartshorne	314
	Objective Idealism (pantheistic): Hegel	315
	Phenomenalism (atheistic): Hume, Ayer, Buddha	319
	Solipsism and the problem of other minds	321
	Metaphysical Nihilism	322
	Why consider "crazy" positions?	323
	Criticisms of solipsism	325
	Desert landscapes and Tropical forests	327
	Affirming, improving, or replacing a worldview	328

PART V PARTING REMARKS   331
Socrates' advice
The difficulties and inconclusiveness of philosophy
The personal importance and intimacy of philosophy
Benefits of philosophy
The spiral of philosophical growth

Index   335

A Preface to Teachers

Dear Colleague:

I am gratified that you are thinking of using or have decided to use *Thinking Philosophically* as a text in your course. In its pages I present many of the basic concepts and positions in philosophy, I engage the reader in thinking dialectically about philosophical issues, and I try to prepare and motivate the reader to engage productively in philosophical discussions.

Thinking Philosophically consists primarily of the lectures I used to give in my Introduction to Philosophy course – though now they are considerably expanded and polished. By putting into written form a great deal of obligatory, foundational material that I used to deliver by lecture, I have freed in-class time to engage students in discussions of that material and to introduce them to primary sources by way of short handouts that we read, interpret, and discuss in class. I frequently present students with opposed primary source handouts on the topic of the day – for example, Aristotle versus Schopenhauer on happiness, Gorgias versus Hegel on human knowledge, Clifford versus James on the ethics of belief, Bertrand Russell versus Carl Jung on religious experience, Socrates versus Thomas Hobbes on conscience. Sometimes a single handout includes opposed ideas, such as Plato's treatment of The Ring of Gyges or the short debate between Socrates and Thrasymachus on justice. On other occasions a single handout from one point of view can be provocative and illuminating, such as Plato's Allegory of the Cave, with which I always begin my Intro course, and to which I then refer at relevant points as the course goes along. The short dialogues of Plato and some of Descartes' *Meditations* also work well as in-class supplements to *Thinking Philosophically*.

In addition to a brief contents at the beginning of *Thinking Philosophically* you will find a detailed contents at the beginning of each chapter. If students read and reflect on the chapter contents before reading the chapter, they should experience less unnecessary confusion, develop a better sense of how different concepts, positions, and topics relate to one another, and achieve a higher level of comprehension when they read the chapter itself. Also, if students use the chapter contents as study guides when preparing for tests – by, for example, turning topics into questions for themselves – they should develop a better sense of what is important and do better on tests than they would otherwise. Finally, a study of the chapter contents should help students get a better sense of where they have been, where they are going, and how the various topics of philosophy connect to one another.

As a technical addendum I should explain two kinds of inconsistencies you will encounter: one regards capitalizations in the contents; the other regards headings that are in the contents but not in the text. Regarding the latter, I have made the chapter contents very detailed, as was just mentioned, to help students be better prepared for what they will be reading, better prepared for tests, and more aware of how the various concepts, parts, and positions of philosophy connect to one another. However, some topics listed in the chapter contents follow so closely on one another in the text or can be located with sufficient ease in the text by looking for key words, which are often italicized, that it seemed excessive and distracting to insert those headings into the text, and so I did not. Key words are italicized so often in the text for two reasons: first, to help the reader understand statements more readily by emphasizing where to focus and which words to group together (as is done in lectures by vocal emphases); second, to help the reader relocate key ideas more easily for review and reflection.

Regarding capitalization in the contents, I usually capitalize the first mention of a position, for example, "Universal Eudaemonism," but lowercase further mentions, for example, "The principle of universal eudaemonism" (see chapter 14, contents, p. 159). When, however, I mention two or more things and am concerned that a failure to capitalize both or all might make it appear that I am favoring one position over another or am suggesting that one thing is less important than another, then I capitalize for the sake of fairness. For example, four of the subheadings in chapter 5 are: "Philosophy and Religion," "Philosophy and Science," "Philosophy and Mathematics," and "Philosophy and History." Strictly speaking, only the first word in each of those headings should be capitalized, but were I to do that, writing "Philosophy and religion," "Philosophy and science," etc., then "Philosophy" would be capitalized in every case and the other disciplines in none – which might give the reader a mistaken, unfortunate impression that I am saying that philosophy is more important than religion, science, mathematics, and the study of history. Hence, I capitalize both disciplines in each case. I also use capitalization to emphasize that each of a series of items is distinct from the others and equally important, for example, "Words, Concepts, Positions, Justifications, and Criticism" (see chapter 15 contents, p. 205).

Whatever approach you take to your own course, I hope you find that *Thinking Philosophically* is sufficiently clear, competent, and comprehensive that it frees you and your students to do good things in class that otherwise you would not have time to do. Insofar as *Thinking Philosophically* needs correction, clarification, trimming, expansion, or other changes, I hope you will let me know so the next edition can be improved.

# Acknowledgments

I feel deeply grateful to the philosophers whom Blackwell Publishers secured to comment on the penultimate version of *Thinking Philosophically*. I was touched by the care which they took in commenting, and I was humbled by their knowledge and insight. They saved me from numerous infelicities and some plain old bone-headed mistakes.

The author and publishers would like to thank Faber & Faber for permission to reproduce an extract from T. S. Eliot, 'Four Quartets' from *Collected Poems 1909–1962*.

Philosophers in
Thinking Philosophically

ALPHABETICAL

Anscombe, Elizabeth (1919–; England)
Anselm, St. (1033–1109; England)
Aquinas, St. Thomas (1225–1249, Italy)
Aristotle (384–322 BC; Greece)
Augustine, St. (354–430; Rome; N. Africa)
Ayer, A. J. (1910–1989; England)
Bentham, Jeremy (1748–1832, England)
Berkeley, George (1685–1753; Ireland)
Blanshard, Brand (1892–1987; USA)
Bradley, F. H. (1846–1924; England)
Clifford, W. K. (1845–1879; England)
Democritus (460–370 BC; Thrace)
Descartes, René (1596–1650; France)
Dewey, John (1859–1952; USA)
Empedocles (ca. 495–435 BC; Sicily)
Epictetus (ca. 50–130; Rome and Greece)
Epicurus (341–270 BC; Athens)
Foot, Philippa (1920–; England)
Frege, Gottlob (1848–1925; Germany)
Hartshorne, Charles (1897–; USA)
Hegel, G. W. F. (1770–1831; Germany)
Heidegger, Martin (1889–1976; Germany)
Hobbes, Thomas (1588–1679; England)
Hume, David (1711–1776; Scotland)
Husserl, Edmund (1859–1938; Germany)
James, William (1842–1910; USA)
Kant, Immanuel (1724–1804; Germany)
Kierkegaard, Søren (1813–1855; Denmark)
Korsgaard, Christine (1952–; USA)
Kripke, Saul (1941–; USA)

Leibniz, G. W. (1646–1716; Germany)
Locke, John (1632–1704; England)
Malebranche, Nicholas (1638–1715; France)
Marx, Karl (1818–1883; Germany)
Mill, John Stuart (1806–1873; England)
Nietzsche, Friedrich (1844–1900; Germany)
Ockham, William of (1285–1349; England)
Paley, William (1743–1805; England)
Pascal, Blaise (1623–1662; France)
Peirce, Charles (1839–1914; USA)
Plato (427–347 BC; Greece)
Popper, Karl (1902–1994; England)
Putnam, Hilary (1926–; USA)
Rawls, John (1921–; USA)
Rousseau, Jean-Jacques (1712–1778; France)
Royce, Josiah (1855–1916; USA)
Russell, Bertrand (1872–1970; England)
Santayana, George (1863–1952; USA)
Sartre, Jean-Paul (1905–1980; France)
Schopenhauer, Arthur (1788–1860; Germany)
Skinner, B. F. (1904–1990; USA)
Socrates (470–399 BC; Greece)
Spinoza, Benedict (1632–1677; Holland)
Swinburne, Richard (1934–; England)
Thales (flourished 585 BC; Asia Minor)
Watts, Alan (1915–1973; USA)
Weil, Simone (1909–1943; France)
Whitehead, A. N. (1861–1947; England; USA)
Wittgenstein, L. (1889–1951; Austria; England)

CHRONOLOGICAL

585 BC	Thales (Asia Minor)
495–435,	Empedocles (Sicily)
470–399,	Socrates (Greece)
460–370,	Democritus (Thrace)
427–347,	Plato (Greece)
384–322,	Aristotle (Greece)
341–270,	Epicurus (Greece)
AD 50–130,	Epictetus (Rome and Greece)
354–430,	Augustine, St. (Rome; N. Africa)
1033–1109,	Anselm, St. (England)
1225–1249,	Aquinas, St. Thomas (Italy)
1285–1349,	Ockham, William of (England)
1588–1679,	Hobbes, Thomas (England)
1596–1650,	Descartes, René (France)
1623–1662,	Pascal, Blaise (France)
1632–1677,	Spinoza, Benedict (Holland)
1632–1704,	Locke, John (England)
1638–1715,	Malebranche, Nicholas (France)
1646–1716,	Leibniz, G. W. (Germany)
1685–1753,	Berkeley, George (Ireland)
1711–1776,	Hume, David (Scotland)
1712–1778,	Rousseau, Jean-Jacques (France)
1724–1804,	Kant, Immanuel (Germany)
1743–1805,	Paley, William (England)
1748–1832,	Bentham, Jeremy (England)
1770–1831,	Hegel, G. W. F. (Germany)
1788–1860,	Schopenhauer, Arthur (Germany)
1806–1873,	Mill, John Stuart (England)
1813–1855,	Kierkegaard, Søren (Denmark)
1818–1883,	Marx, Karl (Germany)
1839–1914,	Peirce, Charles (USA)
1842–1910,	James, William (USA)
1844–1900,	Nietzsche, Friedrich (Germany)
1845–1879,	Clifford, W. K. (England)
1846–1924,	Bradley, F. H. (England)
1848–1925,	Frege, Gottlob (Germany)
1855–1916,	Royce, Josiah (USA)
1859–1938,	Husserl, Edmund (Germany)
1859–1952,	Dewey, John (USA)
1861–1947,	Whitehead, A. N. (England; USA)
1863–1952,	Santayana, George (USA)
1872–1970,	Russell, Bertrand (England)
1889–1951,	Wittgenstein, L. (Austria; England)

1889–1976,	Heidegger, Martin (Germany)
1892–1987,	Blanshard, Brand (USA)
1897–	Hartshorne, Charles (USA)
1902–1994,	Popper, Karl (England)
1904–1990,	Skinner, B. F. (USA)
1905–1980,	Sartre, Jean-Paul (France)
1909–1943,	Weil, Simone (France)
1910–1989,	Ayer, A. J. (England)
1915–1973,	Watts, Alan (USA)
1919–	Anscombe, Elizabeth (England)
1920–	Foot, Philippa (England)
1921–	Rawls, John (USA)
1926–	Putnam, Hilary (USA)
1934–	Swinburne, Richard (England)
1941–	Kripke, Saul (USA)
1952–	Korsgaard, Christine (USA)

But do remember that there are other important philosophers not cited here.

Part I Metaphilosophy

Chapter 1 Introduction

- Three ways into philosophy
- The nature of philosophy
- The three most basic problems in philosophy
- Developing a philosophy of your own

Philosophy is a fascinating subject which is personally relevant to every intelligent human being. I want to tell you why that is so, I want to tell you a great deal about philosophy, and I want to engage you in thinking philosophically. When I speak of philosophy I mean western philosophy as it flourished in ancient Greece, then spread to Europe, Great Britain, and North America. Eastern, or Asian, philosophy is also important – especially Hinduism, Taoism, Confucianism, and Buddhism, but if we were to study eastern philosophy as well as western, that would make this book far too long. However, I do encourage you to study the asian traditions in philosophy later. Because there are certain universal features of philosophy, you will find that *Thinking Philosophically* has prepared you for the study of asian philosophy, as well as for further studies in western philosophy.

Three ways into philosophy

There are three common ways of introducing people to philosophy. One way is to focus on the ancient Greek thinkers who founded western philosophy, especially Socrates, Plato, and Aristotle. Because the rest of the history of philosophy builds on the work of the ancient Greeks, that approach provides students with a sound foundation for further studies in philosophy. However, some students who do not expect to take another philosophy course or do further reading in philosophy find that approach unsatisfying because there are so many other thinkers about whom they learn nothing.

Quite naturally, then, a second approach to introducing students to philosophy is to give them a survey of the history of philosophy. Then they can learn something about most of the giants of western philosophy, starting with the ancient

Greeks, but moving quickly to later eras and other thinkers, such as Augustine, Anselm, Aquinas, Descartes, Leibniz, Spinoza, Pascal, Locke, Berkeley, Hume, Rousseau, Kant, Hegel, Marx, Nietzsche, Kierkegaard, Wittgenstein, Whitehead, and Sartre. That is a wonderful way to learn philosophy because philosophy is, in a very real sense, a 2,600-year-old conversation between individuals and generations. To know how the conversation has proceeded from its beginning, about 600 BC, to the present, is to be well-prepared to enter that conversation oneself. Still, that approach is not satisfying to some people because it does not do what they want to do most: think about and discuss philosophical problems. Consequently, the third and most popular way to teach philosophy is the "problems" approach.

The "problems" approach to philosophy is the one that we will be taking. It consists of identifying, explaining, and attempting to solve philosophical problems – problems that have to do with God, truth, morality, freedom, the mind, and more. As you will see, when discussing those problems I will mention the names and explain the ideas of many great philosophers, but when I do so, it will be to help you understand or attempt to solve some *problem* in philosophy. Consequently, instead of a section on René Descartes, for example, you will see Descartes' name and ideas taken up briefly in different parts of *Thinking Philosophically* in relation to different problems.

In brief, my primary aims are to familiarize you with some of the most basic problems in philosophy, to present alternative solutions to those problems, and to involve you in evaluating those solutions – ultimately leaving you free to make your own decisions about them. The remainder of this introduction is an overview of what is to come. It should give you a broad sense of what we will do and why we will be doing it. As you proceed through *Thinking Philosophically* I encourage you to keep referring back to the tables of contents because they provide you with a quick way to review where we've been, see where we're going, and understand how the many parts of philosophy relate to one another.

The nature of philosophy

The first question we will take up is that of the nature of philosophy. This inquiry into the nature of philosophy is called "meta-philosophy." Some writers prefer to explore specific philosophical problems before asking their readers to consider what philosophy is. There is wisdom in that approach because people can usually better understand a definition of something they have had some first-hand experience with. If a person has never seen or heard of a fish, it would no doubt be helpful to show her a few fish before lecturing her on the question, "What is a fish?" It is not the case, however, that you have no familiarity with philosophy. I am confident that you have thought about some philosophical problems, listened to philosophical discussions about them, and perhaps have entered into those discussions. Maybe no one pointed out that you were listening to philosophy or doing it, but it was philosophy all the same – just as, if English is your native language, you were speaking English long before you learned it was English you were speaking.

To be sure, philosophy, like anything else, can be done well or poorly, in an unsophisticated way or in a highly sophisticated way. The difference between our untutored efforts in philosophy and professional philosophy is at least as great as the difference between amateur football and professional football. Yet just as there is continuity between amateur football and professional football, so there is continuity between our untutored efforts in philosophy and professional philosophy. Consequently, though you may not be familiar with philosophy in its most sophisticated forms, I'm confident you have enough familiarity with it that I need not begin as though you are totally ignorant of the subject.

Still, it would be surprising if you do not have a number of misconceptions about philosophy. For example, few beginners realize how broad philosophy is. Consequently, one of my objectives in the next few chapters is to give you an overview of the many fields within philosophy. Also, many first timers do not realize how rigorous and disciplined philosophy is. Consequently, in addition to explaining what philosophers do, I want to explain how they do it, why they do it, and what their attitude tends to be when they do it well.

Eventually you will want a definition of philosophy, and I will provide one. However, it will have to be *my* definition. There is no official definition of philosophy; I doubt that there can be. Philosophy is an open-ended, pioneering discipline, forever opening up new areas of study and new methods of inquiry. Consequently, philosophy is continually reconceiving itself. As a result, it is impossible to draw four sides around philosophy and say "That's it." Its history is too multifarious and its boldness too daring. However, there is a fairly stable cluster of concerns and questions that have, for the most part, constituted the substance of philosophy over the centuries since ancient Greece. My own conception of philosophy has grown and continues to grow out of my study of the history of philosophy. In my definition I will try to capture what seems to me to have been most distinctive and characteristic of philosophy over the centuries and in the present. Ultimately, however, we must each draw from our rich philosophical heritage and construct our own answers to the questions of philosophy – including the question of the nature of philosophy. That does not mean all answers are equally good. They are not. But it does mean that philosophy itself challenges us to formulate its nature in our own minds and to learn from and improve upon the answers of other people.

The three most basic problems in philosophy

After examining the nature of philosophy we will begin our investigation of what are usually considered to be "the big three" problems of philosophy: the problem of knowledge, the problem of value, and the problem of reality. First we will take up the problem of knowledge. It is fascinating how young children are when they begin asking adults, "How do you know that?" Sometimes they ask that question so relentlessly that it becomes exasperating to the adult. It is unfortunate when we lose the impulse to ask that question. Throughout life it remains one of the most important questions we can ask – of ourselves ("How do *I* know that?"), as well as of others.

The area of philosophy that investigates the nature, sources, authority, and limits of human knowledge is called "epistemology." Clearly, epistemology has a bearing on all that we will be doing, for we will look at various *possibilities* as to the nature of right and wrong, the nature of the good life, the nature of humankind, and the nature of reality. Whenever I present one of these possibilities as true, it is entirely appropriate for you to ask of me, "How do you know that?" When I do not affirm a position as true but simply confront you with alternative possibilities, it is appropriate for you to say, "Okay, that's good, but how can we come to know which of those possibilities is definitely, or at least probably, true?" There are no simple answers to these questions, but there are valuable answers – answers which employ distinctions and insights that have been developed and refined over centuries of time by brilliant thinkers. Among these distinctions and insights, we will look at differences between assertions and arguments, truth and validity, knowledge and belief, and faith and hope. Then we will look into the pursuit of knowledge through ordinary perception, science, and religion. I cannot take you directly to your goal (absolute knowledge), but I can put you aboard the only ship I know of which is headed for that destination, introduce you to its crew, familiarize you with its rigging, and hand you an oar. No one I know of can do more than that.

The next leg of our voyage relates to the fact that the word "philosophy" literally means "love of wisdom." This second leg of our voyage involves two forms of wisdom developed by philosophers: *axiology* and *ethics*. Together these disciplines constitute *value theory*, that is, the general theory of values, both moral and non-moral. Axiology, broadly conceived, is the study of the nature and achievement of happiness. Ethics, broadly conceived, is the study of the nature of moral behavior and character. In addition to examining each of these aspects of value theory, we will ask how they relate to one another. Some people think we must choose between happiness and morality. Some say that if we want to be happy, we must forget about morality; it will only get in the way of our pursuit of happiness. Others say that if we want to be moral, we must forget about happiness; it will only distract us from doing our duty. Perhaps we can find a more satisfying way of relating these two important concerns to one another.

The third and final leg of our journey will take us from the relatively pleasant waters of value theory into the stormy seas of metaphysics. Because value theory is closely related to the experiences and training you have had since childhood, and to the kinds of decisions you have been making for years, you'll probably feel at home in value theory. Metaphysics, by contrast, asks not how we *feel* or what we *want* or what we *ought* to do; it asks how things *are*. What is the nature of reality as a whole and in its parts? To answer that question requires a detachment of mind and a rigor of thought which few people acquire in their first twenty years of life. Consequently, the main difficulty you will probably have in doing metaphysics will be the simple act of trying to really appreciate a radically different way of understanding reality or some part of it.

Because of our pluralistic culture, most of us have been exposed to diverse ways of understanding the world; for that reason, we tend to think of ourselves as liberal and open, but the exposure has usually been superficial. Few of us have ever really had to or tried to understand the world in a way which is radically different from the prevailing way in our culture or subculture. Indeed, cultures, and especially subcultures, often try to protect us and sometimes try to prevent us from seriously examining other ways of understanding reality. They may shield us altogether from these different perceptions, or they may assure us in advance that they're not as good as what we've got (and may even be the work of wicked people or the devil!). In metaphysics, however, it is necessary to do just the opposite: to seek out and clarify all possible basic answers to a metaphysical problem in order that we might compare and evaluate them rigorously and fairly.

Immanuel Kant, one of the great modern philosophers, provides a model for us here. He once said that whenever he set out to find an answer to a philosophical or scientific question (he was a distinguished scientist as well as philosopher), he never settled for the first plausible answer that came to him. Rather, he tried to think of every possible answer to the problem, turning each answer sideways, upside down, and inside out. Why? In order not to become enamored of a wrong answer prematurely or of the right answer for the wrong reason. Consequently, if ABC seemed at first to be the answer to a problem, Kant would go through a process such as the following, unpacking all the possibilities: "Maybe we don't need B and C – maybe A alone is the answer; or maybe we don't need A and C – maybe B is the answer; or maybe C by itself is adequate; or maybe AB is the answer and C is unnecessary; or maybe BC is the answer and A is unnecessary; or maybe AC is necessary but not B. Further, if the order of the elements is important, perhaps the answer is not ABC, but CBA, or BCA, or CAB, or BAC, or BA, or CA, or CB." Clearly, if the answer has to be A or B or C, or some combination thereof, we have just run through all the possibilities and the answer must be among them. Similarly, as we examine each philosophical problem, we shall attempt to "run out the permutations"; that is, we shall try to think of all the basic possible solutions to each problem in order that we might be confident that we have not overlooked anything important when we finally fix on an answer. For example, we will want to consider not only whether there is or is not a God, but also whether God might be different than God is traditionally conceived to be. Perhaps God is finite rather than infinite, or is identical with the universe rather than different from it. But more of that later.

The first metaphysical problem we will take up is that of freedom versus determinism. At that point we will have already talked about morality, which specifies what we ought to do. But isn't talk about what we ought to do based on the assumption that we are free to refuse to do what we ought to do? Immanuel Kant thought so and put his conviction bluntly: "Ought implies can." That is, anything which we ought to do must be something which we can do; it is fitting to hold people responsible for what they ought to do, but it would not be right to hold them responsible for what they could not do. The central issue between libertarianism (which affirms human freedom) and determinism (which denies human freedom) is whether we ever really have a choice as to whether to do something or not do it. Is it the case or not that human actions are as completely determined as the behavior of rocks in a landslide or of geese migrating in fall? We will examine (1) libertarianism, which holds that our actions are not determined and we are responsible for them, (2) hard determinism, which says that our actions are determined and that, therefore, we are not responsible for them, and (3) soft determinism, which says that our actions are determined but we are responsible for them anyway!

Parenthetically, if we had taken up the problem of freedom versus determinism before taking up the problem of value, and had concluded that we are in fact free, then a natural next step would have been to ask, "Okay, now that we believe we have a free choice in what we do, what do we *want* to do with our freedom and – perhaps more important – what *ought* we to do?" Hence, the question of freedom/determinism resolved in the direction of freedom would lead naturally to questions about value and morality. By contrast, if we had decided that human behavior is determined, it would have been natural to ask, "Does that mean that we are not morally responsible?" And, "Do our values *cause* us to behave as we do?" The important point here is to note how these philosophical problems lead to one another. As you'll see more and more, you can begin with just about any philosophical problem and find that it leads eventually to all the others. Hence, which problem you begin with in your study of philosophy is not important. What is important is that you begin and keep going. So let's do.

The second metaphysical problem we will take up is that of the nature of a human. Why is that problem important? Because questions about what we ought to do and whether we are free cannot be answered adequately apart from a consideration of what kind of thing a human is. If a human is a machine of some kind (even a biochemical machine), then it seems unlikely that humans are free, and questions of morality and punishment must be approached in that light. If humans are not like machines, then perhaps they are free in a way that makes them morally responsible. Hence, we shall ask first, "What is a human?" or in more personal terms, "What am I?" In response to that question we shall try to run out the permutations: Am I a body and nothing more? a soul and nothing more? a combination of the two? something else altogether? Your answer to the question, "What am I?", will have important implications for how you should live this life and for whether you should expect life after death.

As I indicated above, nearly every philosophical question and answer has a bearing on all others. When you begin to see those connections, you will have really begun to understand philosophy. Speaking of philosophical connections, we cannot answer the question of the nature of humans without giving attention to the nature of that of which humans are a part: reality as a whole. For example, if reality consists of nothing but matter and space, then a person cannot be a soul or even have a soul. Different implications regarding the nature of humans follow from other positions regarding the nature of reality.

We will look at three very different views of reality. The first, *theism*, has been the most common conception of reality in the western world. It is the belief that reality consists of God and all that God creates. The second position, *materialism*, denies the reality of God, claims that only the physical world exists, and is fast catching up with theism in popularity and influence. Third, and finally, we shall look at *idealism*, which, in one of its forms, says that what should be denied is not God but matter![1]

Consider the possibilities again: Theism says, "Mind and matter are both real." Materialism says, "No, only matter is real." Idealism affirms the remaining possibility: "Matter is not real; only mind is." Because metaphysical idealism is so uncommon in the west, it is difficult for many westerners to appreciate it; however, it is a common position in Asia and in certain western circles, such as Christian Science. Consequently, it is important for us to develop a feeling for its persuasiveness.

Developing a philosophy of your own

Having begun our voyage with the problems of epistemology and worked our way through value theory and metaphysics, I will conclude by encouraging you to begin again – only this time to begin at the end with the problems of metaphysics, then to work your way back through the questions of value theory and epistemology, keeping in mind each step of the way the insights you gained during earlier steps, realizing that a philosophy of life is something to have available for guidance at every moment of life, and yet to continually evaluate, revise, and cultivate.

To be sure, philosophy is an academic subject of study – something that one can keep at arms length to a certain extent; but more deeply philosophy is a way of being in the world – of questioning it, interacting with it, and responding to it. Indeed, humankind is an ongoing dialogue about the topics of philosophy – topics such as good and evil, right and wrong, truth and falsity, appearance and reality. The purpose of this book is to help prepare you to take up your rightful place in that dialogue – to help you become a more appreciative, willing, patient, knowledgeable, insightful, articulate participant in that dialogue.

Note

1 *Metaphysical* idealism, which we will examine toward the end of *Thinking Philosophically*, is a way of understanding the nature of reality. It should not be confused with idealism in the common sense of being strongly committed to lofty ideals. An idealist in that sense may or may not be an idealist in the metaphysical sense. Similarly, a metaphysical materialist may or may not be a materialist in the sense of placing a high value on material possessions.

Reading Further

Plato, "The Allegory of the Cave," in *The Republic*, Book VII, sections 514a–517a.
For an introduction to Asian philosophies read *The Bhagavad-Gita* (a classic Hindu text), *The Teachings of the Compassionate Buddha*, edited by E. A. Burtt (especially "The Dhammapada," pages 51–73), and the *Tao Te Ching*, the masterpiece of Chinese Taoism (see, for example, *The Way of Lao Tzu*, translated and edited with excellent notes by Wing-Tsit Chan, or *The Wisdom of Laotse*, trans. Lin Yutang, which includes the profound and humorous commentary of Chuangtse.

Chapter 2: What is Philosophy?

- Before philosophy
- The historical beginnings of western philosophy
  Thales, Empedocles, Democritus
- The literal meaning of "philosophy"
- The basic problems and areas of philosophy
  Metaphysics, Value Theory, Epistemology
- The interconnectedness of the issues of philosophy
- A definition of philosophy
- Clue words to areas in philosophy
- Sample statements and questions in different areas

"Philosophy" is a word. A word can be looked up in a dictionary. But philosophy itself is not a word; it is something real, so you can no more learn adequately what philosophy is by looking up "philosophy" in a dictionary than you can learn adequately what love or beauty is by looking up "love" or "beauty" in a dictionary. That is why eventually I want not only to tell you about philosophy but also to involve you in it so that you know the thing itself.

Before philosophy

One of the first and most interesting things to note about philosophy is that it had a definite historical beginning. In that way it differs from art and religion, which go back beyond our ability to trace. Philosophy, by contrast, began about 600 BC in the Mediterranean world. What was that new thing which began then and was different from what had gone on before? That is a question which historians are still discussing, but the following point is well-founded and is the one that is important for our purposes. Before about 600 BC, people understood what the world was like and explained how it came about and why many things happened as they did by relating stories about spirits or gods and goddesses and their actions. These stories were passed down uncritically from generation to generation. The Babylonians, for example, understood heaven and earth in terms of a goddess who had been sliced from head to toe into front and back halves. The earth is her bottom half, and the sky is her top half! (A bit gross, but intriguing.)

Another example of the mythical mode of explanation can be found in asian Indian culture. The Hindu *Upanishads* report that the different animal species on earth were created by the ploys of a goddess. She came into existence when a god in his loneliness decided that he wanted a wife to possess. Consequently, he divided himself into a man and a woman. The woman decided she wasn't all that hot to be possessed by a man, so when the man tried to embrace her, she turned herself into a horse. Not to be outdone, he did likewise, and colts were conceived. She, still not happy with her plight, escaped again by turning herself into a deer. He did likewise and – you guessed it – fawns were soon born. You've probably also guessed that the goddess had not made her last attempt at escape by metamorphosis. Indeed, this amorous chase went on and on until all the species of animals on earth had been brought to pass.

Let's take a final example of mythical thinking from another asian country, Japan. Shinto, once the official religion of Japan, held that the islands of Japan came to pass when a god sitting on a rainbow dipped his giant spear into the Pacific Ocean and then lifted the spear from the waters. The drops which fell from the tip of the spear solidified and, presto, the islands of Japan were formed.

These stories from ancient Babylon, India, and Japan, are fascinating in themselves, rich in symbolic significance, and sometimes rich in wisdom. Once upon a time they were taken literally and seriously by millions of people who grew up with them. They are typical of the ways that non-philosophical peoples, western and eastern, today as well as yesterday, think about the world. It should be added that moral codes, as well as world-views, were conveyed and legitimized by stories about the actions and commands of gods and goddesses. See, for example, the god Krishna's statement on the morality of war in *The Bhagavad-Gita* of Hinduism, or the story of the delivery of "The Ten Commandments" in chapter twenty of *Exodus* in the Bible.

The historical beginnings of western philosophy

About 600 BC, however, in coastal cities around the Mediterranean Sea, a few people began trying to understand and explain the world in terms of elements and principles rather than divinities and their actions. Thales, widely recognized as the first philosopher, decided that all things, without exception, are created out of a single element: water. The remarkable thing is how close he came to the truth! Much of the earth is covered by water, and hydrogen (one of the two elements in water) appears to be the most plentiful element in the universe! All Thales needed in addition to water were principles of transformation by which he could explain how water is transformed into things which do not appear to be water, and by which things which do not appear to be water are transformed back into water. Cold, for example, causes water to change into ice; heat causes the ice to change back into water. More heat causes the water to turn into clouds, which then turn back into water. Also, anyone who has seen silt build up at the mouth of a river or has seen sediment settle out of water can imagine how easily one could get the impression that dirt is generated from water. Even more easily one can get the impression that plants are made of water – since when we water them, the water disappears and the plants get larger, and if we cut the plants and squeeze them, liquid oozes out – just as though the plants were made of water. (The human body is pretty squishy, too!)

Thales' explanation was ingenious, but not everyone was satisfied with it. To some of his contemporaries it seemed absurd to say that fire could be generated from water or that water could be generated from fire. Fire and water are "natural enemies." Fire destroys water by boiling it away, and water destroys fire by putting it out. Neither can be generated from the other, so each must be a basic element – neither constituted of the other – and the same holds true for air and earth (dirt). As a result of such reasoning, another early philosopher, Empedocles, concluded that fire, air, earth, and water are the basic elements out of which all other things are composed. These four elements themselves are composed of nothing. Rather, each is a simple, non-composite element.

These basic elements, Empedocles thought, are combined in various ways and proportions by the principle of love, and are separated by the principle of hate. In this way most of the objects with which we are familiar are generated. It is easy to imagine, for example, how the growth of a plant might be explained in terms of fire (the sun), water (rain), and earth (the soil), being joined together by the principle of love, whereas the erosion of soil can be explained in terms of the principle of hate causing air (wind) and water (rain) to carry away the soil. Note that these principles are not portrayed as gods or spirits. They are impersonal forces of attraction (love) and repulsion (hate) which operate inexorably and indifferently, so here we see the beginning of the notion of natural laws.

Empedocles' move beyond Thales was ingenious, but hardly what we would consider a resting place. Democritus rejected the *pluralism* of Empedocles (that is, the belief that there is more than one basic element) and returned to the *monism* of Thales (that is, the belief that all things are constituted of one and only one element). However, Democritus did not agree with Thales that water is the one basic element or substance out of which all things are made. Reason led him to believe that there must be something more basic than fire, air, earth, and water –

something out of which they, too, are made. He concluded that all things are constituted of tiny pellets of matter. He reasoned that if we were able to keep dividing fire, air, earth, and water into smaller and smaller units, we would eventually discover that we were no longer dealing with fire, air, earth, and water, but with tiny little building blocks which are indivisible, indestructible, and invisible to the naked eye.

Democritus called these tiniest parts of the universe "atoms" ("atom" is a Greek word which literally means "not cut" or "not cuttable"). Atoms, according to Democritus, differ among themselves only in shape, size, and the speeds at which they move. They are not, for example, colored or flavored. However, various combinations of atoms cause us to see different colors and taste different flavors when appropriate combinations of atoms stimulate our senses (and, of course, our senses are composed of atoms!). Happily enough, these atoms have little hooks by means of which they are joined and separated as they fall through space eternally. Think of them as tiny velcro balls whizzing through space, sometimes clumping together and sometimes flying apart or being knocked apart. What an amazing fifth-century BC explanation of what reality consists of and how things come to be and pass away. The atomic theory of the universe had its beginning in ancient philosophy! However, most thinkers were convinced by Empedocles' "fire, air, earth, and water" theory, rather than Democritus' atomic theory. Indeed, Empedocles' theory dominated western thought for nearly two thousand years, until the value of the atomic understanding of the universe was rediscovered by modern scientists.

Lest you think that all the first philosophers were materialists, explaining everything in terms of physical elements or atoms, I hasten to add they were not. Socrates, Plato, and Aristotle, the great triumvirate of Greek philosophy, believed that in order to explain adequately what goes on in the universe, it is necessary to postulate the existence of a divine being, as well as of matter. But note: they weren't just saying, "Our forefathers believed in gods, so we do, too." Rather, they *thought* about whether to believe in a God. They presented *arguments* for the existence of God, listened to criticisms of their arguments, considered the merits of those criticisms, and responded as reason seemed to require. That procedure for answering questions about life and the universe was radically different from what had gone on before, and it's still going on today. That "new" way of attempting to answer questions is what philosophy is all about, and it is one of the most distinctive characteristics of western civilization.

What, more exactly, is this thing called "philosophy," which began about 600 BC, is a hallmark of Western civilization, and is still going on? I've given examples of it and hinted at a definition, but I'm reluctant to say, "Well, it's this: A, B, C," because you might be misled into thinking that that's all it is. Philosophy, like love or art, is not a definition or merely an idea, and it cannot be defined in a conclusive way anymore than we can define conclusively what love or art is. Philosophy is a process that exists in humans and between them. Soon I will offer a working definition of philosophy for your consideration, but keep in mind that it is always far more than anyone can say.

The literal meaning of "philosophy"

One helpful thing to note is that the ancient Greek philosophers embodied something of their sense of the nature of philosophy in the very name itself. The word "philosophy" literally means "love of wisdom" (*phil* means "love"; *sophia* means "wisdom"). Note: "philosophy" means "love of wisdom" – not "love of knowledge." That is an important distinction. The philosophers of ancient Greece were not searching for just any kind of knowledge; some things aren't worth bothering to know. For example, how many last names in the Manhattan telephone directory end with the letter "b"? I don't think I'll spend any time trying to gain that knowledge. How about you? Rather, the pioneers of philosophy were interested in the kind of knowledge that enables humans to live a good life – and that kind of knowledge is what we call "wisdom." Yet wisdom has never been the only concern of philosophers, and there is no universally accepted definition of philosophy among philosophers. (Philosophy is not alone in this regard. There are disagreements among the members of most academic disciplines as to how their discipline should be understood.)

Philosophy does, however, have a long and well-documented history, so although I cannot give you an official definition of philosophy, I can give you an overview of the problems which most philosophers have focused on over the centuries, right up to the present and into the foreseeable future. Their methods of approach and their solutions to these problems have varied greatly, but these problems are, nonetheless, the ones that have taken up most of their "working hours" and which characterize philosophical "work." When any of us is thinking about or discussing one of these problems, questions, topics, or issues, we are engaged in philosophy.

The basic problems and areas of philosophy

Nearly all of the many problems dealt with in philosophy can be related to the three basic problems mentioned earlier: the problem of reality, the problem of value, and the problem of knowledge. In figure 2.1 each of these three basic

16 Metaphilosophy

problems is divided into three more specific problems, each of which is associated with one or more sub-areas of philosophy. There are even more areas of philosophy than these, but these should give you an excellent overview of the field. The italicized words indicate topics that are studied in philosophy. The words in parentheses are the names of the parts of philosophy which focus on the topic that is italicized.

Figure 2.1 should help you begin to get a sense of the various parts of philosophy and how they relate to one another, but it should not be accepted rigidly. Other philosophers would include a few areas of philosophy that I have not included, would omit a few areas that I have included, and would arrange these areas somewhat differently. For example, some philosophers would list logic as a basic area of philosophy alongside epistemology, value theory, and metaphysics. Here at the beginning, however, I think it is important for you not only to develop a basic understanding of logic, language, and truth, but also to develop a sense of how they are related to knowledge and our pursuit of it. Now let's take

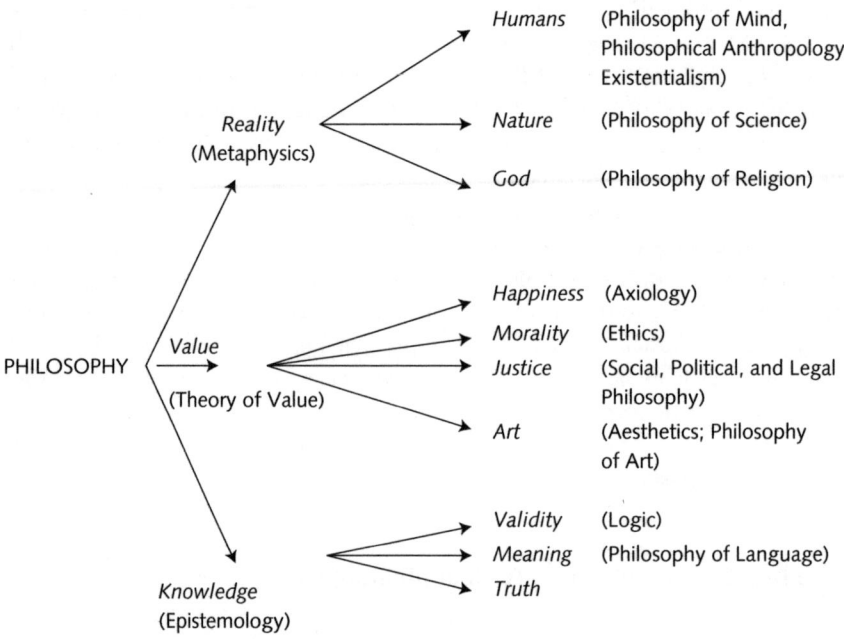

Figure 2.1 Topics and areas in philosophy

a closer look at each of the areas in figure 2.1 in order to get a bird's-eye-view of the whole of philosophy and a sense of how its parts are interconnected.

The first area, metaphysics, takes up the question, "What is the nature of reality?" That question can be sub-divided into three more questions: "What is the nature of a human being?" "What is the nature of Nature?" "What is the nature of God?" Each of these questions, in turn, is concerned with more specific issues. For example, whereas *cultural* anthropologists conduct field studies of human beings in their distinctive cultural settings, from Appalachia to Zambia, *philosophical anthropologists* and *existentialists* (philosophers of human existence) are concerned to understand and clarify the essential and universal traits of human existence, as distinguished from traits which are peculiar to a particular culture or sub-culture. What does it mean to be a human or a person no matter when or where one lives? What experiences do we all have as humans? What responsibilities or situations confront us all as humans?

Philosophy of mind is concerned with such questions as "What is the mind?" "What is the relation between the mind and the brain?" "What is a soul?" "Do humans have souls?" "Can the mind or soul survive the death of the body?" "Is human behavior free or determined?" *Philosophy of science* is concerned with such questions as: "What is matter? space? time?" "What is a natural law?" "Could the laws of nature change?" *Philosophy of religion* asks: "What is the nature of God?" "Does God exist?" "What is the relation of God to moral right and wrong?" By the end of *Thinking Philosophically* we will have examined several answers to most of these questions.

The questions of metaphysics are part of our attempt to understand ourselves and the reality of which we are a part. Yet we are interested not only in how things *are*; we are also interested in how we *want* them to be and how they *ought* to be. Most of us *want to be happy*, think we *ought to be moral*, and believe that neither of these goals can be fully achieved apart from *the right kind of society*. Consequently, philosophers have developed axiology, ethics, social philosophy, political philosophy, and legal philosophy. *Axiology* focuses on questions of value (for example, What makes something valuable or good? How can we distinguish the more valuable from the less valuable? the good from the bad? What are the constituents of a good life? How can we achieve them?). *Ethics* focuses on questions of moral right and wrong (for example, What makes something morally right or morally wrong? Are moral right and wrong relative or absolute? What, if anything, ought we always to do and what ought we never to do as human beings?). *Social, political, and legal philosophy* – three distinct but closely related fields – focus on such questions as: What is justice? What is the nature of a good society? How can we construct it? Are political rights and obligations absolute, relative, or arbitrary? How should individual freedom and social responsibility be related?

The preceding questions relate to our desire to live a good life – "good" in the sense of being, ideally, both happy and moral. To keep this book from getting too long, we will only touch on issues in social, political, and legal philosophy, and we will not at all explore the intriguing issues in philosophy of art.[1] We will, however, explore axiology and ethics at some length. What we find will bristle with implications for the social realm, so I hope the section on value theory will leave you with a desire to explore social, political, and legal philosophy on another occasion. Because art is one of our richest sources of value, I hope the section on axiology will also inspire you to study philosophy of art.

In addition to our interest in reality and value, we are interested in what we as human beings can *know* about reality and value and how we can *achieve* such knowledge. When we do metaphysics and value theory, we will encounter many *claims* as to what is real and what is valuable, but how do we *know* which of those claims, if any, are true? Indeed, what does it *mean* for something to be true? And what does it mean to *know* that something is true? Still more abstractly, what does it mean to mean? As you can see, philosophers rarely stop with the first question asked. Rather, they tend to push toward more and more basic questions until they believe they have reached the most basic ones. Consequently, they write books with such intriguing titles as *Understanding Understanding* and *The Meaning of Meaning*.

When we want to clarify our thoughts about a philosophical issue or communicate those thoughts to someone, we usually do so by means of spoken or written language. The area of philosophy most concerned with trying to understand that kind of meaning (linguistic meaning) is *philosophy of language*.[2] In addition to wanting to understand how language captures and conveys meaning, we want to know what it means for a statement to be *true* and how we can *know* that it is true. This concern with truth and knowledge, as distinguished from meaning, is associated primarily with the area of philosophy known as *epistemology*. When Jesus of Nazareth was on trial before Pilate, he said to Pilate, "For this I was born, and for this I have come into the world, to bear witness to the truth." Pilate replied – perhaps sincerely, perhaps cynically, perhaps wistfully – "What is truth?"[3] That is a most important question, to which philosophers have given centuries of increasingly refined attention. Whatever the correct answer is, it will have profound implications for all of those earlier questions which we asked about reality and value. Indeed, let's next explore some of those implications.

The interconnectedness of the issues of philosophy

What I want to do now is help you acquire a sense of how interconnected the issues of philosophy are. One way to do that is to go back to the diagram of

What is Philosophy? 19

philosophy and its parts (figure 2.1), draw a two-headed diagonal arrow from "Reality" to "Happiness, Morality, Justice" (see figure 2.2) and then ask what bearing specific answers to questions about the nature of reality will have on questions about happiness and morality. For example, on one conception of reality which we will study later, the pursuit of happiness is foolish. Hence, it would seem wise to try to determine whether that conception of reality is true before losing ourselves in the pursuit of happiness. Now ask whether your beliefs about happiness and morality have implications for what you should think about reality. For example, if you believe we have no moral obligations, does that mean you should, logically, also believe that there is no God? If you believe we *do* have moral obligations, does that mean you should also believe that there is more to the universe than just physical processes? Now let your mind roam back and forth along the lines in figure 2.2; ask yourself what implications the answers at one end of an arrow will have for issues at the other end of the arrow. For example, is truth valuable? If yes, why? If no, why not? Is it *morally* wrong to believe things without sufficient evidence to back them up?

From the bird's-eye-view of figure 2.2 it is easy to see that each of the three basic

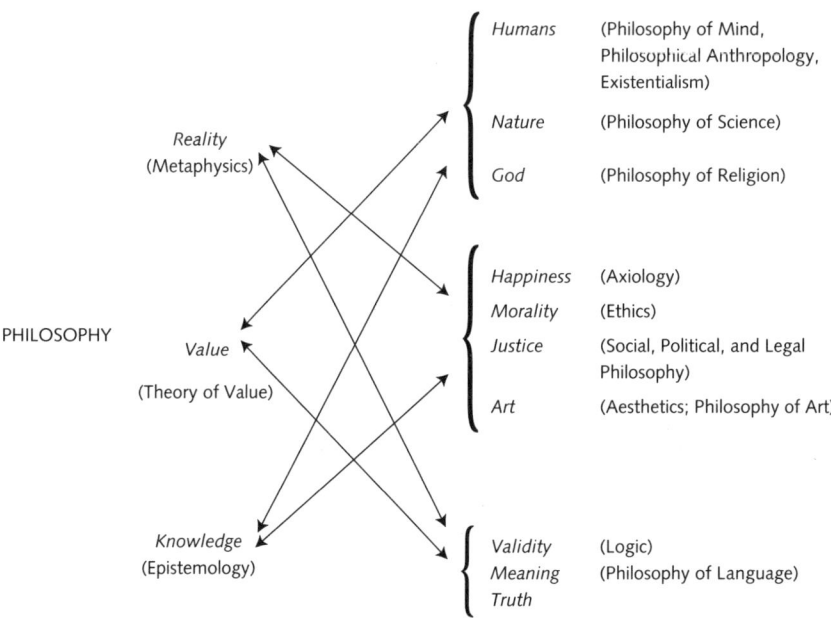

Figure 2.2   Each topic leads to another

areas of philosophy has implications for the other two. Accordingly, although philosophy has numerous distinguishable facets, it is an intimately interconnected whole.

Because of this interconnectedness, no matter what your initial concern in philosophy is – whether it be the issue of reality, value, or knowledge – you will eventually wind up thinking about the other two topics if you just keep going. Questions about *reality* lead to questions about whether we can *know* what reality is like, and if so how. Questions about *whether* we can achieve knowledge lead to questions about whether *reality* is such that we should be optimistic that our minds are capable of knowing the truth about reality, or whether reality is such that we should be pessimistic about our ability to achieve knowledge – especially about philosophical and religious matters. Questions about our *moral duty* raise questions about whether we are *free* to do what we ought to do. Questions about whether we are *free* lead to questions about what we *ought* to do if we are free. Questions about *value* lead to questions as to whether there is anything in *reality* which is objectively valuable, and if so what, and how we can know it to be so. Questions about *reality* disclose that whether or not there are any things in reality that are objectively valuable, there certainly are parts of reality, such as humans, who *value* things. We say we value knowledge, but what *value* is there in *knowing* what *reality* is like? What *value* is there in knowing the *truth*? Isn't ignorance bliss? And isn't bliss (happiness) what we value? But isn't the bliss of ignorance likely to be short-lived because of reality's refusal to be ignored? We will explore answers to these questions in the section on epistemology. For now I just want you to acquire a lively sense of how interconnected all the issues of philosophy are and how one question leads to another until they have all been asked.

A definition of philosophy

To tie the preceding points together, we might say that a human being is a *part* of reality which seeks knowledge of the *whole* of reality, and which seeks to know and live according to what is truly valuable; a human is a seeker of knowledge who values things and a valuer who seeks knowledge. If this description of what it is to be human is even roughly true, then philosophy is a natural, irrepressible part of human existence, for philosophy is *the passionate pursuit of knowledge of the real and the good*.

Each of the major words in this working definition of philosophy calls for explanation. Philosophy is a *passionate* enterprise because, as Jean-Paul Sartre, a twentieth-century existentialist philosopher, might have put it, a person *is* a passion. A person by nature cares very much about his or her existence and his or her world. Indeed, as Martin Heidegger, another twentieth-century existentialist put it, a human being is a caring being; to be human is to care – and two of the things about which we care most are knowledge of what is real and what is not, and knowledge of what is valuable.

Passion alone, however, is not philosophy. I say that philosophy is a passionate *pursuit* because wisdom does not come to us like a wisdom tooth (except insofar as both are uncomfortable!). *Effort* must be expended; *methods* of procedure must be developed, applied, evaluated, and improved upon. Consequently, philosophy cannot be done well without self-conscious discipline. The goal of such discipline is *knowledge*. The philosopher doesn't want just an opinion about reality and value – not even a true opinion. In a pinch, of course, a true opinion is better than a false opinion, but knowledge is even better, since we can have a true opinion without knowing that it's true or why it's true. Let's say, for example, that one of my opinions is that there is intelligent life in galaxies other than our own. My opinion may *be* true, but I certainly do not *know* it is true! Understandably, then, philosophers want more than true beliefs. They want to *know* what is the case. And if we cannot know for certain what is the case about reality and value, then we want to know why that is so and what is the next best alternative to knowledge.

Finally, the philosopher as philosopher pursues knowledge not of just any old thing, but of the nature of reality and value. The philosopher wants to know whether there is a God, a soul, a life after death, where human existence has come from, what its substance and circumstances are, and where it is going. The philosopher wants to know what is truly good, and not only apparently good, and what, if anything, we ought to do with our lives.

Having spoken repeatedly of "the philosopher," I hasten to remind you that we are all philosophers, so if my observations in the last two paragraphs are correct about what "the philosopher" wants, then those should be things that you want, too. Hence, I encourage you to pause, look back over those paragraphs, and ask yourself whether those are things that you, too, want.

To summarize: the many topics and questions we have reviewed regarding reality, value, and knowledge constitute an interconnected enterprise that goes by the name "philosophy". Once we open our minds to any one of these questions or topics, we find ourselves led by intrigue from one topic or question to another until we begin to understand how they are all connected and why a well-rounded philosophy must deal with all of these issues.

Clue words to areas in philosophy

One way to learn to recognize when an issue belongs to one area of philosophy rather than another is to recognize that there are different groups of words that characterize the issues in the different areas of philosophy. Regarding the groups of words below, when you encounter a statement containing one of the words to the *right* of the colon, it may be a clue that you are dealing with a statement that belongs to the area of philosophy to the *left* of the colon. Try to understand *why* the words in each group belong together and how they identify a distinctive area of philosophical concern. That will take you a lot farther than merely memorizing which words belong to each area. Because philosophy deals primarily in concepts rather than facts, it is a general rule in learning philosophy that understanding will take you much farther than memorization.

- Axiology: good, bad, happy, unhappy, pleasure, pain, value
- Ethics: right, wrong, moral, immoral, evil, just, unjust, ought, obligation, right, responsibility, justice
- Epistemology: know, knowledge, true, false, belief, proof, reason, probability, empirical, justification, rational, irrational, probable, improbable, plausible, faith
- Logic: argument, valid, invalid, fallacious, fallacy, imply, implication, infer, inference, deductive, inductive, conclusion, evidence, premise
- Metaphysics: exists, necessary, contingent, possible, impossible, real, unreal, appearance, reality, noumenon, phenomenon

Sample statements and questions in different areas

When you have philosophical thoughts and questions of your own, and when you encounter the philosophical ideas and questions of others, it is important that you be able to identify into which of the following five areas of philosophy – axiology, ethics, epistemology, logic, and metaphysics – that thought, idea, or question belongs. Only then will you be able to respond appropriately to each kind of claim or question as you encounter it in readings, conversations, or reflection (and sometimes, of course, a thought, claim, or question is relevant to more than one area). Below are examples of questions and statements that fall into each area. Again, do not just memorize them. Think about them until you understand why each claim and question belongs where it has been placed rather than elsewhere. (Note: The statements below are only examples. I am not claiming that they are true or that I agree with them.)

- METAPHYSICAL: "Is there life after death?" "There is life after death."
- EPISTEMOLOGICAL: "What evidence would justify belief in life after death?" "If previously dead people came back to life and independently gave

reports that agreed about a life after death, that would justify belief in life after death."
- AXIOLOGICAL: "Would it be good, bad, or indifferent if there is life after death?" "Life after death would be a bad thing. This life is enough."
- ETHICAL: "Ought we to live our lives as though there is life after death?" "We ought to live our lives as though there is life after death."
- LOGICAL: "Is that argument against life after death sound?" "That argument is valid, but it has a false premise, so it is not sound."

Notes

1 Philosophy of art endeavors to understand what art is, what makes something a work of art, and what gives a work of art aesthetic value. It investigates distinctions such as those between nature and art, artistic truth and illusion, and the beautiful and the ugly. It investigates such problems as those of censorship versus freedom of expression, the nature of aesthetic experience, and the criteria for aesthetic evaluation.
2 Verbal language is, of course, not our only vehicle for conveying meaning. In Nikos Kazantzakis's novel *Zorba*, Zorba the Greek often prefers to dance out his meaning. Since works of art, including dance, are often fraught with meaning, philosophy of art, as well as philosophy of language, is very concerned with the nature of meaning.
3 See John 18:33–8 in the New Testament.

Reading Further

Plato, *The Apology*; *The Crito*.
Bertrand Russell, Chapter XV, "The Value of Philosophy," *The Problems of Philosophy*.

Chapter 3: Why We Do Philosophy

- The noetic motive
  Aristotle, Kant
- The cathartic motive
  Socrates, Wittgenstein
- The mystical motive
  Plato, Aristotle, Spinoza, Bradley, Hegel, Heidegger
- The wisdom motive
  Epictetus
- The sport motive

The noetic motive

"Noetic" (no e' tic) comes from the Greek word for "mind." One of our motives for doing philosophy is simply that our minds crave knowledge. Aristotle, who was Plato's most famous student, began his book *Metaphysics* by saying, "All humans by nature desire to know." He was correct. We all desire to know what reality is like. That's just the way we are – creatures full of curiosity, and that curiosity causes us to ask and try to answer philosophical questions.

Immanuel Kant, an eighteenth-century German philosopher, added that we can't help doing philosophy. Why? Because it is as natural as breathing, and like breathing, we do it even when we aren't aware that we're doing it. (Now that you know what philosophy is, you will at times suddenly realize that you are in the middle of thinking philosophically or that you are engaged in a philosophical discussion.) Moreover, when we try to stop doing philosophy, we find it's like trying to hold our breath indefinitely – we just can't do it! The questions involved in philosophy are too fascinating, too perplexing, too urgent, and too tangled up with everyday life for us to be able to resist them indefinitely; so whenever we try to quit, reality itself seems to tease, lure, trick, shove, or enrage us back into doing philosophy.

The cathartic motive

Because of our desire for knowledge and our pursuit of it, we run into puzzles and riddles that haunt us. Consequently, philosophers such as Socrates and Ludwig Wittgenstein have said that doing philosophy is sometimes a matter of catharsis. According to Webster's dictionary, catharsis is "purgation, especially of the bowels." Hence, you might say that according to Webster, philosophy is a kind of intellectual purgative! There's as much truth to that as humor, but in a more serious vein, Socrates believed that doing philosophy frees us from (1) Presumptuousness (especially the presumptuousness of thinking that we know what we do not know), (2) Confusions (resulting from nonsense, vagueness, and ambiguity), (3) Unwarranted Fears (such as perhaps the fear of death), and (4) Harmful Passions (such as the passions for pleasure, possessions, power, and plaudits). Socrates believed that these things encrust the soul, like mud, and prevent it from being filled with the light of truth and goodness. By doing philosophy we cleanse these impediments from our souls, allowing intellectual and spiritual enlightenment to penetrate and illuminate us.

Wittgenstein, a twentieth-century Austrian who did much of his philosophizing in England, spoke of the cathartic function of philosophy in a different way. He said that philosophy gets rid of "mental cramps." Just as the body gets painful cramps, so does the mind. Mental cramps are caused by a misuse of language which causes confusion and perplexing puzzles. For example, is space finite or infinite? You might say, "Well, it's infinite, for if it weren't infinite, there would be something beyond it, but there can't be anything beyond space except more space, so if there's nothing beyond space except more space, then space goes on forever and is infinite!"

Well enough, but then comes the reply: "Insofar as space is, it is what it is. It is complete; there is only so much of it. But since it is complete, it is fixed and definite, and that's what it means to say that something is finite – namely, that it is fixed and definite. Hence, space can't be infinite; it must be finite." Right? Wrong! "For if the universe is finite, then there must be something beyond it, since to be finite is to be limited. But anything which limits it and is beyond it must be included in it, but there can't be anything beyond space except more space! Hence, the universe is infinite."

Well, I'm getting dizzy, and that should be enough to show you what Wittgenstein was talking about when he said that language can cause mental cramps which are quite uncomfortable. Wittgenstein said, picturesquely, that these cramps occur "when language goes on a holiday," that is, when language is used to do things that it is not suited to do.

How does philosophy treat a mental cramp? Sometimes it doesn't give us an answer. It gets rid of the cramp by showing that the cramp is the result of a misuse of language. A follower of Wittgenstein might point out, for example, that whereas the terms "finite" and "infinite" have perfectly good applications in mathematics, they cannot be applied meaningfully to physical space. When they are applied to space, they cause the kind of puzzle we encountered a moment ago. To be sure, we may be bewitched by language into *feeling* that there is an answer to that puzzle about space, but if Wittgenstein is right, there isn't. And that's what philosophy shows us with regard to some questions. It doesn't give us an answer. It makes the cramp go away by showing us that the problem which gave rise to the question is not really a problem – and for that we can be grateful.

The mystical motive

In addition to satisfying our desire to obtain answers and dissolve riddles, philosophy can cause a kind of "consciousness expansion" which transcends our ordinary state of mind in a profoundly uplifting way. This fact is echoed over and over in the writings of philosophers east and west. Plato stated eloquently in his Epistle VII that by means of philosophical investigation we can achieve an occasional flash of illumination in which we are caught up in an ecstasy of understanding which transcends our ordinary level of understanding and cannot be described in ordinary language. That special moment of insight is like a mountain-top experience in which we see and understand things from a panoramic and absolutely convincing point of view. Obviously that moment is an emotional experience, but it is not simply or primarily an emotional experience. The emotions involved are the result of an extraordinary intellectual breakthrough.

Aristotle argues in his *Nicomachean Ethics* that the best and pleasantest life for a human being, the happiest life, is one "which accords with reason . . . as a man's reason is in the highest sense himself."[1] Why is this so? Because, according to Aristotle, reason is the most divine element in a human being, so by our exercise of reason we draw near to the gods and participate in their peace and bliss. Whether Aristotle was correct about the gods, his belief that peace of mind can be achieved by living a life of reason has been confirmed by other philosophers. Benedict Spinoza, a seventeenth-century Jewish philosopher, confesses and argues in his *Ethics* that the surest, least disruptable way to happiness is through the intellectual love of reality.

Still more recently, in the nineteenth century, F. H. Bradley wrote in his *Appearance and Reality*: "All of us, I presume, more or less, are led beyond the region of ordinary facts. Some in one way and some in others, we seem to touch and have communion with what is beyond the visible world. In various manners we find something higher, which both supports and humbles, both chastens and transports us. And, with certain persons, the intellectual effort to understand the universe is a principle way of . . . experiencing the Deity."[2] Some statements, such as this one, speak best for themselves, so I won't comment on it, but you may want to reread it.

Another nineteenth-century philosopher, G. W. F. Hegel, said in his initially enigmatic way that philosophy is "Being aware of itself." What did he mean by that? Hegel believed that the whole of reality is a single, evolving individual which was unaware of itself until the emergence of humans. With the emergence of humans (who are one part of reality), Reality became aware of itself for the first time. By means of humans doing philosophy, Reality is examining and becoming more aware of itself. We are the eyes and the brain cells of Reality. Alan Watts, a twentieth-century thinker, captured this idea beautifully when he wrote, "Each one of us is therefore an aperture through which the universe knows itself, but not all of itself, from a particular point of view . . ."[3] Perhaps these ideas from Hegel and Watts mean nothing to you now, but when their shells crack and their meanings flow forth, you will have a strange, expansive feeling as you think of yourself and reality in an entirely new way.

Finally, I mention Martin Heidegger, a twentieth-century German existentialist, who in the spirit of Hegel said that philosophy is Being tolling unto being. Here "Being" (with a capital "B") stands for the whole and the depth of reality; "being" (lowercase) stands for the individual – for you and for me. Philosophy is the result of Being tolling like a bell for our attention, whispering our names, calling us, luring and teasing us out of our individual selves, out of our socialized conceptions of reality and into a direct encounter with Being itself. Clearly, such an encounter will disrupt, override, perhaps shatter, one's ordinary consciousness; just as surely, the lure and impact of such an experience is one of the reasons that people do philosophy.

The wisdom motive

The "wisdom motive," like the "noetic motive," is based on a desire for knowledge. But whereas the noetic motive is based on profound curiosity about everything, the wisdom motive is more focused. It is concerned with understanding and living life well. Philosophy, as we have seen, literally means "love of wisdom." Wisdom is insight regarding the nature and living of *human* existence. Epictetus, an ancient stoic philosopher, put the matter this way: "Philosophy does not promise to procure any outward good for man; otherwise it would admit something beyond its proper theme. For as the material of a carpenter is wood and of a statuary, brass; so of the art of living, the material is each person's own life."[4]

The function of philosophy, Epictetus was saying, is to teach us not the art of carpentry or metalwork, but the art of living. In each art there is a specific "stuff" to work with. The carpenter works with wood and the artisan with metal. What does the philosopher work with? His or her own individual life. In what sense does he or she do this? In something like the following sense. Your life is the raw material out of which it is your task to create something beautiful and satisfying. Someone once said, "What you are is God's gift to you. What you make of yourself is your gift to God." There is truth in that statement whether you believe in God or not. The truth is that you are a bundle of possibilities waiting to be actualized. When you become self-conscious and able to take charge of your life, you are a partially begun but unfinished work of art which can only be finished properly by yourself. How to go about "finishing" oneself (or perhaps better, "actualizing" or "fulfilling" or "developing" oneself) is not obvious. It doesn't happen naturally. It's something one must think about and work at. One of the most important things that philosophy does is help us find wisdom for the shaping and living of our lives.

The sport motive

Finally, people do philosophy for the sheer fun of it! In certain respects, philosophy is like a game which is fascinating in and of itself. There is no end to it, and any number can play. It can be carried on in a competitive way or a cooperative way. It has all the thrill of a chase, and its quarry – truth – is as beautiful, awe-inspiring, and elusive as any prize that people have ever hunted. Moreover, there's no danger of "overhunting" the species and making it extinct. The more sophisticated you become as a philosophical hunter, the more numerous, subtle, and fascinating will be the questions you see to pursue answers to. If you don't take yourself too seriously, many a good laugh is in store as you see your quarry elude you once again in a totally unexpected way – just when you thought you had it for sure! Such a disappointment is no more discouraging to the mature philosopher than it is to the bear-hunter who has developed the greatest admiration, and even affection, for crafty ol' slewfoot – who always escapes by some clever maneuver. And remember, the objective of the philosopher, figuratively speaking, is not to slay ol' slewfoot, but to befriend or master him.

A comedian once confessed that he had taken a philosophy course in college. He added that it was really going to come in handy – if he were ever committed to solitary confinement! That may seem like a put-down, but anything we can enjoy even in solitary confinement is a heck of a valuable thing to know! Indeed, one of the most unappreciated values of studying philosophy and the other liberal arts is that they enable us to enjoy ourselves in solitude and with one another in inexpensive ways, by thinking, reading, discussing, and perhaps writing or creating art.

Blaise Pascal thought such an ability to be of enormous social, as well as personal, importance. "Sometimes," he mused, "when I set to thinking about the various activities of men, the dangers and troubles which they face at Court, or in war, giving rise to so many quarrels and passions, daring and often wicked enterprises and so on, I have often said that the sole cause of man's unhappiness is that he does not know how to stay quietly in his room. A man wealthy enough for life's needs would never leave home to go to sea or besiege some fortress if he knew how to stay at home and enjoy it."[5] Along the same line an anonymous author has said, "People usually quarrel because they do not know how to argue." One of the primary objectives of *Thinking Philosophically* is to help you learn how to argue (engage in philosophical discussion) in a positive, cooperative way rather than in a negative, quarrelsome way. Chapter 6, "Methods for Doing Philosophy", is devoted entirely to helping you achieve that objective.

Notes

1   Aristotle, *Nicomachean Ethics*, Book X, 1178a8, any edition.
2   F. H. Bradley, *Appearance and Reality* (Oxford: The Clarendon Press, 1930), p. 5.
3   Alan R. Watts, "Philosophy Beyond Words," *The Owl of Minerva*, ed. Charles J. Bontempo and S. Jack Odell (NY: McGraw-Hill, 1975), p. 196.
4   Epictetus in *Great Traditions in Ethics*, ed. Albert, Denise, and Peterfreund, 4th edn. (NY: Van Nostrand Reinhold Co., 1980), p. 101.
5   Blaise Pascal, *Pensées*, tr. A. J. Krailsheimer (NY: Penguin Books, 1966), p. 67, #136.

Reading Further

One valuable cross-section of opinions among philosophers regarding the nature of philosophy can be found in *The Owl of Minerva*, ed. Charles J. Bontempo and S. Jack Odell. A more recent collection can be found in *After Philosophy: End or Transformation?*, ed. Kenneth Baynes, et al.

Chapter 4

The Two Most Basic Causes of Philosophy

- Ambiguity
    - Erik and his grandparents
    - Two children at a zoo
    - Thomas Hobbes
    - The given and interpretation of the given
- Curiosity
- Vagueness, ambivalence, and ambiguity

Ambiguity

Now I want to identify what I think are the two most *basic* causes of our doing philosophy. The first of these causes is *the ambiguity of human experience*. For something to be ambiguous is for it to be subject to two or more interpretations when it is not clear which interpretation is correct. For example, someone might point to an automobile and say, "That car is hot." The speaker might mean "That car travels at very high speeds" or "The engine of that car has overheated" or "That is a stolen car" or some combination of these. Without some context clues or clarification from the speaker, we would not know exactly what to think about that car.

When we think of ambiguity, we most often have ambiguous sentences in mind. However, an experience, as well as a sentence, can be ambiguous – that is, can be open to more than one plausible interpretation without any clear indication as to which interpretation is correct. For an example, here is a true story. A young woman once took a seminar with a distinguished philosopher. She noticed during meetings of the seminar that whenever she spoke, the professor turned away from her. After a while she began to be annoyed by this and felt that she was being snubbed. This situation went on until she told another student about it, and that student brought the situation to the attention of the professor. He was chagrined to find out how his behavior had been interpreted. He explained to his student that he had a significant hearing loss in one ear, so he turned away from her not as an expression of disregard but as an expression of high regard. He wanted to be sure that he heard everything that she said, so when she spoke he turned his eyes away from her in order to direct his good ear toward her!

Here is another true story of the ambiguity of experience. A mother named Kay Haugaard reported that when her middle son, Erik, was a child, he one day asked her, "How can there be ground up in the sky?" At first she was utterly baffled by his question. Then insight came, and she thought she understood, so she told him about planets and about how the Earth, which is made of ground or dirt, is held in the sky by gravity. Erik seemed satisfied and left. The next day, however, his mother overheard him say to a playmate, "No, it isn't. Oregon is up in the sky!" Then his mother realized that she had badly misunderstood his question about how ground could be up in the sky. The only times that Erik had gone to visit his grandparents in Oregon had been by airplane, by flying up into the sky and then landing on Oregon. Erik had surmised that he went to Oregon much like a bird flies from the ground up into a tree rather than like a bird that flies from the ground in one yard to the ground in another yard. On his trips, he could not tell what was happening just by looking out the window of the plane, so to him the most plausible understanding of what had happened on his trips was that he had flown up into the sky to visit his grandparents and then flown back down to earth – a fascinating and plausible interpretation. The charming end to this story is that before Erik's mother explained air travel to Erik, she told him that his grandparents were coming for a visit from Oregon in their car. Erik, she reports, "wrinkled his nose and asked, 'How can their car go through the air?'"[1]

As children we all, like Erik, struggled to make sense of our fragmentary experiences of the world. Sometimes we were, from the adult perspective, hilariously or intriguingly wrong. I was once at the National Zoo in Washington, D.C. In the reptile house I paused before a water tank full of frogs. I was fascinated by the perspective because the surface of the water in the tank was several feet above my head, so I was looking up through the water at a number of frogs suspended motionless just below the surface of the water, their heads touching, but not breaking, the surface tension of the water. Two preschool children next to me were also fascinated by those frogs. The little boy asked the little girl, "Why are they like that?" I'm quite confident that the little girl had never seen anything like that before, yet she replied with barely a pause, "They must be stuck to the ceiling." What a wonderful example of the ambiguity of experience and of the human urge to resolve that ambiguity. Mistaken though she was, the little girl disambiguated the situation in a clever, charming way. It did indeed look as though the motionless frogs were glued to the surface and hanging down, though the truth of the matter was quite the opposite; they were being buoyed up from below rather than held from above.

What I want to emphasize here is that although the ambiguity of human experience is partially overcome by the time we are adults, it is never entirely overcome, and especially with regard to philosophical issues. Consider, for example, a statement by Thomas Hobbes, an important seventeenth-century British philosopher. Hobbes once said, "Someone who says that God has spoken to him in a dream has said no more than that he dreamt that God spoke to him." Well – maybe, and maybe not. What is indisputable to the individual is that the dream did occur. What is not clear is whether the person merely dreamt that God spoke to him or whether God really did speak to him in his dream.

Consider also the ambiguity of near-death experiences in which people whose hearts have quit beating have the experience of being outside of their bodies looking down at them, or seem to travel to another realm and meet people whom they know to have died. Those experiences can be interpreted as evidence that humans have a soul which survives the death of the body, or they can be interpreted as illusions which the brain generates when blood quits flowing through it. Consider also feelings of guilt. Do they mean that there really are things that are morally wrong? Or do they only mean that those who rear us can condition us to feel that certain things are morally wrong? (We will return to all of these issues at more length later.)

Human experience can, then, be interpreted as though there is a God or as though there is not, as though there is life after death or as though there is not, as though some actions are morally wrong or as though they are not, as though human beings have freedom of choice or as though they do not, as though history is meaningful and going somewhere or as though it is not, as though life is just a dream or as though it is not. Philosophy is, to a significant extent, an endeavor to identify these ambiguities in experience, to distinguish what really is in our experience from what has been injected into it by interpretation (such as the little girl's seeing the frogs as stuck to the ceiling), to determine how many interpretations of our ambiguous experiences are plausible, and to evaluate and choose among those interpretations. Hence, because of the ambiguity of human experience we need to try to distinguish between what is *given* to us in experience and how we should *interpret* the given. (It is also important to realize that art and literature deal with these same ambiguities in their own ways.)

In brief, experience, as well as language, can be ambiguous. Moreover, many of our experiences are ambiguous. If our experiences were not ambiguous but, rather, had one clear meaning, then philosophy would not have arisen. We would not have needed it because all of the answers to our philosophical questions would either be clear or clearly unavailable. However, the answers to our philosophical questions are not immediately obvious. More than one answer is often compatible with our experiences. It is not clear which of those incompatible answers is true, and we cannot decide between them by using scientific methods. That is why we need to do philosophy.

Curiosity

Ambiguity by itself, however, would not be sufficient to cause us to do philosophy. If experience were ambiguous but we didn't *care* which interpretation of it was the truth, we wouldn't do philosophy. Similarly, if we cared about the truth, but just looking at our experiences made the truth immediately evident, we would not need to do philosophy to find the truth; we could just look at our experiences and see the truth – just as when we want to know how much money we have with us, all we have to do is get it out and count it up; we don't have to do philosophy. But getting the answers to our deepest questions is never that easy, whether in science, history, or philosophy. Hence, it takes both ambiguity and curiosity to get us going philosophically.

Vagueness, ambivalence, and ambiguity

In order to do an effective job of disambiguating ambiguities it is important to be able to distinguish *ambiguity*, *ambivalence*, and *vagueness* from one another. They are not the same thing – though they all have to do with things not being clear. *Ambiguity* pertains to multiple *meanings*, as when a sentence or an event has two or more possible meanings, sometimes quite different meanings. For example, say you are on your way out of a building and meet someone coming in. You ask, "Is it still raining?" The other person, who commutes a long way to work, says, "It's all over." That could mean "The rain has stopped," or it could mean "The whole region is covered by rain." Obviously, if you get the wrong meaning for this sentence, you could get a surprise when you go out. Similarly, a crash downstairs in the middle of the night could mean a burglar has broken in or it could mean your cat is horsing around. Obviously, these two different meanings for the same sound call for very different responses on your part. Hence, it is important that we correctly *disambiguate* ambiguous sentences and events, that is, untangle and distinguish their different possible meanings.

Ambivalence pertains to *feelings*. An individual is ambivalent toward something when she has mixed feelings about it; for example, when at one and the same time she both likes a certain idea and doesn't like it, or wants to do a certain thing and doesn't want to do it, or likes a certain person and doesn't like that person.

Ambiguity should also be distinguished from *vagueness*. Something which is *ambiguous* has two or more clear interpretations; the problem is that, at least initially, it is not clear which interpretation is correct. Something which is *vague* has *no* clear interpretation. A blurry image in a microscope ("I can't make out what it is") or an unclear response (a shrug of the shoulders in response to the question "How long will you be gone?") is vague, not ambiguous. Sometimes that which is vague can be clarified (by focusing the microscope or asking more questions, for example), but in the case of vagueness, clarity is achieved by making something more precise, not by choosing between multiple interpretations of it.

Sometimes philosophically relevant experiences are vague, not ambiguous. For example, we may have an intense, unusual experience which seems to resonate with philosophical significance, yet we may not know what to make of it or how to understand it. Also, we may have an ambivalent reaction to a philosophically relevant experience. For example, we may have an ambivalent reaction to an experience which suggests that there are no moral rights and wrongs: on the one hand we might be horrified by that thought because it would mean that nothing is morally evil; on the other hand we might like that thought because it would mean there are no moral rules which we ought to obey.

Note

1   *Unitarian-Universalist World*, 1, 15 (November 1, 1970), p. 5.

Reading Further

On the topic of ambiguity, see Ludwig Wittgenstein, *Philosophical Investigations*, part II, section xi, and especially the duck/rabbit illustration. Also, John Hick, "Seeing-as and Religious Experience," in *Faith*, ed. Terence Penelhum (published first in *Philosophy of Religion: Proceedings of the 8th International Wittgenstein Symposium, 1983*, ed. Wolfgang L. Gombocz). For a fuller discussion by Hick see, "Part Two: The Religious Ambiguity of the Universe," *An Interpretation of Religion* (plus chapters 8, 9, and 10).

Chapter 5: Reason, Philosophy, and Other Disciplines

- An expanded definition of philosophy
- Philosophy and Religion
- Philosophy and Science
- Philosophy and Mathematics
- Philosophy and History

An expanded definition of philosophy

In chapter 1 I defined philosophy as "the passionate pursuit of knowledge of the real and the good." Now I want to expand that definition to say: *Philosophy is the passionate pursuit **by means of reason** of knowledge of the real and the good.* I have added "by means of reason" to my earlier definition to make clear the authority by which philosophy proceeds, namely, reason. Reason is, of course, not employed only by philosophy. It is employed in every area of human life, but for different purposes and in different ways. Hence, it should help make the nature of philosophy clearer if we compare and contrast philosophy's use of reason to how it is used in some other areas of intellectual activity. Religion, science, mathematics, and history provide especially illuminating comparisons and contrasts.

Philosophy and Religion

Philosophy and religion are similar in that they are both concerned with the nature of reality and the way to human fulfillment. However, in addition to using *reason* to find answers to questions about the nature of reality and the way to human fulfillment, religions often also turn to what they believe to be a *divine revelation* contained in a sacred scripture. Jews turn to the Torah; Christians turn to the New Testament; Muslims turn to the Koran; Hindus to the Bhagavad Gita; etc. The orthodox members of each of these religions believe that their scriptures are a gift from a divine being who in them has revealed important truths that we could not discover on our own – truths that are critical to understanding the nature of ultimate reality and the way to human fulfillment.

Philosophy, by contrast, recognizes no authoritative scripture. Within philosophy every claim that anyone makes is considered to be an appropriate target for critical investigation by means of reason. What is reason? What does it mean to investigate something by means of reason? That is difficult to say, but it means something like this: taking nothing for granted and asking of every claim, "What evidence is there for believing it? What arguments are there to support it? Are those arguments strong enough to justify believing it?" Philosophy examines every position and asks what reasons there are for accepting or rejecting it; philosophy allows nothing to be sacrosanct and beyond the pale of rigorous investigation – not even reason itself! In brief, philosophy is the attempt to see what we can know just by depending on ordinary human experience and the powers of the human mind.

Please keep in mind that although philosophy can come into conflict with religion, the two are not necessarily incompatible. There is a rumor that if one studies philosophy, one will cease to believe in God. That simply is not true. There are many first-rate philosophers who are deeply religious. There are many others who are not. Philosophy has led some people to abandon their religious beliefs, and sometimes even to become hostile to religion, but it has led others to become more appreciative of religion. Moreover, a philosopher can be very interested in seeing how much we can know by means of human reason, yet also believe that there are truths that human reason cannot reach but which God could reveal to us.

Whether a particular philosopher will be open to the possibility of learning truths by means of divine revelation will depend, of course, on what that individual believes he or she knows by means of reason about the nature of reality. If a person believes that reason shows that there is nothing in reality beyond the physical world, then that person will believe that there is not a God; therefore that person will also believe that there are no revelations from God. However, if an individual believes that reason shows that there *is*, or *probably is*, or *may be* a God, then he or she will be open to the possibility of learning truth through divine revelation. There is, then, no predicting what kind of religious impact philosophy will have on a particular person. Moreover, sometimes its effect is to first cause individuals to abandon the beliefs with which they grew up, and then years later to cause them to return to at least some of those beliefs on a more mature, sophisticated level of understanding.

Philosophy and Science

Now let's look at similarities and differences between philosophy and science. Philosophy and science are alike in that both want to see what truths can be learned by human effort alone, and both proceed in disciplined, systematic ways. However, there are at least two significant differences between philosophy and science.

First, philosophy is concerned with basic questions about value and morality, that is, with questions about what is really good, as distinguished from what merely appears to be good but is not, and with questions about what we ought and ought not to do. By contrast, pure science, as distinguished from applied science, is limited to questions about *how things are in actuality*, rather than how they ought to be or what would make them better.

Second, whereas philosophers, as well as scientists, are interested in finding out how things are in actuality, science is limited to *empirical* questions about the way things are. That is, science, by virtue of its distinctive methods of investigation, is limited to questions that can be answered by "looking to see," that is, by using our senses. In order for an hypothesis to be subject to investigation by science, it must be subject to confirmation or falsification by means of our senses – though we may aid our senses by means of telescopes, microscopes, sophisticated electronic equipment, etc. If a claim can be evaluated adequately in these ways, then it is a claim that belongs to science, not philosophy.

If a claim *cannot* be investigated in these ways, that is, if it is not an empirical claim, then it may well be a claim that calls for philosophical investigation. Claims regarding the number of planets in our solar system are clearly claims that can and should be investigated by science, not philosophy. However, claims regarding what is morally right and morally wrong cannot be resolved by empirical procedures, so in the division of labor between philosophy and science, moral claims fall to philosophy. Also, claims regarding the nature and existence of God cannot be resolved by empirical procedures, so they, too, fall to philosophy. Whether capital punishment is morally wrong, whether an awesomely intelligent, nonphysical being created the universe, are questions which cannot be answered by looking through telescopes or microscopes, mixing chemicals in test tubes, or doing mathematical analyses of scientific data.

Whether a problem should be taken up by science or by philosophy is not always clear. Sometimes it seems that a problem is a scientific problem, yet after investigation it proves to be philosophical in nature. Sometimes a problem that seems to be a philosophical problem is discovered, after awhile, to be a problem that can be investigated scientifically. However, philosophy and science together take up nearly every basic question that can be asked about the nature of reality. In fact, I like to think of philosophy and science as the two legs of the human pursuit of basic knowledge. That is why it is important that we keep in touch with both of them, at least in terms of what their leading practitioners are discovering. Keep in mind also that there is nothing other than the limitations of time and energy to prevent a philosopher from also being a scientist, or a scientist from being a philosopher.

I have characterized *scientific* problems as empirical, that is, physical. Philosophical problems, by contrast, are non-empirical, that is, *metaphysical*. "Metaphysics" literally means "after physics." Nearly three centuries after Aristotle's death, Andronicus of Rhodes undertook the enormous task of organizing Aristotle's many writings on many topics. Part of what Andronicus did was group shorter writings together into volumes on similar topics. He grouped together some writings on the generation, change, and deterioration of physical entities and titled those writings *Physics*. Then he grouped together some writings on God, essence, and the nature of existence, and he titled those writings *Metaphysics* ("After Physics"), simply to indicate where they were located among the many volumes of Aristotle's writings. Today we associate the word "metaphysics" with the kinds of topics discussed in the *Metaphysics*, rather than with where the volume on those topics was placed by Andronicus' in his arrangement of Aristotle's writings.

Because of the literal meaning of "metaphysics," it still makes sense to think of metaphysical topics as topics which come after physics in the sense of going beyond topics that can be dealt with physically, that is, empirically. Hence, if you think of the area of philosophy call "metaphysics" as dealing with non-empirical questions about what *exists*, and think of the other areas of philosophy (such as ethics) as dealing with *other* kinds of metaphysical questions, that is, other questions that cannot be answered by empirical methods, then I think you will find it illuminating to think of *science* as dealing with physical questions and *philosophy* as dealing with meta-physical questions – that is, important questions that still need to be answered after we have answered the questions that can be answered empirically.

Another revealing difference between philosophy and science is that most scientists (especially "natural" scientists as distinguished from "social" scientists) have some kind of special equipment, and sometimes special clothing, to help them carry out their investigations. By contrast, no laboratory or physical tools or special clothing would help us do our philosophical work better. That is because philosophical work consists for the most part of thinking about ordinary experience, formulating concepts and distinctions (such as those between appearance and reality, justice and injustice) for organizing and interpreting our experience of the world, constructing hypotheses about what, if anything, exists beyond the physical world, and deciding how we ought to live in the world as we think it is. Special equipment seems to be of no help in doing those things.

Parenthetically, people in physics, chemistry, and biology receive a Doctor of Philosophy (Ph.D.) degree as their highest graduate school degree. Why don't they receive a P.D. (Physics Doctorate) or C.D. (Chemistry Doctorate)? Because once upon a time philosophy (which, remember, literally means "love of wisdom") included all of the humanities and sciences. To study philosophy once meant to study the basic truths about everything that was thought to be worth studying – and that, of course, included study of the physical world, as well as of things metaphysical. In that era the study of the physical world was called "natural philosophy," and topics that could not be studied by sensory observation and physical experimentation were grouped under "metaphysics" (when the topics were primarily religious they were usually labelled "theology" – which literally means "the study of God"). Teachers of the liberal arts typically studied and taught about both of those aspects of reality. René Descartes, for example, founded analytic geometry and wrote important works on physics, as well as on God, knowledge, and the mind. Blaise Pascal made important contributions to probability theory and the study of atmospheric pressure, as well as to philosophical anthropology. Immanuel Kant made important contributions to astronomy, as well as to ethics and philosophy of mind.

Beginning roughly with the mid nineteenth century, however, the explosion of knowledge in the sciences and the extensive training required for mastery of the increasingly mathematical and experimental methods of the sciences limited fewer and fewer people to knowledge of both natural philosophy and metaphysical philosophy. As a consequence, what we now know as science gradually became separate from philosophy. However, in some universities science is still called "natural philosophy," that is, philosophy of nature. That name for science still has merit today. It helps make clear that although different methods have to be used to pursue different types of knowledge about reality, ideally the pursuit of knowledge is comprehensive, ignoring nothing important. Hence, it is important to read newspapers, magazines, journals, and books to keep up with natural philosophy (science) as well as metaphysical philosophy.

Philosophy and Mathematics

Let's continue trying to get a clearer understanding of philosophy by comparing and contrasting it to another discipline with which you have had some experience, mathematics. Philosophy and mathematics are quite similar, as numerous students have noticed and told me, because they both put great emphasis on rigorous reasoning, and they both ideally want to *prove* the truth of the position that they think is true. Hence, in both fields you will encounter rigorous efforts to prove or demonstrate conclusively that a position is true or is false.

A crucial difference between math and philosophy is that pure mathematicians (but not applied mathematicians) need not be interested in whether the mathematical truths that they discover have any application to the world we live in. Pure mathematicians love the intellectual beauty of mathematical reasoning and the clever, sometimes awesome, insights to which it leads; they (as mathematicians) are in love with abstract truths, not concrete truths about the world. Indeed, to some pure mathematicians it is a source of pride and pleasure that their mathematical investigations and discoveries have no known application to our world. (But please note: what has *begun* as pure mathematics has sometimes been discovered later to have enormous practical uses!)

By contrast, the philosopher is interested primarily in concrete truths, that is, truths that apply to our understanding of reality and our living in it. Perhaps that is why "philosophy" literally means "love of wisdom" rather than "love of knowledge." Wisdom is that part of knowledge which pertains to the living of a good life in this world. Please do not think, however, that I am claiming that mathematics is somehow inferior to philosophy. Mathematics is unsurpassed as a means of personal intellectual growth, and it continues to provide conceptual tools that enable scientists to break new ground. Moreover, mathematical concepts and reasoning have been very important and highly regarded in the history of philosophy. The most dramatic example of this comes from Plato. He thought it was so important that his new students in philosophy first have considerable experience with mathematical reasoning that he posted the following message over the entrance to his philosophical academy: "Let no one enter here who knows not mathematics." Furthermore, many great western philosophers have made significant contributions to mathematics as well as to philosophy, for example, René Descartes, Blaise Pascal, G. W. Leibniz, Gottlob Frege, A. N. Whitehead, and Bertrand Russell. Such contributions continued in the late twentieth century with mathematician/philosophers like Hilary Putnam and Saul Kripke.

Philosophy and History

Now let's think a bit about philosophy and the academic discipline called "history." The philosopher's ideal is to be able to prove what he or she claims to be true. More often than not, however, the philosopher winds up with a case like that of the historian – a case that makes a specific belief appear plausible, and perhaps even probable, but not certain. This is because both the philosopher and the historian ordinarily operate with incomplete data or evidence. As a consequence, historians create what they think is the most plausible explanation or interpretation of their incomplete data – staying open to the possibility that a different, even more plausible interpretation of the data might emerge, or that new data might be discovered that makes the old explanation become implausible, thereby calling for a new explanation that accommodates the new data better. Similarly, philosophers are trying to interpret data that is usually far from complete and which, at least in some cases, perhaps can never be complete. Hence, the philosopher, like the historian, is forced by the incompleteness of his or her data to be rationally creative, to think up an explanation of the data that is plausible, on the one hand, and that, on the other hand, seems at least as plausible as any other explanation that has been offered, and which, ideally, is more plausible or probable than any other explanation that is available. This kind of reasoning is sometimes called "reasoning to the best explanation."

The big contrast between history and philosophy is more obvious than with religion, science, and mathematics. The discipline of history limits itself to the discovery of truths about the past of the physical world of ordinary experience. Furthermore, history, like science, limits itself to empirical methods for evaluating whether an historical claim or an interpretation of history is true or false, probable or improbable, plausible or implausible. Its data are physical documents, artifacts, and testimony.

Philosophy, by contrast, leaves to the sciences (natural and social) problems that can be dealt with adequately by empirical methods. Also, unlike history it is primarily interested not in how things *were* but in how things *are* – but not in how things are with regard to the physical aspects of reality (that is the turf of the sciences), but, rather, in how things are with regard to those aspects of reality that cannot be investigated empirically.

Whether we *can* gain knowledge about anything that cannot be investigated empirically is an important question (and a philosophical one); *empiricists* say we cannot; *rationalists* say we can. Nonetheless, certain questions that cannot be investigated empirically – such as how to live a good life, what to think about death, and whether there is a God – are so important that surely we do not want to ignore them, and just as surely we do not want to simply think what someone else tells us to think about them. There are many people who would be glad to tell you what to think about these topics, but those people disagree with one another (sometimes wildly), so surely you want to think about these matters for yourself. To be sure, we have a philosophical obligation to seek out and listen to what other people have to say about philosophical issues, but having done that, I think we have a subsequent obligation to think about and decide on these issues for ourselves.

Reading Further

For introductory articles on philosophy of art, philosophy of law, philosophy of science, etc., see *The Blackwell Companion to Philosophy* or *The Cambridge Dictionary of Philosophy*.

Chapter 6: Methods for Doing Philosophy

- The Socratic Method
  Clarification
  Evaluation
- Running out the permutations
  Descartes
  Benefits of
- Rational Dialogue
  Assertion
  Justification
  Comprehension
     Descartes
  Evaluation

The goal of philosophy is knowledge of the truth about reality and value. The human situation motivates us toward this goal because it (the human situation) is shot through with ambiguities that invite, intrigue, confuse, challenge, and trouble us. As I will explain more fully later, in order to pursue the goal of philosophy effectively, we need to have an appropriate attitude and to be highly motivated. However, even high motivation and a right attitude are not enough to ensure progress toward the goal of philosophy. We need in addition to know *how* to focus and carry out our ambitious good intentions. Specifically, we need some effective methods for carrying on the business of philosophy. Consequently, let's look at three widely used methods of doing philosophy: the Socratic method, running out permutations, and rational dialogue.

The Socratic Method

The Socratic method is named in honor of its greatest practitioner: Socrates of ancient Athens. Socrates' most famous student, Plato, wrote many philosophical dialogues in which he depicts Socrates engaging in discussion according to the following pattern: he asks a series of related questions that build upon one another in order to help individuals *clarify and evaluate* their beliefs about a philosophical topic. The first objective of these questions is simply to help someone state clearly and to their own satisfaction *what* they believe about something, such as the nature of love or justice or knowledge.

I am sure that at times you, as I, have attempted to state one of your ideas to someone but have found yourself unable to do so. Sometimes in these situations we say something like, "I know what I mean, but I can't seem to put it into words." In such situations Socrates would try to help by stating his best guess at what the other person was trying to say. If the other person said, "No, that's not what I'm trying to say," then Socrates would make another guess or try to get the other person to try again. This process would continue until the other person was satisfied that his or her position had been stated satisfactorily.

We, too, can use this method of asking questions and proposing statements to help an individual clarify his or her thoughts about a topic. When the other person agrees that their idea on the topic under consideration has been stated adequately (perhaps with help from you), they may say something like, "Yes, that states exactly what I've been trying to say." At that point, and only then, we should move to the second phase of the Socratic method. In that phase we use a second series of related questions to help the other person *evaluate* the idea they have just clarified. Note that we, as philosophers, are only half through with our task when we have helped the other person clarify their belief. That is because we want to know not only *what* the other person thinks but also whether it is true.

It was because of this second phase of the Socratic method that Socrates got into more and more trouble with some of the citizens of Athens until finally he was jailed and executed. Those people were perfectly happy for Socrates to help them *articulate* their ideas, but they did not like it when he insisted (as he always did) that they should then proceed to *evaluate* their ideas. That was especially disturbing to them because it was often in public places that they made the statements which Socrates would then question. Frequently their ideas did not stand up well under Socrates' probing questions, so they felt embarrassed or tricked and got angry at Socrates – though he had done them the favor of helping them see that one of their beliefs had a problem of which they had not been aware before he questioned them about it. In Plato's dialogue *Theatetus*, Socrates tells young Theatetus that sometimes such people got so angry with him that they seemed positively ready to bite him! (For a brilliant, moving account of Socrates' defense of his role as "the gadfly of Athens," read Plato's short dialogue *Apology*.)

We, too, should be prepared for the possibility of hostile reactions from people when we encourage them to evaluate their beliefs and when we ask them questions for that purpose. This half of the Socratic method should be carried out

Methods for Doing Philosophy 45

with appropriate sensitivity to the other person, but surely it should not be avoided because of the risk that someone might take offense. We want to know whether the other person's idea is a good one or not; surely they, too, should want to know. The only way to know such a thing (apart from a direct experience of the matter) is to ask what reasons there are for thinking that their belief is true or false and to consider how strong those reasons are.

Before leaving the Socratic method, let me emphasize that we should be willing at appropriate times to be the person questioned, as well as at other times to be the questioner. Keep in mind also that you can be both the questioner and the questionee. With regard to any topic, you can ask yourself: What do I believe about that? Then you can ask yourself questions of clarification until you get your position stated fully to your own satisfaction. Next you can ask yourself questions of evaluation until you have as clear a sense as you think you can achieve at that time of your reasons for thinking that your belief is true, and of how conclusive or inconclusive, strong or weak you think those reasons are.

It can be very helpful to write out all of your thoughts as you apply the Socratic method to yourself. Indeed, this can be a very useful procedure for developing a paper on one of your beliefs: clarify what your belief about some topic is by explaining what you do not believe as well as what you do believe about that topic; think about ways in which what you say might be misunderstood by others or be unclear to them, and then clarify your meaning; state your reasons or evidence for your belief; state actual or possible criticisms of your belief; state alternatives to your position; state your criticisms of those criticisms and alternatives; finally, assess your reasons for your belief; that is, state whether you think your reasons are conclusive, strong, moderately strong, weak but better than the alternatives, etc., and why. This process involves risk. It may lead you to modify a belief that you have held, or to give it up altogether. It may also, however, encourage you to keep your belief and to hold it more strongly than ever. We cannot know in advance to what conclusion the process of rational evaluation will lead us, but there seems to be no other leader more worth following.

Running out the permutations

Another method of doing philosophy is what I call "running out the permutations" on a philosophical problem, that is, trying to identify all of the basic possible solutions to a problem. To understand what this means, think about the use of "permutations" in probability theory. When explaining permutations, math teachers often illustrate them by means of a pair of dice, each with six sides numbered one through six. When such dice are thrown on a smooth flat surface, there are many combinations that can turn up, but those combinations are strictly limited in number. You have to get a 1-1, a 1-2, a 1-3, a 1-4, a 1-5, a 1-6, a 2-2, a 2-3, and so on. Don't bet any money on a 2-7! That combination is not possible with such dice.

The value of running out the permutations before placing a bet is to ensure that the combination you bet on is a possible winner. That was what René Descartes was getting at when he prescribed the following rule of thought for himself: "Make enumerations so complete and reviews so general that I would be sure of having omitted nothing."[1] Similarly, when we are thinking about a problem in philosophy, it is helpful to try to figure out all of the basic possible solutions so as to help ensure that we are not overlooking a possible solution. After all, any possibility that we overlook, and therefore do not consider, may be the correct solution.

Recall young Erik. Erik's interpretation of the given (his experience of flying to where his grandparents lived) was that his airplane had gone to Oregon like a rocketship would go to the moon. He probably did not think of the possibility that his airplane went to Oregon like a bird flies from one yard to another. If Erik had thought of that possibility, presumably he would have thought that it made more sense than the other interpretation – especially when his mother told him that his grandparents were coming from Oregon in their car!

Clearly, then, it is important to resist the inclination to limit our attention to those answers which first pop into our heads or which are the most common answers in our family, religion, or society. Sometimes it is the uncommon answer, and sometimes even the counterintuitive or bizarre answer, that turns out to be true. Hence, to the extent that we ignore or are ignorant of possible solutions, to that extent we run a risk that the truth is not among the alternatives that we are considering. For example, in the western part of the world, when we think about what a human being is we tend to think in terms of two alternatives: a human is a combination of a body and a soul or a human is just a body. But obviously there is another possibility: that a human is just a soul, a spiritual thing, and that the body (rather than the soul) is an illusion. To most of us that seems like a crazy idea that is not worthy of equal consideration along with the other two possibilities. To many people in Asia and some in the west, however, it is an idea that does not seem crazy and even seems true. Hence, when thinking about what a human being is, we should consider all three of these possibilities together with the fourth possibility that a human is neither a body nor a soul nor a combination of these two, but something else altogether.

Consider another example: we tend to think either that death is the end of human existence or that there is everlasting life after death. There is, however, at least one more possibility: there is life after death, but it will not last forever, so that although we shall continue to live after death, we shall eventually perish forever. Surely that is a possibility that should be considered along with the usual two, plus any additional equally basic possibilities that there might be. (For a wide-ranging discussion of such possibilities, see Plato's *Phaedo*, which also contains an account of Socrates' last hours of life, which he spent in vigorous discussions of death with his friends and other visitors.)

To summarize, disciplining ourselves to run out the permutations on a problem (1) increases the probability that the truth is among the possibilities that we are considering and (2) helps make us more objective in evaluating the various permutations, as it becomes clear quickly that our preferred alternative is only one possibility among others, all of which deserve full and fair consideration. Moreover, we should keep it ever in mind that some of the most important and valuable ideas in history seemed obviously wrong and sometimes even crazy to most people when they were first proposed – ideas such as: the earth goes around the sun, not vice versa; the sun is much larger than the moon; physical objects are mostly empty space; the species of animals that currently exist evolved from predecessors which were quite unlike them; many illnesses are caused by invisible, impersonal entities (bacteria and viruses). To be sure, most ideas that seem obviously wrong or crazy are wrong or crazy. However, it has happened often enough in human history that an idea which initially seemed obviously wrong or crazy turned out later to be perfectly true and very important, so we cannot afford to overlook, ignore, or dismiss such ideas, and especially when they come from appropriately experienced or educated people.

Rational Dialogue

A third method for doing philosophy is rational dialogue. In philosophy, rational dialogue consists of dynamic interaction between two people who are cooperating to try to get at the truth about some topic. There is a definite structure to this process. The first stage of rational dialogue is *assertion*. If no one says anything, we are not likely to make progress toward the truth because if no one says anything, then there is nothing to discuss; there is just silence. Hence, someone must assert something – even if it is 'just for the sake of argument,' that is, just to have something to think about, to get started with, to evaluate, and perhaps to try to improve upon.

The second stage of rational dialogue consists of the person who has made an assertion *providing a justification* for it, that is, stating a reason or reasons for thinking that the assertion is true. This is terribly important. In philosophy we are very interested in *what* people believe, but we are equally interested in *why* they believe what they believe. That is because we want to consider whether to adopt the other person's belief for ourselves. If the other person gives us no reason for her belief, then we have no reason for adopting that belief (unless we can think of one on our own). To be sure, we do not want to commit the fallacy of *argumentum ad ignorantiam* (the argument from ignorance). That fallacy consists of thinking that if a person is ignorant of (cannot give) a good reason for thinking what she thinks, that therefore what she thinks must be false. It is possible for a person to have a true belief without being able to state a good reason for holding it. However, if we hold a different belief on that topic, or no belief at all, then the speaker has given us no reason to adopt her belief instead of ours. Hence, mere assertion does not take us to the goal of philosophy, which is to gain *knowledge* of the truth. We do not want to accept a position as true unless we have a good reason for thinking that it is true.

The preceding responsibilities (assertion and justification of a position) are the responsibilities of the *speaker* in a rational dialogue. The *listener* in a rational dialogue also has two responsibilities. The first responsibility of the listener is *comprehension*, that is, to understand correctly what the speaker has said. After all, if the listener does not understand correctly what the speaker said, then the listener has no right to agree with what the speaker said (because that would be agreeing not with what the speaker said, but, rather, with a misunderstanding of what the speaker said). Similarly, the listener would have no right to criticize what the speaker said (because that would not be criticizing what the speaker said; it would be criticizing a misunderstanding of what the speaker said).

For several reasons, such misunderstandings are much more common than we usually realize. First, each of us (including me) wants to think that he or she is a great understander of what people mean, and so we tend to assume that we have understood what the other person has said – even though we may not have. Second, when we think we have understood someone else correctly, it feels as though we have, and we tend to think that that feeling means that we *have* understood the other person correctly – even though that feeling might be mistaken. Third, the different backgrounds that we bring to philosophical discussions, and the abstractness of philosophical concepts, make it quite easy for discussants to think they have the same thing in mind when in fact they do not (when, for example, they talk about God or justice or love). Fourth, we like to think that when *we* speak, we speak clearly and other people understand us, so we tend to assume that when other people speak, they speak clearly and we understand them. Often, however, that is not so.

René Descartes, "the father of modern philosophy," often found himself the victim of such well-intentioned misunderstandings and consequently wrote the following: "Although I have often explained some of my opinions to very intelligent people who seemed to understand them very distinctly while I was speaking, nevertheless when they retold them I have noticed that they have almost always so changed them that I could no longer accept them as my own".[2] Many years of teaching have revealed to me case after case in which students thought they agreed (or disagreed) with one another when – after I encouraged further discussion between them – they discovered that they had misunderstood one another. Hence, the importance of the comprehension phase of rational dialogue cannot be overemphasized.

One of the most effective ways to help ensure that we really have understood what someone has said is to repeat back to them what we have understood them to mean, and to ask them if they agree with our summary. If they do not, then we should ask them to explain their position again, and (yes) then we should repeat what we now (after the second explanation) understand them to mean. The most common mistake here is to skip this step because we assume that we could not possibly misunderstand someone after they explain their position to us a second time, but it is very possible and not unusual, especially in philosophical discussions.

This means that great patience is called for on the part of both the speaker and the listener. The speaker should not be impatient with a listener who is trying to understand but does not do so the first or second time, and the listener should not be impatient with a speaker who has to explain his or her meaning several times before being understood. The speaker should be grateful to have such a serious listener, and the listener should be grateful to have such an accommodating speaker. Tedious as it may seem, it is worth the time it takes to reach the point where we can state the speaker's position to his or her satisfaction. Only then can we rightly proceed to the final stage of rational dialogue.

The fourth and final stage of a rational dialogue is *evaluation*; it is similar to the second half of the Socratic method. (In the Socratic Method, Socrates' aim was to get the other person to evaluate their own position. In rational dialogue the listener's primary objective is to evaluate the other person's position for him or herself.) In our philosophizing we are interested in what other people think, and why they think what they think, but because our ultimate goal is to gain knowledge of the truth, we cannot just stop with finding out what other people think and why they think it. Rather, once we are clear on their what and their why, we need to evaluate or critique as vigorously as we can what has been claimed and what the reasons are that have been put forward to support it. The point of such evaluation is to enable us to decide whether or not to accept the other person's assertion and justification as our own. (In chapter 12, on logic, you will receive more specific guidance about how to evaluate a claim and the reasons given for it.)

Note that if we do not evaluate the other person's claim and reasons, then we have no reason to adopt their position (unless we know or can think of on our own some adequate reason to hold that position). Surely we should not accept someone else's philosophical position simply because they believe it and have reasons for it. Their reasons may be weak or irrelevant, and their belief may be false.

If after evaluation you are persuaded by the other person's position and the reasons given or thought of for it, then that particular dialogue is over. You have been rationally persuaded into agreement with the speaker. However, if you do not agree with the other person's position, then guess what? You and the other person should change positions – you should become the speaker in the dialogue and the other person should become the listener. You, as the new speaker, should then assert that you do not accept the other person's position and give your reasons for not accepting it. The other person (the new listener, the old speaker) should first endeavor to understand you correctly, and, second, should evaluate your reasons for disagreeing. If the listener is persuaded by your reasons, then the two of you have reached agreement, and the dialogue on that point is over. If the listener is not persuaded by your reasons, then the two of you should change places again – the other person becoming the speaker again, and you becoming the listener again. The dialogue should seesaw back and forth like this until, ideally, you reach agreement with one another. Sometimes, of course, you will have to suspend such a discussion in order to attend to other things (like going to work or classes!). Othertimes, even if you do not have to stop, if you find that the discussion has reached a standstill because neither of you has anything new to say, yet neither of you has been persuaded by the other, then it is probably time to put the matter aside until one of you discovers or thinks of something new on the topic.

To summarize, there are four phases to a rational dialogue: two phases (assertion and justification) belong to the speaker; two phases (comprehension and evaluation) belong to the listener.

Without any one of these four phases, we do not have philosophy. Without an assertion by someone, there is just silence. If the silence is broken by an assertion but no justification of it is given, then there is no reason for us to think that the assertion is true. If we are given both an assertion and a justification, but we have misunderstood the other person's assertion or justification, then we do not have a right to proceed to agreement or disagreement. If we have understood correctly but have not evaluated the assertion and the reasons for it, then we are not justified in accepting it as true or rejecting it as false (assuming that we do not have any additional reasons of our own for thinking the position true or false, and assuming that the other person is not an expert while we are novices regarding

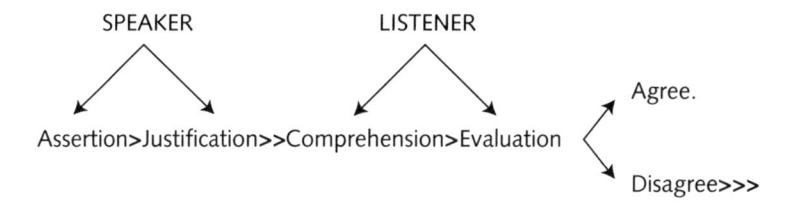

```
        SPEAKER            LISTENER
         ╱╲                 ╱╲                  ↗ Agree.
        ╱  ╲               ╱  ╲               ⟨
Assertion>Justification>>Comprehension>Evaluation
                                                 ↘ Disagree>>>
```

Figure 6.1 The dynamics of rational dialogue

the matter under discussion; if someone is an expert with regard to the matter under discussion, that is a good reason for accepting their position rather than that of a novice).

A rational dialogue is usually carried on between two people, but keep in mind that a rational dialogue can also be carried on among three or more people, or by one person with him or herself. As to the latter, with regard to any topic of interest, you can play the speaker to yourself by asking what you think about that topic and what your reasons are for thinking that way. Then you can play the listener by asking whether what you have said is clear to you and is likely to be clear to others. If it is not likely to be clear to others, you should work at reformulating your position and your justification for it until you are satisfied with its clarity. In the stage of evaluation, you can play devil's advocate to yourself by rejecting your position and then becoming the new speaker, stating the strongest case you can think of against the position that you originally stated as speaker. Then, as the new listener, you should see how effectively you can answer the criticisms and questions that, as devil's advocate, you have posed to yourself. Parenthetically, this is a very effective way to develop a paper or presentation on a topic. Developing the mental agility to carry on this kind of rational dialogue with yourself will be very valuable to you throughout your life. However, you should not limit yourself to solitary dialogue about your ideas, as I will explain at the end of chapter 9.

Notes

1 René Descartes, *Discourse on Method*, tr. Donald A. Cress (Indianapolis, IN: Hackett Publishing, 1981), p. 10; or any edition, Part Two.
2 René Descartes, *Discourse on Method and Meditations*, tr. Laurence Lafleur (NY: Bobbs-Merrill Co., 1960), pp. 50–1.

Reading Further

Plato, *Euthyphro*, 12d–13e, is a splendid example of the socratic method at work. *Meno* illustrates the power of the socratic method and rational dialogue. *Theatetus*, 148e–151d; Socrates portrays himself as a philosophical midwife. *Republic*, 338a–339e; an excellent example of the four stages of rational dialogue.

René Descartes, *Discourse on Method*; *Rules for the Direction of the Mind*.

Chapter 7: Things Philosophers Do

- Exposit
    Becoming aware of personal convictions
    Expressing beliefs
        Socrates as Mid-wife
    Philosophy of "X": nature of, presuppositions of, objectives of, limits of "X"
- Analyze
    Meanings of concepts
    Relations of concepts to one another
- Synthesize
    The enormity and growth of empirical data
- Describe
    The subjective realm
    The phenomenological method
- Speculate
    Is there more?
    The whole and its parts
- Prescribe
    Words
    Arguments
    Behavior
- Criticize
    Popper
    Poking holes
    Proposing alternatives
    Truth or comfort?

In the last chapter we examined methods that philosophers use to do the kinds of things they do. But what kinds of things do philosophers do? We will examine seven such things that should give you a good, but not exhaustive, sense of the nature and range of philosophical activities. You will be more attracted to some of these activities than others, as are professional philosophers, and you may conclude, as have some professional philosophers, that not all of these activities are appropriate for philosophy. I shall treat them as though they are all legitimate; certainly each has played an important part in the historical development of philosophy.

Exposit

Philosophers exposit. Expository philosophy endeavors to lay bare what is covered up, to make conscious the unconscious, to make explicit the implicit. It does this both for individual people and for fields of study. Regarding individual people, philosophers endeavor to help them make conscious their unconscious assumptions about the nature of reality, value, and knowledge. Karl Popper, an important twentieth-century philosopher, speaks in the following statements to the need for expository philosophy: "All men and all women are philosophers, though some are more so than others."[1] Moreover, "If they are not conscious of having philosophical problems, they have, at any rate, philosophical prejudices. Most of these are theories which they unconsciously take for granted, or which they have absorbed from their intellectual environment or from tradition."[2] Popper adds that "Since few of these theories are consciously held, they are prejudices in the sense that they are held without critical examination, even though they may be of great importance for the practical actions of people, and for their whole life."[3]

For us the important point in Popper's statements is that most of us are not fully aware of the assumptions about reality, value, and knowledge which are operative in our lives and guide our actions – and some of those assumptions we are not aware of at all. Expository philosophy is an effort to become clearly aware of the vague and unconscious assumptions that guide our lives. Until we become aware of them, we cannot assume responsibility for our lives, because until we know what our beliefs are, we can neither evaluate them, nor affirm them, nor reject them, nor modify them. We can only be driven blindly by them.

Socrates, the greatest of expository philosophers, described his role in Athenian society as that of a mid-wife of ideas.[4] Out of a conviction that "the unexamined life is not worth living," he devoted himself to helping his fellow Athenians give birth to the ideas growing within them, helping them find the words by which to transform vague notions into clear thoughts. No doubt we have all had the experience of not being able to find the right words (or any words at all!) to say what we want to say, so most of us can appreciate the value of someone who plays philosophical midwife when we are at a loss for words.

As well as helping the individual clarify his or her own ideas to him or herself, expository philosophy facilitates communication between people, since until we are able to articulate our beliefs clearly, we are unable to communicate them clearly. Along this line, a biochemist once told me of a party at which he and some fellow scientists became embroiled in a debate that began to bog down. Fortunately, he said, a member of the group had studied a bit of philosophy and helped the discussion keep moving forward by clarifying the points and presuppositions of the various disputants whenever they seemed to be speaking unclearly or misunderstanding one another.

In brief, in its expository role philosophy serves individuals (1) by helping them clarify to themselves the presuppositions on which they have been operating vaguely or unconsciously, and (2) by helping them clarify their points and assumptions to other people.

Philosophy also plays an expository role in relation to other fields of study. Philosophy of law, philosophy of religion, philosophy of science, and philosophy of art, have already been mentioned; the list could go on: philosophy of mathematics, philosophy of language, philosophy of history, philosophy of social science, philosophy of psychology, etc. Indeed, there can be and ought to be a "philosophy of" every major field of inquiry because such a philosophy is an endeavor to clarify and examine the basic concepts and presuppositions of that field, its potential contributions to human knowledge, the limitations of what it can contribute, and its theoretical relations to other fields of study.

As Adam Schaff, a contemporary Marxist philosopher, pointed out, the concepts of truth and falsehood, cause and effect, necessity and chance, are primary concerns of philosophy. Consequently, "since no branch of science (or everyday thinking either) can do without such concepts as truth and falsehood, cause and effect, necessity and chance, *none* can do without its specific philosophy."[5] Schaff goes on to say that it is unimportant whether philosophical reflections about mathematics, law, psychology, and so on, come "from professional philosophers or from philosophizing representatives of the specialized discipline in question" – as long as the non-professional philosopher is philosophically competent and the professional philosopher is knowledgeable in the relevant field outside philosophy.

"Philosophy of X" is important because intellectual progress in various fields has more than once had to wait upon the resolution of a conceptual confusion or the exposure of a seemingly necessary assumption as unnecessary after all. Philosophizing – whether by professional philosophers or philosophically competent members of the relevant fields – has contributed to such progress and continues to contribute by its expository work in these fields.

Analyze

Philosophers analyze. Whereas expository philosophy helps individuals and disciplines lay bare the concepts and assumptions by which they operate, the focus of analytic philosophy is on the concepts themselves. When we do analytic philosophy we lift out and investigate important concepts which we use everyday yet know so little about. Some of these concepts are: appearance and reality, truth and falsity, cause and effect, right and wrong, good and evil, space and time, mind and body. We have each used all of these concepts. Yet what do they mean? That's a very difficult question to answer – though until we try it, it seems like it should be easy.

St. Augustine spoke for most of us when he said of the concept of time, "What, then, is time? If no one asks me, I know; if I want to explain it to someone who does ask me, I do not know."[6] We can, of course, look up the word "time" in a dictionary, and that is a good thing to do, but it is only a first step on a long journey. A sophisticated understanding of the concept of time requires far more than that – a book at least! And it is philosophers – though not only philosophers – who write articles, essays, and books on time and many other concepts. In fact, if you want to find an article or book on the concept of love, death, time, meaning, justice, faith, probability, science, God, or just about anything else, you need but browse in the philosophy section of a library.

Philosophical analysis, as one philosophical activity among others, performs the "microscopic" function within philosophy. It involves carefully lifting out a specific concept and putting it under the magnifying lens of rational analysis. This enables us to see more clearly the internal complexities of the concept (perhaps revealing facts which weren't evident at all before), and it enables us to discern more accurately the logical relations of that concept to related concepts. In recent years, for example, philosophers have given a good deal of attention to the individual concepts of knowledge, belief, faith, and hope, and to how they relate to one another. We will analyze these concepts in the chapter on epistemology.

Synthesize

Philosophers *synthesize*. What do they synthesize? The findings and theories of the natural and social sciences. Why do they do this? In order to construct a picture of the physical universe which keeps current with the latest positions of the sciences and which shows how the sciences relate to one another. Scientists have produced a vast amount of literature on their discoveries, and they continue to generate new information at an overwhelming rate. Yet the individual scientist is normally too busy doing research on specific problems in his or her own field to have time to keep abreast of what is going on in other fields, much less to relate the developments in all scientific fields to one another. Moreover, a research scientist may not be interested in or have the gifts for this kind of wide ranging, synthesizing task. The research scientist is interested in isolating problems which can be resolved clearly. His or her gift is usually that of intense concentration on a single point.

Some scientists, to be sure, also have the gift of global synthesis; some of them, such as Newton, Darwin, and Einstein, have made profound philosophical contributions. But scientists need not be interested in or gifted for a broad survey and synthesis of the findings of the sciences in order to do successful and important research in their areas of specialization. Nor do philosophers interested in the integration of scientific findings have to be interested in or gifted for laboratory research. Normally they work from the research of others; if they did not, they would hardly have time for the task of synthesis. They have no laboratory, no sophisticated machines, or electronic devices. Their special tools are those of language, logic, and discernment. Given these tools, if they keep current with the most interesting and important discoveries and hypotheses in the sciences, then they will be fitted to reflect on the relations of the sciences to one another and on the ways in which the claims and hypotheses of the sciences might be woven into a consistent and plausible view of the physical universe.

It should be noted that reflection on scientific findings and integration of them with one another is not a recent interest of philosophers. From Aristotle to Kant, philosophers *made* important scientific contributions, as well as studied them. Indeed, as noted earlier, until about AD 1800, what we call "science" was part of philosophy; it was called "natural philosophy". However, the explosion of scientific research from about 1800 forward made it impossible for most individuals to do natural philosophy (science) in addition to doing axiology, ethics, logic, epistemology, and metaphysics, so a division of labor developed. Those people who do natural philosophy are now called "scientists". Because non-scientist philosophers want to develop the most accurate and plausible understanding of the universe that they can, they remain very interested in the findings of their natural philosopher colleagues.

Describe

Philosophers *describe*. That might seem redundant, since description has come into play in each of the preceding sections, but there is a difference. When one is doing expository philosophy, one is describing what someone believes; when one is doing analytic philosophy, one is describing a concept. In descriptive philosophy, one is attempting to describe neither a concept nor what someone believes, but the world itself. Yet isn't that what synthetic philosophy does – describe the world? Yes, but with an important difference. Synthetic philosophy limits itself to the findings of the sciences; descriptive philosophy does not. To be sure, descriptive philosophy (as I am defining it here) limits itself to ordinary space and time, but ordinary space and time may contain things that are not entirely subject to scientific analysis – things such as intentions and emotions, and mental states and processes such as feeling, hoping, thinking, intending, and dreaming.

Note, I said that ordinary space and time *may* contain things that are not entirely subject to scientific analysis (and perhaps some that are not subject to empirical study at all). Whether such things do exist is a controversial matter. If everything in ordinary space and time can be thoroughly investigated by scientific analysis, then there is no need for descriptive philosophy in addition to synthetic philosophy. However, if everything in ordinary space and time cannot be analyzed and understood satisfactorily by means of scientific methods and categories, then something more is needed, something that goes beyond the methods and findings of the sciences yet limits itself to ordinary space and time as they are revealed through sensation and introspection. That "something more" is what I am calling "descriptive philosophy".

Within the vast realm of ordinary space and time, descriptive philosophy helps us see more clearly and understand more adequately the world as it is presented to us by our senses and introspection – the world that has been there all along. It's a bit like having a local historian conduct us through our own home town in such a way that our awareness of the "old" things is enhanced, and we become aware for the first time of things we must have seen a hundred times before but never noticed. This phenomenon was captured well by the poet and playwright T. S. Eliot when he said: "We shall not cease from exploration / And the end of all our exploring / Will be to arrive where we started / And know the place for the first time."[7]

More abstractly, philosophy in its descriptive role endeavors to discern the most fundamental categories of reality as they are present in ordinary human experience. Consequently, it works at boiling down to a minimal but comprehensive set of categories the bewildering diversity of experience and language.[8] To put the point another way, philosophers in their descriptive role try to put together a "relief map" of reality as it presents itself in human experience – a map which accounts for everything in experience by means of a minimal but adequate set of categories.

The primary tool of descriptive philosophy is *the phenomenological method*.

This method, pioneered by Edmund Husserl in the nineteenth century, requires that we try to apprehend things by allowing them to reveal themselves to us as they are. This means that we should refrain from interpreting them – lest by interpretation we distort our understanding of them. That may sound easy, but it is extremely difficult because we are predisposed from infancy to interpret things according to the beliefs and values formed in us by our parents, peers, religion, and culture. We can come to see things "as they are in themselves," that is, as they are apart from interpretation, only by learning to "bracket out" our prejudices – that is, our pre-judgments. For example, a child reared in a racist family or society can only with difficulty come to see a person of another race the way he or she really is, without automatically imputing negative characteristics to that person. (Archie Bunker, a television comedy character, was a wonderful example of someone desperately in need of a strong dose of phenomenology. Though the women, blacks, and Polish people in Archie's life were intelligent and virtuous, he couldn't see through his stereotypes and prejudices to their intelligence and virtue. The contrast between the way those people were (as the viewer saw them) and the way Archie saw them made him a laughable buffoon.)

Most of us learn somewhere along the line, usually by hard experience, to make a distinction between the way things are apart from interpretation and the way they seem as a result of interpretation. Phenomenology is a highly self-conscious, disciplined attempt to perfect methods and skills that enable us to apply that distinction to anything and everything. By means of it we work toward the ideal of an awareness of the world that is undistorted by presuppositions.

What I have said about phenomenology does not imply that it denies the existence of things that are not given in direct experience, such as perhaps God, matter, the soul, and the minds of other people. The point of phenomenology is not to deny or affirm the existence of anything; its point is to provide an accurate description of what is *given* in experience. Once we know what is given and what is not, then we can decide whether to go beyond the given by affirming that which is not immediately given, such as God, matter, the soul, and other minds. These are issues that we will explore much later, in the section on metaphysics. (For a brilliant and haunting story about the difference between the given and interpretation of the given see Plato's "Allegory of the Cave" in his dialogue *Republic* at the opening of Book VII.)

Parenthetically, existentialism, which was mentioned as part of philosophical anthropology, utilizes the phenomenological method in order to search for the essential and universal features of *human* existence. Descriptive philosophy has the broader objective of providing a set of categories in terms of which *everything* in the realm of ordinary experience can be accounted for.[9]

Speculate

Philosophers speculate. The inherent limitations of synthetic philosophy and descriptive philosophy point beyond themselves to speculative philosophy. Speculative philosophy "lets out all the stops." It stops at neither the limits of scientific investigation nor those of ordinary experience. To do speculative philosophy is to ask whether there is anything *more* to reality than that which is revealed by science and ordinary experience. The objective of speculative philosophy is to comprehend the *whole* of reality by means of *reason*, so if reason requires that speculative philosophers go beyond the limits of science and ordinary experience in order to account for the whole of reality, then they do so.

The reverse of this attitude was exhibited by an ancient Roman named Cotta. He stated, "I have always defended and will always defend the traditional ceremonies of religion, . . . if you, as a philosopher can justify my belief on rational grounds, good but I am bound to believe our ancestors, even though they give no reason."[10] Cotta's attitude toward reason is anathema to the philosopher. To be sure, philosophers generally agree that not every true proposition can be justified by reason, but they also agree that what *can* be comprehended and justified by reason *should* be.

On philosophical topics philosophers are not willing to accept on bare authority what they can comprehend for themselves. This is not to say that they never accept a philosophical claim on the basis of human or divine authority. It is to say, first, that they want to be convinced that a purported authority is worthy of trust before trusting him or her, and, second, that when it is possible for them to know by rational insight the same thing that they are asked to accept on authority, they endeavor to do so. As was pointed out earlier, philosophers want, whenever possible, to *know* what is true – both for the sake of the satisfaction of knowing, and in order to escape the doubt that can haunt us when we accept things on mere authority.

Reason, then, is the primary tool of speculative philosophy. By means of it we try (1) to determine whether there is or may be more to reality than is given in ordinary experience, and (2) to comprehend the whole of reality – analyzing it into its parts and then comprehending those parts in their dynamic relations to one another, even as the student of a combustion engine must know not only the assembled whole but also the individual parts, and not only the individual parts but how they function in relation to one another in order to constitute a running engine.

In fact, we might say that "the whole and its parts" is the motto of the speculative philosopher. Descriptive philosophy, by contrast, limits itself to that which is given in direct experience. Consequently, the question as to whether there is anything in existence which is not present through ordinary experience does not arise in descriptive philosophy. Indeed, when one is doing descriptive philosophy, one self-consciously stays away from the "Is there anything more?" question in order to focus entirely on what is directly given.

Still, the question as to whether there is anything more to reality than is revealed in ordinary experience or is theorized by science is an intriguing and important question. Consequently, you can be sure there is a branch of philosophy which gives attention to it, namely, speculative philosophy. The speculative philosopher, in his or her desire to account for the *whole* of reality, investigates the extent to which reason does or does not require that we postulate the existence of things such as God, matter, other minds, souls, and dimensions of reality other than those revealed by sensation and introspection. For example, William James, a great American philosopher and psychologist, concluded that

> normal waking consciousness, rational consciousness as we call it, is but one special type of consciousness, whilst all about it, parted from it by the filmiest of screens, there lie potential forms of consciousness entirely different. We may go through life without suspecting their existence; but apply the requisite stimulus and at a touch they are there in all their completeness . . . No account of the universe in its totality can be final which leaves these other forms of consciousness quite disregarded. How to regard them is the question – for they are so discontinuous with ordinary consciousness . . . At any rate, they forbid a premature closing of our accounts with reality.[11]

James' statement is commendable both for its insistence that we acknowledge radically different forms of consciousness and for its cautiousness about the metaphysical conclusions we should draw from them.

Speaking metaphorically, whereas descriptive philosophy looks at the surface of things, speculative philosophy asks whether there is anything beyond the surface except more of the same kind of thing that is seen on the surface, and if so, what. Whereas description and analysis perform *microscopic* functions in philosophy – examining experiences and concepts in great detail, synthesis and speculation perform *telescopic* functions – helping us understand reality more comprehensively and connectedly than is ordinary, and helping us understand the possibility or impossibility, the plausibility or implausibility, the necessity or lack of necessity, of realities which are not directly perceived. Some of the great works in speculative philosophy are by Plato, Aristotle, Lucretius, Plotinus, Augustine, Spinoza, Berkeley, Hegel, Marx, and Whitehead.

Prescribe

Philosophers prescribe. Philosophy is prescriptive as well as descriptive. It goes beyond saying how things *are* to saying how they *ought* to be, and it does this with regard to several things. First, it proposes standards for the use of language *so that meaning might be clear*. Earlier we saw the importance of clarity of meaning for self-understanding and communication. Consequently, it is only natural that when philosophers encounter – often as a result of their expository or analytic work – confusions or mistakes in the use of a word, they recommend how the word ought to be used in order to avoid those problems.

Here are two examples of unfortunate uses of words. A newspaper article reported that a citizen in an inner-city neighborhood told two policemen that a man was trying to rape a woman in a nearby alley. The policemen hurried to the alley, spotted the attempted rape, and shouted a warning at the man. The man shoved the woman away and fired three shots at the policemen. The policemen fired back and killed him. A police official said he was greatly relieved when the incident did not lead to a violent mob scene. He told a reporter, "I've worked this area for 19 years. With a police-involved shooting, a lot of people head to the scene, and they can get enraged and try to start trouble by spreading facts that aren't true." The problem here, and I've heard it elsewhere as well, is the notion that facts can be false. The word "fact," according to the Oxford English Dictionary, means "truth; reality; a thing known for certain to have occurred or be true."

If we use the word "fact" to mean things that are false, as well as things that are true, then we deprive ourselves of a simple way of saying that something is known for certain to be true. If we say that facts can be false, we delude ourselves into thinking or feeling that nothing is true or is a demonstrated truth, and surely that is false. To be sure, we can *think* that something is a fact when it is not, but that just means we can be mistaken about what we think is a fact. That does not mean that there are no facts (truths or demonstrated truths). Surely it is a fact that I exist whenever I think I exist (how could I think I exist if I did not exist?), and surely it is a fact that the area of a circle is greater than the area of any triangle constructed within it.

A second example comes from a friend. My friend worked for an institution which got into financial difficulty, so its president decided to downsize the staff, that is, to cut expenses by eliminating employees. When people are "downsized" they may be excellent employees; indeed, they often are. However, their jobs are considered to be eliminable (though often the various parts of their jobs are parceled out to those workers who remain). After it was announced that my friend's job was being eliminated due to downsizing, someone came up to him and said, "I'm so sorry to hear that you've been fired!" To be fired means to have one's job terminated because one's boss is unhappy with one's performance or behavior. Consequently, for two reasons my friend was stunned by his friend's statement. First, he was not happy about losing his job, but he had been at peace with it because he had been assured by his boss that his performance had been excellent and his elimination was purely a financial decision. Now, however, he had to wonder, "Does this person know something I don't? Was I really let go because of unsatisfactory performance?" Hence, he began to doubt himself when he needn't have. Second, he was concerned that if people talked about his elimination in terms of being "fired," then even though his performance had been excellent, word might spread to potential employers that he had been fired, even though he had not been, so they might be reluctant to hire him lest the rumors of his having been fired be true. In light of a subsequent newspaper article, his concern was well-taken. That article consisted of professional advice about how to interview for a new job after being fired from an old one. The article said: "Never, never tell anyone you were fired!"

For the kinds of reasons illustrated above, the following observation by Socrates is as relevant now as it was in ancient Athens: "Misstatements are not merely jarring in their immediate context; they also have a bad effect upon the soul." Or, in a different translation: "To express oneself badly is not only faulty as far as the language goes, but does some harm to the soul."[12] Hence, prescriptive philosophy with regard to language remains important.

Second, philosophers propose standards of logic *so that reasoning might be valid*. Knowledge of such standards enables us to construct valid arguments, to recognize invalid ones, and to understand why they are invalid. Fallacious arguments mislead us only when we do not spot them. Valid arguments can serve us only if we can recognize them and construct them.

Third, philosophers propose standards for scientific research *so that empirical truth might be secured*. As you may have inferred from earlier chapters, contemporary science is built on centuries of conceptual clarification by thinkers going back to ancient Greece. Recall, for example, the monism of Thales, the pluralism of Empedocles, and the atomism of Democritus. Clearly, much progress has been made with regard to what we ought and ought not to mean when we speak of empirical knowledge, and with regard to methods for securing it. Much remains to be done, however, and philosophers are among those who are examining and refining the concepts and methods that are fundamental to the sciences.

Fourth, philosophers propose standards for human conduct *so that action might be moral*. To be sure, philosophers are not unique in making moral pronouncements. Anyone can make them, and most people do. Nor do philosophers have a special pipeline to the truth about morality (or anything else). As a group, professional philosophers are distinguished only by the enormous amount of time and the rigor of attention that they give to morality and other topics. Their reasoning and conclusions merit our attention not out of courtesy but out of probability. Why? Because the more extensively the members of an informed, intelligent group of people have studied a topic – whether it be morality, science, logic, or anything else – the more likely it is that they have something valuable to say about it. It is of no importance that the people within the human community who give extraordinary attention to the topic of morality are called "philosophers". It is of great importance that there be within the human community a group which gives intense attention to this very important topic.

Criticize

Philosophers criticize. Recall Karl Popper's statement that all men and women have "philosophical prejudices," that is, have beliefs about reality and morality that have been absorbed uncritically from their social environments and that have a profound influence on how they live their lives. This fact itself, according to Popper, is sufficient justification for the existence of philosophy. Using the word "apology" in the sense of "a good reason", and referring to the universality of philosophical prejudices, he wrote:

> It is an apology for the existence of professional philosophy that [people] are needed who *examine critically* these widespread and influential theories.
> This is the insecure starting point of all science and of all philosophy. All philosophy must start from the dubious and often pernicious views of uncritical common sense. Its aim is enlightened, critical, common sense: a view nearer to the truth, and with a less pernicious influence on human life.[13]

Popper, not an elitist, went on to insist that although professional philosophers are needed to evaluate the views of uncritical common sense, their views, too, should be evaluated. No one and no thing is exempt from philosophical evaluation.

Philosophy carries out its function of criticism in two basic ways: by *poking holes* and *proposing alternatives*. Regarding the first way, philosophy pokes holes in propositions and arguments. Less figuratively, it examines propositions to see if they are meaningless, vague, ambiguous, or false; it examines arguments to see whether they are valid or invalid, sound or unsound. More generally, philosophers test assertions and arguments to see how strong they really are. Appearances in this regard are often misleading – as most of us have learned through hard experience with advertisements, sales pitches, and promises.

Indeed, it seems a fair reading of history to say that philosophy was born as the polar opposite of *sophistry*, the art of appearance. The sophists of ancient Greece made money by teaching success-oriented young men how to use language in the marketplace, in the court of law, and in the political forum to achieve what they wanted. The object of the sophists' teaching was not truth, but the appearance of truth. By contrast, the object of philosophy is truth. In a conversation with a sophist, Socrates expressed the attitude of philosophy in these words about himself: "And what kind of man am I? One of those who would gladly be refuted if anything I say is not true, and would gladly refute another who says what is not true, but would be no less happy to be refuted myself than to refute, for I consider that a greater benefit."[14]

A second way in which philosophers criticize theories is by proposing alternatives to them. This function of philosophy is important because of the limitations of our individual experience and imagination. Sometimes we ascribe greater strength to a theory than it deserves because we know of no alternative. We reason to ourselves, "I can think of no alternative, so I suppose it is true" – only to have someone later present an alternative. Whether such a "present" is embarrassing or not, it should teach us the importance of considering whether we have really examined all the possibilities (run out all the permutations) and weighed them fairly.

As we saw in the last chapter, one of the jobs of philosophers is to work at clarifying all the basic types of solution to each philosophical problem. For example, we must consider not only whether a human being is a body only or a combination of a body and a soul, but also whether a human being is a soul only or something other than either a body or a soul. By figuring out and setting forth these basic alternatives, philosophy helps ensure that we do not commit ourselves prematurely to a position, and that we do have the correct answer among the positions that we are considering.

Why do philosophers poke holes in propositions and propose alternatives to whatever is claimed? Not because they are quarrelsome people, but in order to advance our pursuit of truth.[15] Such things as these must be done if we are to determine the true strength of a claim or an argument. It's a bit like testing a girder before using it in a bridge that we must travel over, or like comparing girders to one another in order to choose the stronger one. More abstractly, a proposed solution to a problem must be critiqued in order that we might determine (1) whether it is an *adequate* solution to the problem, (2) whether it is the *only* adequate solution or one of two or more adequate solutions, and (3) if the latter, whether it is *superior, inferior, or equal* to the other solutions. Clearly, our problem solving efforts will become more effective and satisfactory when asking these questions becomes second-nature to us.

As I mentioned earlier, philosophy omits nothing from critical analysis. Consequently, philosophy can be anxiety inducing as we see our own most precious beliefs submitted to evaluation. Often we don't realize how emotionally attached to a belief we are until it comes under attack. That makes for one of the interesting by-products of doing philosophy: we discover unexpected emotional reactions to certain positions, such as theism or atheism, when they are attacked or defended. Indeed, two of the most certain things I can say are (1) that your study of philosophy will introduce you to ideas that will strike you as among the strangest or most obviously wrong things you have ever heard, and (2) that what *you* think is strange or obviously wrong, some other people – sane, intelligent, and educated – will see as familiar and obviously true!

Parenthetically, I once visited a psychiatric institution and was told not to discuss three topics with the patients: religion, sex, and politics. Philosophy wades into all three topics in a "no holds barred" manner. Indeed, we might define philosophy in its critical role as *an unflinching analysis of all beliefs, including the ones that are most precious to us.* For this reason the nineteenth-century German philosopher Friedrich Nietzsche pointed out that pursuit of truth is not always compatible with kindness to self. "The seeker of knowledge," he wrote,

> operates as an artist and glorifier of cruelty, in that he compels his spirit to perceive *against* its own inclination, and often enough against the wishes of his heart: – he forces it to say Nay, where he would like to affirm, love, and adore; indeed, every instance of taking a thing profoundly and fundamentally, is a violation, an intentional injuring of the fundamental will of the spirit, which instinctively aims at appearance and superficiality, – even in every desire for knowledge there is a drop of cruelty.[16]

Nietzsche's language may be too strong, but his basic point is correct. We must choose between the ideal of truth and the ideal of comfort. We cannot always have both.

Notes

1 Karl Popper, "How I See Philosophy," in *The Owl of Minerva*, ed. Charles J. Bontempo and S. Jack Odell (NY: McGraw-Hill, 1975), p. 42.
2 Ibid., p. 48.
3 Ibid.
4 See Plato's dialogue *Theatetus*, 148c–51d, any edition.
5 Adam Schaff, "What Philosophers Do," *The Owl of Minerva*, p. 181.

6   St. Augustine, *The Confessions of St. Augustine*, tr. John K. Ryan (Garden City, New York: Image Books, 1960), Book 11, Chapter 14, p. 287.
7   See the last stanza of Eliot's poem "Four Quartets."
8   Language is a depository in which we create and save labels for the distinctions that we experience: red, blue, space, time, tortoise, turtle, mental, physical, star, planet, moon, etc.
9   Some people think that by means of the phenomenological method we get at the way things truly are. The objective of the phenomenological method, for these people, is to allow things to reveal themselves to us as they really are, without interference or distortion from us.

   Other people think it is impossible to experience anything without interpreting it. They think that an experience is always the product of an interaction between the object that is experienced and the dispositions that the individual brings to that experience. Just as we cannot see anything except through our eyes (which cause us to see an object in a human way, rather than in the way that a fly or other organism sees it), they say we cannot experience anything except through the predispositions (physical and psychological) that we bring to an experience. See, for example, John McDowell's *Mind and World* (Cambridge, MA: Harvard University Press, 1994), esp. pp. 3–23.
10 James Thrower, *A Short History of Western Atheism* (London: Pemberton Brooks, 1971), pp. 48–9.
11 William James, *The Varieties of Religious Experience* (NY: The Modern Library, 1902), pp. 378–9.
12 The translations are, respectively, by Hugh Tredennick, *The Collected Dialogues of Plato*, ed. Edith Hamilton and Huntington Cairns (NY: Pantheon Books, 1961), pp. 95–6, and G. M. A. Grube, *Five Dialogues* (Indianapolis, IN: Hackett, 1981), p. 153.
13 Karl Popper, *The Owl of Minerva*, p. 48.
14 Plato, *Gorgias* (458a), in *The Collected Dialogues of Plato*, ed. Hamilton and Cairns, p. 241.
15 St. Paul speaks in the New Testament of people who have "a morbid craving for controversy and for disputes about words" (I Timothy 6:4). That description might fit some philosophers – flawed creatures that we are – but obviously it would not be part of a healthy philosophical attitude, the nature of which we will explore in the next chapter.
16 Friedrich Nietzsche, *Beyond Good and Evil*, sec. 230, tr. Helen Zimmern, in *The Philosophy of Nietzsche* (NY: The Modern Library, 1954), p. 536.

Reading Further

Expository philosophy: E. A. Burtt, *In Search of Philosophic Understanding*.
Analytic philosophy: Alan R. White, "Conceptual Analysis," *The Owl of Minerva*, pp. 103–18, or John Hospers, *An Introduction to Philosophical Analysis*, 4th edition or later.

Synthetic philosophy: see the *Foundations of the Unity of Science*, a series published by The University of Chicago Press.

Descriptive philosophy: Richard Zaner, *The Way of Phenomenology*. John Wisdom, "Epistemological Enlightenment," *The Owl of Minerva*, pp. 245–58.

Speculative philosophy: Alfred North Whitehead, *Process and Reality*, Chapter 1, section 1.

Chapter 8

A Healthy Philosophical Attitude

- Caring rather than indifferent
- Courageous rather than timid
- Open rather than closed
- Grateful rather than resentful
- Assertive rather than passive

In order to do a thing well, a person needs to be highly motivated, to have an effective method for doing it, and to have the right attitude as she or he goes about doing it. These are points that have been emphasized by nearly every successful athletic coach, music teacher, trainer of sales personnel, etc. They are also important in philosophy. We have already examined motives and methods, so now let's think about the *attitude* that is most likely to help us do philosophy well. We will focus on five facets of that attitude: caring, courage, openness, gratitude, and assertiveness.

Caring rather than indifferent

First, in order to do philosophy well we must *care* about what the true answers are to the questions of philosophy. If we are indifferent to what the truth is, our attitude will not sustain us through the demands and difficulties of the philosophic process. Fortunately, there are both personal and practical reasons to care about what the truth is. The primary personal reason, I think, is to achieve the dignity of having sought the truth in spite of trepidations about what it might turn out to be. It is a matter of caring about one's character – of wanting to be intellectually tough and courageous rather than intellectually weak and cowardly. It is a matter of honoring and satisfying the natural aim of the mind, which is to seek the truth.

Practical concern for the truth can be cultivated by noting that because our beliefs guide our actions, and our actions have consequences, therefore our beliefs have consequences for our lives and the lives of others. In every-day-life it is painfully obvious that beliefs have consequences. The child believes the bleach is good to drink. The college student believes she can't get AIDS from having unprotected sex just once. The middle-aged man believes he can drink heavily yet drive safely. They are all mistaken and can suffer devastating consequences.

Philosophical beliefs also have consequences. At the very least they have consequences for how we should live our lives, and how we live our lives may have serious repercussions for our happiness and destiny. For example, if death is the end of personal existence, then there will be no *personal* repercussions for us after death because of how we live our lives (though, as Buddhists emphasize, there may be ongoing *social* repercussions about which we should care). However, Socrates believed and Hindus believe that we each have an eternal, indestructible soul that inescapably suffers in another life the consequences of any and all bad behavior in this life.

If we follow the belief of Socrates and Hinduism and live as though there is life and punishment after death when there isn't, then we may misdirect our lives and waste enormous amounts of time. Yet, if we reject their belief and live as though there is no life or punishment after death when there is, then we may bring horrific misery upon ourselves. Surely from a practical point of view such an issue is worth caring about, of trying to get right. So, for other reasons, are numerous other philosophical issues, such as whether life is meaningful or not, and whether humans have a free will or not. To the extent that we see the practical importance of the truth about such matters, we will care about attaining it.

Courageous rather than timid

Second, we should care more about what the truth really is than we care about whether the truth is what we now think it is or whether it is what we would like it to be. It is usually easier and less embarrassing for the truth to turn out to be what we think it is, for then we do not have to admit that we were wrong and do not have to change our thinking. Moreover, it is natural to want the truth to be as we would like it to be. Yet if we are going to seek truth most effectively, we must not let our feelings and desires interfere with our recognition of what the truth really is, or probably is, and we must pursue the truth to the end even when it begins to appear that it will be something other than we would like it to be.

M. Scott Peck, M.D., best-selling author of *The Road Less Traveled*, has emphasized these points because of his experience as a psychiatrist and his concern for the emotional health and spiritual growth of people. "Our view of reality," Peck says, "is like a map with which to negotiate the terrain of life." The more of our beliefs that are true (the more accurate our map is), "the better equipped we are to deal with the world." Yet we shy away from revising our maps (beliefs). Why? Because "The process of making revisions, particularly major revisions, is painful, sometimes excruciatingly painful. And herein lies the major source of many of the ills of mankind." Peck concludes this way:

> Truth or reality is avoided when it is painful. We can revise our maps only when we have the discipline to overcome that pain. To have such discipline, we must be totally dedicated to truth. That is to say that we must always hold truth, as best we can determine it, to be more important, more vital to our self-interest, than our comfort. Conversely, we must always consider our personal discomfort relatively unimportant and, indeed, even welcome it in the service of the search for truth. Mental health is an ongoing process of dedication to reality at all costs.[1]

That statement should remind you of Nietzsche's statement at the end of the last chapter – which it would be good to stop and reread at this point.

In brief, the truth is what it is. To be philosophical in the best sense is to be determined to get at the truth whatever it is – even if the truth turns out to be very different from what we currently think it is, and even if it turns out to be very different from what we would like it to be. Such determination requires philosophical courage.

Open rather than closed

Third, as a consequence of the preceding points, we should endeavor to be as open as possible to (1) criticisms of what we currently believe and (2) alternatives to what we currently believe. It is natural for us to think that the truth about something is the way it seems or feels to us, but still we should listen to and examine criticisms and alternatives as fairly and vigorously as we can because we are all *fallible*, that is, we are all capable of being wrong; none of us is infallible, that is, incapable of being wrong.

Indeed, we should go beyond merely being open to criticisms and alternatives. We should seek them out. Charles Peirce, a great American philosopher, captured this aspect of the philosophic attitude when he said: "No blight can so surely arrest all intellectual growth as the blight of cocksureness." Consequently, Peirce added, the first step toward finding out the truth is to acknowledge that we do not already satisfactorily know it.[2]

We should not, however, confuse openness with (1) having no opinion or (2) thinking that incompatible opinions can all be true. Openness does not require that we have no opinion about what the truth is. In order to move forward in life, we need opinions about many things about which we do not and perhaps cannot have proven knowledge. Moreover, if we do not know for certain what the truth is, then your opinion as to what it is may be correct! What openness requires is that you thoughtfully consider criticisms of and alternatives to your opinions; you need not abandon your opinions unless you encounter a good reason to do so. Openness also does not mean thinking that all opinions are correct or equally good. Genuinely opposed opinions cannot all be correct. If one person believes there is intelligent life on other planets and another person believes there is not, then (assuming they mean the same things by "intelligent life" and "other planets") they cannot both be right – even if we do not *know* who is right.

Grateful rather than resentful

Fourth, we should never be resentful toward those who criticize our beliefs or present alternatives to them. We should be grateful to them. After all, those who present alternatives to our current beliefs may be presenting us with the possibility of replacing our current beliefs with better ones – and surely we would prefer better ones if they are available. Furthermore, if their alternatives do not in our best judgment prove better than what we currently believe, then we have lost nothing significant, and we have gained a richer sense of our own position and how it compares to alternatives.

The other person may, of course, present us not with an alternative to our position, but only with a criticism of it. Ordinarily such a criticism claims either that our assertion is false or that our justification for it is inadequate, or both. If we are persuaded that a criticism is correct (and if it is correct!), then our critic has saved us from holding a false belief or from mistakenly thinking that our justification for our belief is adequate when it isn't. In either case, we have gained something important and therefore should be grateful to those who are willing to take the time to criticize our beliefs or present alternatives to them.

Before going further, however, I want to sound two notes of sympathy. First, submitting our ideas to criticism by others can be anxiety inducing because if our critic convinces us that our belief is false or probably false, and we are not presented with a satisfying alternative to our old belief, then if our old belief was something from which we drew comfort and guidance (which is often the case with philosophical and religious beliefs), then we may feel empty and disoriented when we feel compelled to give up our old belief. That is always a risk when we do philosophy – a risk that recalls the issue with which we began this section, namely, whether we care more for the truth than we care for emotional comfort. Philosophy calls us to put the dignity and courage of intellectual honesty ahead of emotional comfort.

My second note of sympathy arises from the fact that I realize that sometimes people who present criticisms or alternatives to what we believe do so in an offensive manner. I'm not suggesting that we should be grateful for their obnoxious attitude. Such an attitude – often dogmatic or condescending – is unfortunate and makes it difficult to pay attention to *what* the person is saying. Keep in mind, however, that a dogmatic or arrogant person may be correct in what he or she is saying. If such a person is right and we are wrong, then surely we want to benefit from what they are saying, and surely we should be glad that we can benefit from them in this way. Meanwhile, we should keep in mind that an arrogant or dogmatic attitude creates static in the process of philosophical communication and sometimes drowns it out altogether. Still, it would be shortsighted and wrong to prefer to possibly remain in error rather than to be delivered of it because the person who could deliver us is arrogant, rude, insulting, or dogmatic.

Assertive rather than passive

Fifth, in order to do philosophy well, we should be assertive rather than passive; we should *express* our thoughts about philosophical issues – whether we are confident of the value of those thoughts or not. If we are not confident, we can say that we are just proposing the idea for consideration; we can ask, "What about this possibility?"

Assertiveness is so important because the objective of philosophy is to get at the truth about things; its objective is not just to avoid being wrong. What is the difference? Consider the following statement: "If we get the truth, we have avoided being wrong, and if we avoid being wrong, then we have gotten at the truth." Right? Not entirely. The first half of the statement is true. If we get the truth, then we have avoided being wrong. However, if we avoid being wrong, it is not necessarily true that we have gotten at the truth. Consider an analogy. I can play a whole basketball game without missing the basket once. How? By not taking any shots! But the point of the game, as a competitive sport, is to win – and no one can win who does not take a shot! To be sure, every time I take a shot, I risk missing the basket, but that is a risk that must be taken in order to possibly win.

Similarly, if no one asserts (or even thinks) anything in response to philosophical questions, then no one will be wrong, but neither will there be any possibility of someone being right – or of eliminating wrong answers or of replacing weaker answers with stronger answers and thereby getting closer to the truth. Hence, it is important that each of us take the risk of being wrong by making assertions for consideration. We should be grateful to those who take that risk, and we should assume our fair share of it – difficult as that is for many of us to do.

We should not try to excuse ourselves by thinking, "I don't have anything of value to say, so I won't say anything." Speaking up helps *us* think our thoughts more clearly, and others usually do benefit in some way from what we say – though they may not tell us so and may not even realize it until later. Moreover, *we* cannot benefit from the criticisms and alternatives of other people unless they know what our positions are so that they can respond to them. Hence, silence, rather than assertiveness, is doubly unfortunate – it deprives us and others of philosophical benefits.

To summarize, it seems better to run the risk of feeling uncomfortable in order to increase our chances of getting at the truth than to choose comfort over truth. The pursuit of truth is moved forward more effectively by us stating our opinions to others and then having those opinions enriched or examined by our listeners than it is moved forward by keeping our opinions to ourselves. After all, the ultimate objective of philosophizing is not merely to avoid error. Its ultimate objective is to gain knowledge of the truth. Movement toward that objective will be most effective if we state our beliefs and ideas to one another and evaluate them together in ways that promote continuing, candid, rational dialogue.

Notes

1 M. Scott Peck, M.D., *The Road Less Traveled* (NY: Simon & Schuster, 1978), pp. 44–5, 50–1.
2 Charles S. Peirce, *Philosophical Writings of Peirce*, ed. Justus Buchler (NY: Dover Publications, Inc., 1955, p. 4.

Reading Further

Plato, *Apology*; *Crito*; *Gorgias* (457c–458b).
Charles Peirce, *Philosophical Writings of Charles Peirce*, ed. Justus Buchler, pp. 1–4, 5–22.

Chapter 9: Alternatives to Philosophy

- Neglect
  Why neglect? Boring. Hard. Other attractions.
  Impossible to avoid
  Ought not to be ignored
- Skepticism
  Why skepticism?
  An epistemological position
  Criteria of a justified belief
    Coherent, comprehensive, resilient, useful, fruitful, simple
- Dogmatism
  What it is
  Two problems
    Fallibility
    Equals disagree
  Values of inconclusive investigations
- Solitude
  Why people choose it: embarrassment; conflict
  Two problems
    We need one another
    Better effort when observed
      Descartes and Socrates, Jaspers and Dewey
- Philosophy is important and inescapable
- We are responsible for our beliefs
- Philosophy not adversarial

Maintaining high motivation, keeping a proper philosophical attitude, and carrying out the methods of philosophy for the sake of getting at the truth is very demanding. Sometimes it is frustrating and discouraging. Consequently, the following question arises: "Why should I go to all this trouble?" That question presupposes that there are alternatives to doing philosophy, so let's examine some of the more common proposals.

Neglect

One common reaction to philosophy is to say, "Thanks, but I'm not interested. I've got better things to do." If you were to say that, you would not be the first person (or the last) to do so. But before you say it, let's think about an assumption that lies behind it, namely, the assumption that philosophy is not important enough to keep among our more important things to do. In reaction to this position I want to say, first, I am convinced that it is probably impossible for a person not to philosophize, that is, not to think about questions regarding morality, the meaning of life and death, the relation of the mind and the body, whether criminals act freely, and so on. To be sure, we do not have to think about such things all the time, but sooner or later life forces us up against them: we find ourselves having to make a decision that involves the welfare of other people, or we find ourselves plunged into grief and depression by the death of someone we love very much, or we achieve our goals but find they are not satisfying and so begin to wonder whether life is meaningless, or someone tells us that there is a God who loves us when we very much need and want to believe that but cannot believe it until we are convinced that it is true, and so on. In brief, life keeps throwing us into situations that call for philosophical decisions, and it seems impossible that anyone can escape. If I am correct, then neglecting philosophy altogether is not a real alternative. We are going to philosophize whether we like it or not, whether we want to or not. Hence, since we cannot help but do philosophy, and since philosophy pertains to some of the most important issues in our lives, we should do it self-consciously and as well as we can.

Second, even if we cannot neglect philosophy altogether, to try to do so, or to try to minimize the time we spend on it, seems wrong, or at least regrettable. Perhaps the reason why was captured best by Socrates when he said, "An unexamined life is not worth living for a human" – for a turtle perhaps, but not for a human (*Apology* 38a). Blaise Pascal, a seventeenth-century French philosopher, also spoke strongly on this point: "Man is obviously made for thinking. Therein lies all his dignity and his merit; and his whole duty is to think as he ought."[1]

So far as we know, we humans are unique in all of the universe because we alone, so far as we know, are able to conceive and think about the questions of philosophy. Our minds are part of the glory of the universe. It would be a shame for us to let that capacity lie unused. We might as well have been cattle or lizards if we are not going to use our capacity for philosophical reflection. A popular slogan states, "A mind is a terrible thing to waste." It is also terrible to waste *part* of a mind – and especially the part that is capable of doing philosophy. Hence, those who feel disinterested in philosophy should ask why they feel that way and whether it is a result of misunderstanding what philosophy is. Philosophy deals with urgent issues that pertain to all of us. It does not, then, seem realistic or right to neglect philosophy, so let's look at another proposed alternative.

Skepticism

Some people say, "It's not that I'm not interested in philosophy. I am. However, I do not think that the human mind is able to answer the questions of philosophy. Philosophers have been arguing over the same questions for 2,500 years without getting any closer to answering them! The questions are too deep and the mind is too short. Hence, doing philosophy is a waste of time."

The preceding is not an uncommon reaction to philosophy. I can sympathize with it to some extent. However, it involves a delusion that needs to be exposed – the delusion that the speaker is rejecting philosophy. In fact, the speaker is not rejecting philosophy. Rather, the speaker is taking a philosophical position. More specifically, the speaker is taking an epistemological position regarding what the human mind can and cannot know. To take such a position is legitimate, but it should be acknowledged for what it is: a philosophical position, rather than a rejection of philosophy.

Skepticism about philosophy is a philosophical position, and those who hold that position should join in rational dialogue with those who think that the human mind can accomplish more than skeptics think it can. G. W. F. Hegel, for example, was profoundly optimistic about the powers of the human mind. In a speech at the University of Heidelberg he said:

> Man, because he is mind, should and must deem himself worthy of the highest; he cannot think too highly of the greatness and power of his mind, and, with this belief, nothing will be so difficult and hard that it will not reveal itself to him. The being of the universe, at first hidden and concealed, has no power which can offer resistance to the search for knowledge; it has to lay itself open before the seeker – to set before his eyes and give for his enjoyment, its riches and its depths.[2]

To be sure, Hegel may have been wrong and the skeptics may be right, but in our current state of knowledge the skeptics have no right to stand outside the circle of philosophical dialogue and act as though their position is a matter of obvious fact when it is not. Rather, skepticism is only one position among many, all of which need to be compared and evaluated.

When considering the merits of skepticism it is also important to note that there are at least *six* criteria by which we can judge that one theory is *preferable* to another even when we do not know for certain which theory is true. *One* mark of a good theory is that it is *coherent*; that is, it does not contradict itself or imply anything absurd; its various parts cohere (fit consistently) with one another, and its implications are consistent with known facts. If one theory consists of nothing but beliefs that are consistent with one another and none of which implies anything absurd, and an alternative theory consists of beliefs some of which contradict one another or imply something absurd, then (all other things being equal between the two theories) the coherent theory is preferable to the incoherent theory because when two propositions contradict one another, they cannot both be true (because the truth of the one implies the falsity of the other). For example, "I was born before the end of World War II" and "I was not born before the end of World War II" cannot both be true of me. Hence, at least one proposition in a self-contradictory theory must be false, which means the theory is false (that is, not totally true).

Here is a philosophical example of how this might work. Jean-Paul Sartre argued that atheists who believe that there are absolute values are contradicting themselves. That is, Sartre thought that the two propositions "There is no God" and "There are absolute values" contradict one another because if there is no God, then there is no one capable of establishing values with absoluteness, that is, in such a way that they apply to everyone and do not change. Yet there are some philosophies, such as the humanism of the eighteenth century, which hold both of these beliefs. If Sartre was correct about these beliefs being contradictory, then any atheistic humanism which involves belief in absolute values must be rejected because we know by virtue of the contradiction between two of its beliefs that at least one of its beliefs is false (even if we are not sure which one it is). Hence, *logical self-consistency* is one mark of a true position. To be sure, just because a position is self-consistent, and not self-contradictory, does not ensure that it is true. Self-consistency is a necessary but not sufficient condition of a position being true. Nonetheless, all other things being equal, a self-consistent theory *might* be true whereas a self-contradictory theory cannot be true.

Similarly, if a theory implies something absurd, then it does not cohere (fit) well with something that seems to be an established fact, but a good theory will cohere with (be compatible or consistent with) all facts. Hence, a theory which does not imply anything absurd should be preferred to one which does. For example, if a theory which you are thinking about implies that you do not exist, then it has an absurd implication and should be rejected – since if you did not exist you could not be thinking about that theory, but you are. In brief, a good theory is coherent – both with itself and with established facts.

A *second* mark of a good theory is *comprehensiveness*. If one theory accounts satisfactorily for all of human experience to date, whereas a second theory does not account satisfactorily for the evolution of animal species or the vividness and value of religious experience, then the less comprehensive theory is less satisfactory than the more comprehensive theory. After all, our intellectual ideal in speculative philosophy is to discover or construct a theory that explains everything! Hence, the more a theory explains, the better it is. Consequently, a theory that explains more is better than a theory that explains less (all other things being equal between the theories, such as neither having a contradiction within it). To be sure, the more comprehensive theory may not be true, as it might (without us realizing it) involve a false explanation of something. Still, all other things being equal, a more comprehensive theory is to be preferred to a less comprehensive theory because the ultimate aim of the human mind is a correct understanding of everything.

A *third* mark of a good theory is that it *can survive rational dialogue*. The more successfully a theory can survive widespread, vigorous examination and criticism, the more likely it is that it is true, since presumably the true explanation of things is the one that will best survive the gauntlet of criticisms and alternatives. Of course this does not mean that the theory that has best survived open debate to the present moment is definitely true – but it is a point in its favor.

A *fourth* mark of a good theory is that it *proves useful* in ways that one would expect if it is true. One of the reasons that truth is so valuable is because when we act according to it, the consequences of our actions are what we expect them to be. By contrast, when we act on beliefs that we think are true but are actually false, the consequences of our actions are not what we expect, and may be inconvenient, harmful, or tragic. Hence, two of the main reasons why we want to know the truth is so we can avoid nasty surprises and achieve our goals. In brief, if a philosophical theory with practical implications is true, then it should be useful in appropriate situations – such as an ethical crisis or a logical confusion. If it does not prove useful in such situations, that is evidence against it being true. If it does prove useful, that is evidence in favor of it being true. It is not proof, however, as sometimes action on a false belief is followed by a desired result.

Consider a hypochondriac who always feels better after taking the medicine which his doctor prescribes for his bouts of malaise. The hypochondriac believes it is the medicine which helps him, whereas in fact what the doctor has prescribed is not medicine but a placebo. Hence, the hypochondriac's belief is false. What he takes does not help him. What helps him is his taking something that he believes will help him.

A *fifth* mark of a good theory is that it is *fruitful* rather than sterile; that is, it excites new ideas in the minds of those who hold it, and those new ideas prove to be good ideas. Isaac Newton's theory of gravitation enabled us to understand not only why apples fall from trees but also why planets orbit around the sun and how the moon causes low tides. By contrast, to say of everything that has occurred, "It was fate," would provide us with few, if any, further insights.

A *sixth* criterion is *simplicity*, which will be discussed more fully in the chapters on metaphysics. For now we need to note that if one theory is simpler, less complicated, than another, then all other things being equal between the two theories, the simpler theory is preferable. It is not necessary that the simpler of two adequate theories will always be the true one, but it does seem clear, all other things being equal, that the burden of justification should be on the person who claims that we need to make a theory more complicated than it seems we need to. Why do that extra work? Isn't that a waste of energy?

Recall, for example, that in ancient Greece Empedocles concluded that air, fire, earth, and water are the basic elements of all physical things. To say that something is a basic element is to say that it is not a compound of other things; air is nothing but air; fire is nothing but fire; etc. By contrast, mud is a compound of earth and water. Empedocles' position prevailed for more than a thousand years in western thought. In the eighteenth century, however, scientists began to discover gases such as hydrogen, nitrogen, and oxygen. At first many scientists continued to believe that air was an element in addition to those newly discovered gases. Gradually, however, they began to realize that air was not something in addition to those newly discovered gases but was, rather, the combination of those gases. It was simpler, then, and just as adequate to think of the atmosphere as consisting of nothing but those newly discovered gases and to give up belief in air as a gas in addition to those other gases. Consequently, you will not find air in the table of chemical elements, but you will find hydrogen, nitrogen and oxygen.

Similarly, some people say that the existence and the nature of the universe, and the events which happen in it, can be explained adequately without belief in God; therefore, by the principle of simplicity, belief in God should be dropped. Here, however, we begin to see how the different criteria of a good theory can come into tension with one another. Some people argue that although a worldview without God would be simpler, it would not be adequately comprehensive; that is, it would not adequately explain certain things about the universe. Hence, they conclude, we need to postulate the existence of God in order to provide an adequately comprehensive explanation of the universe. (We will explore this debate at length in the chapters on theism and materialism.)

Unfortunately, none of these six criteria is a proof that a claim is true, nor do all of them together constitute a proof. Truth is not so easy a quarry as that. Consider, for example, geocentrism, the theory that the earth is the center of the universe. That theory was not self-contradictory; it gave a comprehensive explanation of astronomical phenomena known at the time; it was of practical value to sailors who navigated by the stars. Moreover, most of the best minds once thought that the sun and stars revolve around the earth. They were all wrong.

What led to the overthrow of geocentrism was the fact that it was a dreadfully complicated theory which did not explain certain things satisfactorily. Consequently, a few stubborn souls kept trying to figure out an explanation which would, like geocentrism, be self-consistent, comprehensive, and useful, but would be simpler, account for troublesome data more satisfactorily, and survive rational scrutiny with fewer objections and unanswered questions. Finally, especially due to the work of Copernicus, Kepler, and Galileo, support for the geocentric theory began to collapse and the heliocentric theory came to be universally accepted as a more satisfactory explanation of the relation of Earth to the Sun. It is important to realize that people like Copernicus began to shift from geocentrism to heliocentrism long before they had anything that could be considered a proof of heliocentrism. They were attracted by how the heliocentric theory was so much less complicated than the geocentric theory yet could account for all the same phenomena equally well.

To summarize, a good explanation of things will (1) be coherent, (2) be comprehensive, that is, account for all that needs to be accounted for, (3) be hardy, that is, survive the keenest and most widespread scrutiny to which we can put it, (4) be of practical value in appropriate situations, (5) be fruitful, giving rise to additional illuminating insights, and (6) be as simple as we can make it without violating the preceding criteria. Moreover, it is reasonable to think that the more of these qualities a theory has, the closer it is to the truth. We should, then, keep these qualities in mind as guiding lights while we search for the truth. But even when these demands seem satisfied by a theory which is widely accepted, we should be willing to examine it again if it is in any way unsatisfactory. Seemingly small problems can indicate the need for a profound change of mind – as in the changes from geocentrism to heliocentrism, and from Isaac Newton's theory of gravitational attraction to Albert Einstein's special theory of relativity.

In brief, even if we cannot know for certain the truth about reality and value, it seems plausible to believe that a theory which is coherent, comprehensive, useful, hardy, fruitful, and simple is more likely to be true than a theory which does not meet these criteria. Hence, a failure to discover for certain which theory is true need not paralyze us or drive us to skepticism when we are confronted with competing theories. We have rational criteria for deciding which theories are better.

Dogmatism

A third alternative to philosophy is that of dogmatism. The dogmatist says, "Contrary to people who are not interested in the issues of philosophy, I am quite interested in them. Contrary to the skeptic, I believe that the human mind can know the answers to the questions of philosophy. However, we do not need to 'do' philosophy to figure out the answers to its questions. All we need to do is look around us, and it quickly becomes evident what the answers are. Philosophy is for people who don't want to face the facts because they don't like them." In brief, the dogmatist thinks that doing philosophy is unnecessary because it is obvious what reality is like.

The dogmatist is right that when we do philosophy, sometimes our desires and aversions improperly determine our presuppositions or conclusions. However, that kind of criticism is a two-edged sword: If it cuts against those who do philosophy, it also cuts against those who think the answers to philosophical problems are obvious. Their needs and feelings might be influencing the way *they* see things. Perhaps they cannot tolerate uncertainty and need to feel that what they believe is true is unquestionably true, and so are blinded by emotion to the plausibility of ideas that differ from their own. Or perhaps they lack the background or imagination to realize that there are serious alternatives to the way they see things. Everyone, then, needs to be sensitive to the fact that some of his or her beliefs may be a function of emotion rather than evidence or reason, but that need is no greater for the philosophical person than it is for the dogmatic person.

Moreover, the dogmatist has two other problems. First, dogmatism does not honor the fact that we are all fallible, that is, capable of being mistaken, even about things that seem obvious to us (for example, the sun seems to go around the earth, and it seems smaller than the moon). Experience has shown us so many times – over the centuries and in our private lives – that we can be mistaken in some of our most confident beliefs, that it is hard to believe we are not all more humble and open to criticisms and alternatives than we are. But we are not, and that fact is an indication of just how entrenched the problem of dogmatism is – perhaps not with regard to all of our beliefs, but almost always with regard to some of them.

A second problem with dogmatism is that it does not account adequately for the fact that equally intelligent, sincere, and educated individuals can disagree profoundly with one another over the answers to the questions of philosophy. This fact seems to indicate clearly that our experience of reality is ambiguous, subject to different interpretations, each interpretation having some plausibility, yet no interpretation being obviously true. For example, there are intelligent, sincere, educated people who think it is obvious that there is *no* God, and there are other such people who think it is obvious that there *is* a God. If by "God" they mean the same thing, then it is obvious that they cannot both be right, and it seems equally obvious that they should use the methods of philosophy to engage in discussion of their disagreement and to try to reach the truth. Hence, because of human fallibility and the disagreement of sincere, intelligent, educated people with one another, dogmatism about philosophical beliefs is not warranted.

To be sure, the beliefs of a dogmatist might be true, so the problem with their believing is not with *what* they believe but with *how* they believe. If their belief is *true*, their dogmatic attitude may alienate other people from finding out the truth. (Who wants to believe that someone who is arrogant is right? Who even wants to listen to them?) If the dogmatist's belief is *false*, then his or her attitude tends to prevent him or her from finding out the truth, since he or she refuses to take seriously ideas other than those that he or she already believes. Ironically, the more dogmatic a mistaken person is about knowing the truth, the less likely he or she is to find the truth! (This point is illustrated in a haunting way in Plato's Allegory of the Cave; see *Republic*, Book VI.)

Before turning to a fourth alternative to philosophy, we need to consider an important fact to which skeptics often appeal when they dismiss the value of philosophy. The fact of which I speak is the fact that the philosophical problems of ancient Greece are still with us nearly 3000 years later. Doesn't this failure to reach solutions indicate clearly that doing philosophy is a dead-end activity, a waste of time?

I share the skeptic's frustration with the fact that philosophy, unlike science, has not solved any of its basic problems in a decisive, universally agreed upon way. Still, I believe there are values to philosophy which save it from the bleak conclusion that it is a waste of time. The values of which I speak are at the heart of what a liberal arts education is all about. The liberal arts, insofar as they are successful, liberate us from ignorance – and I hope I need not convince you that ignorance is *not* bliss, at least not usually. More often than not, ignorance leads us to make mistakes that hurt ourselves or others, and it can leave us unaware of valuable alternatives that we would have preferred if only we had known about them. Hence, ignorance can impoverish and harm us, and often does.

To be sure, philosophy does not liberate us from ignorance by giving us knowledge of the solutions to its problems. Rather, it helps liberate us from ignorance by exposing us to alternative ways of thinking about reality and living our lives, and by helping us learn to think of yet other ways. It was in order to help free us from philosophical narrowness that I emphasized the importance of running out the permutations on each philosophical problem. The more alternative ways we have for thinking about reality and living our lives, the more likely it is that we will find a satisfying philosophy among those alternatives. Generating such alternatives – and rational criteria for choosing among them – has been one of the most important contributions of philosophy to the pursuit of truth and the living of life.

To be aware of a larger number of philosophical alternatives is, of course, not to *know* which one of them, if any, is true. But the more alternatives we are aware of, the more likely it is that the truth is among them, and when we do not know which theory is true but must choose among them for practical reasons, it seems better to have a wide variety of positions to choose from than to be limited to the positions of our parents or peers.

Let me try an analogy. Think of being very gifted musically and of loving music, yet of being reared in a town where the only musical instrument played or known is the tuba (weird but imaginable!). So far as you know, the tuba is the only musical instrument in all of the world; you love it and excel at it. Then one day a friend takes you to a neighboring town where you discover that in addition to tubas, there are trombones, trumpets, French horns, reed and wind instruments, string instruments, pianos, and organs! That experience would, I think, be a wonderful discovery for you. You might still decide to stick with the tuba for your own musical instrument; it is, after all, a wonderful instrument in its own right. Surely, though, you would be all the richer for knowing of those additional instruments, being able to study them, and being able to join with those who play them.

Similarly, the ways in which philosophy enriches our awareness of options makes it worth the confusion, anxiety, and frustration involved in discovering and studying them, even if it does not tell us which alternative is correct. Moreover, the seeming inability of philosophy to determine the truth about certain issues can be seen as a good thing because it means that we get to choose the alternative that we prefer – within the limits of rationality – rather than having an alternative forced upon us from the outside. (Plato's "Allegory of the Cave" speaks powerfully to the way that philosophy expands our horizons. See *Republic*, Book VII. See also Plato's *Euthyphro* 9e.)

Solitude

The fourth alternative I want to discuss is characterized by the following attitude: "I'm definitely interested in the questions of philosophy, and I certainly don't think that the answers to them are obvious. Furthermore, I do believe that the human mind can make progress with them, but I don't want to engage other people in dialogue about them. It's too stressful, and sometimes embarrassing. So, I'll just think about the issues of philosophy by myself, thank you."

Perhaps this fourth position is not an alternative to philosophy. Perhaps it just consists of doing philosophy by oneself, in one's own head. However, whether such solitary thinking is truly philosophy or not, it is not preferable to doing philosophy in the full-blooded sense which includes rational dialogue with other people. It is true that many of us experience distress or embarrassment at the mere thought of expressing our philosophical ideas and trying to justify them to other people. That is a reaction with which I can sympathize. Too many times, unfortunately, such anticipatory distress is the result of having been treated rudely by others when we have spoken up in the past. But whatever its cause, I believe there are at least two good reasons for each of us to suppress that reaction, be courageous, and speak up anyway (as we saw in the chapter on a healthy philosophical attitude).

First, each of us is limited in experience, insight, and imagination. Consequently, we need one another in order to have as wide a range of ideas to choose among as possible. Other people will know or think of alternatives that we will not know or think of. Also, we need one another in order to adequately evaluate those ideas that we currently believe, plus any ideas that we are considering for adoption. If we do not speak of these ideas with other people, we lose the benefits they could have given us, and they lose the benefits that we could have given them.

Second, we are not likely to think as clearly and rigorously as we can if we only think privately. We usually rise to our best when we are communicating or preparing to communicate with other people. The Bible makes this point in Proverbs 27:17, "As iron sharpens iron, so one man sharpens another" (New International Version). René Descartes realized this truth after reflecting on his own philosophizing. "Doubtless," he said, "we always scrutinize more closely that which we expect to be read by others than that which we do for ourselves alone, and frequently the ideas which seemed true to me when I first conceived them have appeared false when I wished to put them on paper."[3] It is likely, then, that if we limit our philosophizing to private reflections, we will not formulate our ideas as clearly and rigorously as we otherwise would – and as Descartes noted, *writing* out our thoughts usually motivates us to be even more careful than when we merely speak them out. If we each limit our philosophizing to the privacy of our own mind, we are all too likely to succumb to self-delusion and self-pampering.

Socrates made the preceding point long ago when he said that the greatest good for humans is to discuss virtue and other such topics everyday (*Apology* 38a). For Socrates the concept of virtue was not some kind of prudish moral notion; rather, it had to do with that way of life that would bring true happiness. Yet what he said is initially puzzling. To be virtuous is to *act* in a certain way, so why didn't he say that the greatest good for a human is to *act* virtuously every day?

I think he did not put it that way because he observed that our actions follow from our beliefs, and our beliefs generally follow from our thinking about things. Hence, before we act, we should think; specifically, before we attempt to act virtuously, we should try to make sure that we know which kinds of actions are virtuous.

In that case, though, why didn't Socrates say that the greatest good for a person is to *think* about virtue everyday? After all, isn't thinking what philosophy is all about? Yes, but discussing a topic does involve thinking about it (or at least it should!), so by discussing virtue everyday, rather than just thinking about it, we do think about it – plus we get the benefits of hearing other people's experiences, ideas, and reactions (and they get the benefit of hearing our experiences, ideas, and reactions). Hence, rational dialogue is much more likely to help us progress toward satisfactory positions on virtue, justice, love, the mind, God, and other such topics than is solitary reflection (or safe discussions with people who agree with us or don't like to disagree).

In brief, to philosophize well we need others, especially those others who disagree thoughtfully with our beliefs. As Karl Jaspers put it: "The truth needs two." Philosophy at its best, as John Dewey put it, is "a cooperative enterprise." At its best, then, rational dialogue calls for two or more participants, each of whom is willing to take the risks of venturing ideas, is willing to listen intently and fairly, is willing to be candid in reaction to the ideas of others, and is sensitive to how easily the precious web of communication can be destroyed by indifference, dogmatism, contentiousness, insincerity, or disdain.

Philosophy is important and inescapable

There seems, then, to be no worthy alternative to doing philosophy in the form of rational dialogue, and life forces us to deal with the problems of philosophy whether we want to or not. Hence, the question is not whether to get involved with philosophy; the question is whether philosophy is important enough to do well. For all the reasons given above, I think the answer is "yes." The better part of wisdom is to enter into philosophy as fully and effectively as we can. That does not mean becoming a professional philosopher or majoring in philosophy in college. Rather, it means acknowledging the everyday relevance and importance of philosophy, taking a good philosophical attitude toward the pursuit of solutions to its problems, and becoming skilled in the concepts and methods of philosophy so as to be able to think and communicate effectively about philosophical problems.

We are responsible for our beliefs

Finally, I would like to comment on the fact that we are not responsible for the philosophical beliefs that we had as children. Those beliefs were taught to us by people whom we trusted, and we probably knew of no serious alternatives.

Now, however, having reached a level of intellectual maturity, *we* are responsible for retaining, modifying, or replacing those beliefs. That is an awesome responsibility, but an exciting one as well. Surely in our new freedom we want to have the best beliefs that we can find or conceive. Just as surely, we do not want to turn over to anyone else the task of deciding our beliefs for us. Answering the questions of philosophy is a very personal responsibility that we should let no one take from us or take over for us. Yet our answers should be sought through dialogue with others, and retained, modified, or rejected in dialogue with others.

Because we are going to have philosophical beliefs whether we like it or not, surely it is better to have clear, specific beliefs that we can act on and then evaluate – rather than to muddle along in a mental fog, not clearly aware of the beliefs by which we are, in fact, living our lives. When Saul Bellow, a Nobel prize-winning novelist, came to realize this point, he said, "With increasing frequency I dismiss as merely respectable opinions I have long held – or thought I held – and try to discern what I have really lived by, and what others live by."[4] Bellow's distinction between what we think our beliefs are and what they really are is an important one. He is suggesting that we find out whether what we think our beliefs are is what our beliefs really are by looking at our actions and inferring from them the beliefs by which we are guided. The ideas that actually guide our actions are our real beliefs – whether we are aware of them or not.

If you unearth your real beliefs, compare them to what others believe, evaluate them all, and then decide that your beliefs are as good as, or better than, any others around, then great! If, by contrast, you discover that some of your beliefs are not satisfactory, then you have a chance to improve upon them – and surely that is better than continuing to hold a wrong or inferior belief.

We seem, then, to have no honorable alternative to dealing with the problems of philosophy by means of private reflection and public dialogue – presenting our beliefs and sentiments to one another, listening and responding sensitively and respectfully. By means of such attitudes and procedures, progress in philosophy is *dialectical*, that is, philosophical progress results from the conflict of opposing ideas, neither of which is entirely satisfactory, but the clash of which often produces a third idea which improves upon both of the first two.[5] By virtue of the dialectic of philosophy over the centuries and in our own time, we have a better understanding of the implications, strengths, and weaknesses of old ideas, and we have generated new ideas for consideration and comparison. That, I submit, is a real and important kind of progress that philosophy has made. Hence, a skeptic would not be correct in thinking that no progress has been made in philosophy.

Philosophy not adversarial

Because we "argue" with one another over philosophical issues, it is somewhat natural to think of philosophy as an adversarial process like that of two lawyers (the prosecution and the defense) arguing against one another in a court of law. For two reasons, however, this comparison of philosophers to lawyers is not a good one. First, each lawyer's objective is to win the case for their client – whether their client is innocent or not. By contrast, the objective of two arguing philosophers is to get at the truth – whatever it is and whoever is right. As a student put it, quite eloquently, the objective of each lawyer is to win, whether he or she has the truth or not, whereas the objective of each philosopher is to get at the truth, whether he or she wins the argument or not. (Philosophers being human, they are not always so disinterested, but that is the ideal.)[6]

Second, in a trial proceeding there is an authority (the judge or the jury) whom the lawyers are trying to convince and who finally determines whether the accused person is innocent or not. In philosophy, by contrast, there is no judge or jury with the authority to make one position wrong and the other right. There is just us. To be sure, some of us are more learned than others about philosophical matters, and the better part of wisdom is to heed and learn from people who know more than we do, but when it comes time for decisions to be made regarding a philosophical issue, we are all equally responsible for our beliefs, so we should each make our own decisions.

Notes

1 Blaise Pascal, *Pensées*, tr. A. J. Krailsheimer (NY: Penguin Books, 1966), p. 235, #620.
2 J. Glen Gray, ed., *G. W. F. Hegel on Art, Religion, Philosophy* (NY: Harper Torchbook, 1970), pp. 20–1.
3 René Descartes, *Discourse on Method and Meditations* (Indianapolis, IN: Bobbs-Merrill, 1960), p. 48, or sec. 66 in the French edition or any translation thereof.
4 *The Ithaca Journal* (Ithaca, NY), Dec. 13, 1976, p. 28.
5 G. W. F. Hegel called the parts of this process the thesis, the antithesis, and the synthesis. The thesis is what is asserted. The antithesis is an alternative which is critical of the thesis. The synthesis is a creative effort to improve upon and go beyond both the thesis and the antithesis.
6 My comment should not be interpreted as a criticism of criminal lawyers. It is because the truth as to the guilt or innocence of the defendant in a criminal trial is not known

that it is appropriate for each of the opposed lawyers to try to win the case for their side. It is hoped that this process of clashing opinions will result in the truth coming out, but that does not change the fact that, ordinarily, the objective of each lawyer is to win the case, not to get at the truth.

Reading Further

On misology (the distrust of reason) see Plato's *Phaedo*, 85c–d and 89d–91b, and the opening of the third dialogue in George Berkeley's *Three Dialogues Between Hylas and Philonous*.

For negative assessments of the value of traditional western philosophy see D. T. Suzuki's "The Sense of Zen" in his *Zen Buddhism*, ed. William Barrett, pp. 3–26, and Alan Watts's "Philosophy Beyond Words" in *The Owl of Minerva*, ed. Charles J. Bontempo and S. Jack Odell, pp. 191–200.

Part II Epistemology

Chapter 10: What is Truth?

- Non-epistemic uses of "true"
- The kind of thing that is true or false
- The nature of truth
- What makes an assertion true or false
- Competing conceptions of truth
- Why truth is important
- Three laws of thought
  - The Law of Identity
  - The Law of Non-Contradiction
  - The Law of Excluded Middle
- Six sources of truth
  - Sensation, introspection, reason, intuition, dreams, revelation
  - They are not infallible

Now we begin doing philosophy rather than looking at it from the outside. Earlier we saw that epistemology, one of the basic areas of philosophy, investigates the nature of knowledge, sources of knowledge, methods for achieving knowledge, and methods for evaluating knowledge claims. We also saw that in philosophy one thing leads to another. Because knowledge is a form of belief, we must ask what belief is if we are going to understand what knowledge is; because belief consists of thinking that something is true, we must understand what truth is in order to understand adequately what belief is; because the goal of belief is knowledge of the truth, we must ask when it is that we are justified in thinking that we have the truth. Let's begin our exploration of these epistemic concepts by asking, "What is truth?"

Philosophers often ask awkward questions – at least they seem awkward in comparison to the questions that we usually ask – but they are important questions, and they usually cannot be formulated in less awkward ways. For example, we often ask "What time is it?", but we rarely ask, "What is time?" We often ask "Is that statement true?", but we rarely ask "What is truth?" Socrates thought these "What is X?" questions to be terribly important because if we do not know what X is, how can we tell whether something is an example of X? If we do not know what a polygon is, how can we tell whether something we see is a polygon? If we do not know what truth is, how can we tell whether a statement is true or not?

Non-epistemic uses of "true"

Before trying to answer the question, "What is truth?", let's ask an even more awkward question: "What kind of a thing is it that is true or false?" There are many different kinds of things that are true or false; that fact is one cause of our confusion over the nature of truth. "True" can be an adjective. For example, "His teeth were false, but his love was true." Also, we distinguish true gold from fool's gold (iron pyrites, which resemble gold in sparkle and weight). "True" functions also as an adverb: the arrow flew straight and true; cattle from the same pure stock will breed true. "True" even serves as a verb: the carpenter trued the door frame (made each of its angles 90 degrees); the mechanic will true your bicycle wheel (adjust the tightness of the spokes so that the wheel doesn't wobble when it spins).

The kind of thing that is true or false

It is the adjectival function rather than the adverbial or verb function of "true" and "false" that we are most interested in in philosophy, and the kind of thing the truth or falsity of which epistemologists are most interested in is *the assertion*. An assertion is a statement or claim that something is the case. Often the word "*proposition*" is used by philosophers to refer to *what* is being asserted or stated or claimed. For example, in writing the words, "It is raining," I mean to assert the proposition that it is raining. I assert the same proposition in French and German, respectively, by writing, "Il pleut" and "Es regnet." Hence, a proposition is *the meaning* of an assertion, claim, or statement. Obviously, the same proposition can be asserted in different languages (and even in different words in the same language; for example, "Precipitation is occurring"). In ordinary language, however, these four terms – assertion, claim, statement, and proposition – are roughly synonymous with one another, so we will say that the kind of thing that is true or false is an assertion or claim or statement or proposition. (A more advanced treatment would disclose subtle differences between these concepts.)

The nature of truth

Now that we have an idea of the kind of thing that is true or false, let's ask, "What is truth?" Given that an assertion is the kind of thing that is true or false, it turns out that truth is the property of an assertion which says that things are a certain way when in fact they are that way. If, for example, I assert that Socrates died before Aristotle was born, then if that's the way things were, namely, if in fact Socrates died before Aristotle was born, then my assertion has the property of truth. If Socrates did not die before Aristotle was born, then my assertion has the property of falsity. Hence, falsity is the property of an assertion which says that things are a certain way when in fact they are not that way.

What makes an assertion true or false

Now let's ask: What makes an assertion to be true or false? What conveys upon an assertion the property of truth or falsity? According to *the correspondence theory of truth*, the answer can be captured in one word: REALITY! What makes an assertion to be true or false is the way reality is, or "the way things are" (or the way things were or will be, depending on the tense of the verb). An assertion is always about the way reality is in some respect – using "reality" to include not only the physical universe but all else that there might be, such as angels and numbers. Therefore, if reality is the way an assertion states that it is, then reality makes that assertion true. If reality is not the way an assertion says it is, then reality makes that assertion false. Aristotle made this point in Book IV of his *Metaphysics* when he said that a statement is true when it says of what is that it is and of what is not that it is not, and is false when it says of what is that it is not and of what is not that it is. Bertrand Russell, one of the great twentieth-century philosophers, said that "beliefs (a) depend on minds for their *existence*, (b) do not depend on minds for their *truth*"; "a belief is true when there is a corresponding fact, and is false when there is no corresponding fact."[1]

Some people react negatively to this depiction of truth as hard fact. They object, "But how do you *know* what reality is like?" That is an important question. We will focus on it soon. For now, however, notice that whether we know for certain what reality is like is usually *irrelevant* to whether an assertion is true or false. What makes an assertion true is not whether we think it is true. For example, the truth of the proposition, "*Pi* is an irrational number," does not depend on what I think; it doesn't even depend on whether I've ever heard of *pi*. (Exceptions occur, of course, in those instances in which our assertions are claims that we believe something. If I say, "I believe I am funny," my claim is not a claim about what other people think about me; it is a claim about what I think about myself. Therefore, whether my claim is true or false is determined not by whether I really am funny; it is determined by whether I really think I am funny or whether I have misspoken or am lying.)

In some instances, then, it is the reality of what we think or do not think that makes our claims true or false. Most often, though, our claims are about things other than ourselves and our states of mind. For example, some people think there is intelligent life on other planets; other people think there is no intelligent life on other planets. Because these two groups contradict one another, one group has to be right; the other group has to be wrong. We do not know now which group is right, but one group is, and it is the way reality is that makes that group right – even though we do not know now which group is correct.

Think also about this. The pursuit of truth, such as science is engaged in, would not make sense if it were not true that reality is one way rather than another before we find out how it is. A scientific *hypothesis* is a scientist's best guess as to how reality is in some specific respect. Scientific *research* consists of taking steps to discover whether an hypothesis is true or false – but the hypothesis is either true or false *before* scientists find out which it is. Before the research begins, reality either is or is not like the hypothesis asserts that it is. If that were not the case, scientists could not *discover* how things are, as things would not be a certain way, and so there would be nothing to discover. But scientists do discover various facts about reality, so reality must be that way before they discover it.

Competing conceptions of truth

The preceding conception of truth is called a "correspondence" conception of truth because it says that a belief is true when it corresponds to the way reality is. Some thinkers reject the correspondence theory of truth because, they say, it is unverifiable. We cannot, they say, check whether a belief is true by looking to see whether reality really is the way the belief says it is. To think we could do that would be to *assume* that reality is the way we experience it to be, but how could we ever know that reality apart from us is the way we think it is? To be sure, we can check one experience against another experience (as when we move closer to something to see if it still looks the way it looks when we are farther away, or as when we feel something to see if it feels the way it looks). We cannot, however, get outside of experience to see whether reality really is the way our experience leads us to think it is. Hence, to say that a true belief is one that corresponds to the way reality really is is a useless conception of truth.

The pragmatic theory of truth seizes on the fact that our beliefs guide our actions and our actions have consequences. Pragmatists, such as the American philosophers William James and John Dewey, say that what it means for a belief to be true is that it works or that it guides us successfully when we act on it. A false belief is one that does not turn out the way we expect it to when we act on it. That much, say pragmatists, we can know through experience, and that is all we need to mean or can verify when we say that a proposition is true. Whether our beliefs correspond to reality as it "really" is is something we cannot know, and so it must not be what we mean when we say that a belief is true.

Defenders of the correspondence conception of truth reply that the pragmatic conception of truth has a fatal flaw: A false belief can work. Physicians, for example, use placebos because sometimes false beliefs work. A patient's false belief that a placebo is real medicine which will help him can cause him to get better. Hence, we need to make a distinction between a true belief and a belief that works. To be sure, the correspondence theorist agrees that a true belief will work, but it works because it is true; it is not true because it works. It is true because reality is the way it says it is, and that is what it means for a statement to be true.

Other thinkers who are dissatisfied with both the correspondence theory of truth and the pragmatic theory of truth argue that truth consists of a certain kind of coherence of a belief with other beliefs. That is, what it means for a belief to be true (and what makes a belief true) is that it fits coherently with other beliefs that are solidly established, whereas a false belief is one that does not cohere with other beliefs that are well-established. For example, the claim that I am ten thousand years old is false because it does not cohere with the testimony of my parents, or with the date on my birth certificate, or with what we know of the life-span of humans, etc. By contrast, the claim that I was born on July 8, 1940, is true because it coheres with what my parents say, what their friends say, what the attending physician says, what the hospital records say, what my birth certificate says, etc. As was mentioned earlier, we cannot get outside our experience to see whether I was "really" born on 7/8/40. We can only check whether a belief coheres with other well-established beliefs; when it does, it is true.

The most serious problem with the coherence theory of truth, according to its critics, is that truth, as conceived by the coherence theorist, seems to have no connection to reality. Its only connection is to others' beliefs, and it is conceivable that a large system of beliefs that cohere with one another could be an illusion. For example, although most of our beliefs about living in a world of people and animals, trees and flowers, etc., fit quite nicely with one another, it could be that they don't correspond to reality. Perhaps most of our beliefs are false because we are really nothing but lumps of matter floating in space and our experiences are being caused by gaseous beings who, for their own entertainment, are bombarding us with streams of photons, or some such thing.

To be sure, says the correspondence theorist, a true belief will cohere with other true beliefs, and it will "work" when we act on it. Moreover, if a belief works when we act on it and if it coheres with other beliefs which we have good reason to think are true, then that is evidence in favor of the truth of that belief. However, what *makes* a belief true, and what it *means* for a belief to be true is not that it works or coheres with other true beliefs. What makes it true is that reality is the way it says it is. In the final analysis, says the correspondence theorist, the notion that what it means for a belief to be true is that it corresponds to reality is difficult, and perhaps impossible, to get away from.

Why truth is important

Having examined what truth is and what it is a property of, now let's ask, "Is truth important?" Most people seem to think it is. Scientists, mathematicians, historians, and journalists, among others, are constantly working to get at the truth. In light of the preceding paragraphs, now we can understand why truth is considered to be so important. Truth in our beliefs is important because it means that our beliefs are in line with reality; it means that reality is the way we think it is. That is important because our *actions* are guided by our beliefs. When we act on true beliefs, we get the results we expect; when we act on false beliefs, we do not. Many a parent has given aspirin to his or her feverish teenage child because the parent *believed* it would help relieve the child's fever and wouldn't hurt even if it didn't help. We now know reality isn't that way. There is such a thing as Reyes Syndrome, which is a serious affliction that can be triggered in teenagers by aspirin. Tragically, some teens have died and others have been seriously harmed because their parents' beliefs about reality were not true.

True beliefs enable us to take better advantage of the beneficial features of reality and do a better job of avoiding its harmful features. False beliefs put us on a blind collision course with reality, and in a head-on collision with reality, it is clear who is going to lose. Perhaps that is why Marilyn vos Savant, listed in the *Guinness Book of World Records* under "Highest IQ," answered as she did in her column, "Ask Marilyn". When Larry Cates asked her, "What is the most powerful concept, and what makes it powerful?", her answer was: "Truth. The reason it's so powerful is that, whether we like it or not, there isn't a darned thing we can do about it."[2]

In addition to the practical importance of true beliefs, we noted earlier (in the section on courage in chapter 8) that self-respect is an important ingredient of a deeply satisfying life, and intellectual honesty is an important ingredient of self-respect. We may not always be happy with what we discover the truth to be, or with what it seems probably to be according to the best evidence we can find, but we can always be happy with the integrity we show by seeking the truth with firm resolve and acknowledging what we find. Furthermore, once we know what the truth *is*, if we do not like what it is but it can be changed, we are in a better position to improve what it will be. Reinhold Niebuhr, a twentieth-century theologian, incorporated some of these points into a popular prayer:

> God grant me the serenity to accept the things I cannot change,
> The courage to change the things I can,
> And the wisdom to know the difference.

Three laws of thought

Now let's examine three laws or rules of thought that we must follow or else fall into confusion regarding truth and falsity. First there is the Law of Identity. This law is deceptively simple. It states, "If an assertion is true, then it is true." You might be thinking, "Well of course it is! Why even say so!" But trust me – this principle is violated much more than you might think. People contradict themselves unintentionally more frequently than they realize. (Sometimes they contradict themselves intentionally – usually in order to deceive someone.)

Part of the practical significance of the Law of Identity is that if someone says that something is true, then they should *continue* to say that it is true. For that reason the Law of Identity is at the heart of the process of cross-examination in a trial. In the cross-examination procedure in a trial, what is the prosecuting lawyer trying to do? She is trying to see if she can catch the defendant (or a witness) in a self-contradiction or in a contradiction of the testimony of some other witness. Such contradictions are important because they suggest that the defendant (or a witness) is lying or does not remember the events as well as she originally claimed. For the same reasons, contradictions between the testimonies of witnesses other than the defendant are also important.

Here is where the second law, the Law of Non-Contradiction, comes in. It states: "An assertion cannot be both true and false." This goes beyond the Law of Identity. The defendant who contradicts himself at a trial might try to defend himself by saying, "Yes, earlier I said I was *not* at the scene of the crime, and I wasn't, but I also was." The Law of Non-Contradiction says that such a thing is impossible. The truth cannot contradict itself. If an assertion is true, then it is true and only true; it cannot *also* be false. After all, part of what it *means* to say that an assertion is true is that it is *not false* – just as to say that an assertion is false is to say that it is *not true*. Hence, it cannot be the case that the defendant both was at the scene of the crime and was not.

Our defendant might now try to wiggle out of his self-contradiction by saying, "Actually, I was confused in what I said earlier. You see, it's not true that I was at the scene of the crime, but it's also not false that I was at the scene of the crime. The truth is a third possibility. After all, just because a car isn't red doesn't mean it has to be blue; there are other possibilities. Similarly, just because it's not true that I was not at the scene of the crime does not mean that it has to be true that I was at the scene of the crime."

Now, however, the defendant has violated our third and last law of thought, the Law of Excluded Middle. This Law states: "An assertion must be true or false." The point of this law is that for an assertion there is no third alternative between being true or false. Yes, there are colors other than red and blue for cars, but there are no epistemological properties other than true and false for propositions. The possibility of a middle ground between true and false is excluded by the fact that either reality is or is not the way an assertion says it is. Hence, an assertion cannot be neither true nor false. It must be one or the other (Law of Excluded Middle); it cannot be both (Law of Non-Contradiction); and whichever it is, that is what it is, and not anything else (Law of Identity).

Assertions can, of course, be vague rather than clear, suggestive rather than explicit, indirect rather than direct, subjective rather than objective, and complex rather than simple, making it difficult and sometimes impossible to understand and evaluate with confidence what has been said. For example, "He is rather handsome" or "My love is like a summer's day." Moreover, if what is uttered or written is meaningless, it is neither true nor false because it does not assert anything; it is not a genuine assertion; for example, "The glibble is petch." (Philosophy of language gives extensive attention to linguistic meaning, vagueness, analogy, and metaphor.)

Six sources of truth

Now that we have a sense of what truth is and why it is important, let's ask how we can find it. What are the sources of truth that are available to us? There seem to be numerous sources of truth, including sensation, intuition, introspection, reason, dreams, and revelation. Let's consider each of these sources, though not necessarily in their order of importance.

First, if a friend says there is milk in the refrigerator, but we think there is not, we can use *sensation* to find out the truth of the matter; we can look inside the 'fridge – or if we are blind, we can feel and smell or taste.

A second source of truth is *introspection*. In the midst of a very busy day someone might say, "I have a mild headache. What about you?" You won't find out whether you have a headache by looking into a mirror or listening to your heart with a stethoscope or sniffing your foot. You will pause and do what we call "introspecting"; that is, you will attend to what you are experiencing inwardly – and you may discover that you had been so busy that you had not noticed that, like your friend, you, too, have a mild headache.

Reason is a third source of truth. By means of reason we discover, for example, geometrical truths, such as that the internal angles of a plane triangle must add up to exactly 180 degrees, and that the circumference of a circle must be more than three times as long as its diameter. We can, of course, empirically measure the angles of numerous triangles and see that in each case they equal 180 degrees. However, we cannot empirically measure the internal angles of every possible triangle; they are infinite in number, so it is only by means of reason that we can understand that the internal angles of *every* plane triangle must equal 180 degrees.

A fourth source of truth is *intuition*. Sometimes the truth comes to a person unexpectedly and apart from any conscious reasoning process. Nearly all of us seem to have true intuitions; a few people seem to have an unusual number of true intuitions. How intuition works is not clear. Perhaps it is a reasoning process that goes on unconsciously and then flashes an answer on the screen of consciousness; perhaps it is a result of extrasensory perception; perhaps it works different ways at different times. However it is caused, the person who experiences it feels that his or her intuition is true.

Wilfred Benitez, an outstanding boxer, fought for years, became champion of the welterweight division, and was undefeated until he fought Sugar Ray Leonard, who stopped him by a technical knockout in the 15th and last round of a close, exciting fight. After that agonizing loss, Benitez beat his next opponent impressively and called for a rematch with Leonard. Anticipating the rematch, Benitez said, "I know I'm going to knock him out. That's safe to say because that's the truth." If it was the truth, it was certainly safe to say. Unfortunately, the rematch never happened, but Benitez's pronouncement illustrates well the confidence of intuition.

Dreams are a fifth source of truth. I read about an unemployed California carpenter who dreamt that he was going to "hit it big." The dream was so powerful that he woke up in the middle of the night and could not go back to sleep. Early in the morning he packed and traveled to Reno, Nevada. He went to a casino, couldn't play the dollar slot machine that he wanted to play (someone else was playing it), so he started playing a quarter machine. After about only $30, he hit the jackpot for $67,261! Note: I read this in an Associated Press report, not in a supermarket tabloid! Besides, it is well known that one of the most important chemical discoveries of the twentieth century, the shape of the benzene ring, came to its discoverer in a dream.

Finally, though my list is surely not exhaustive, *revelation* is a sixth source of truth. I may have no way of finding out what you did on your sixth birthday, but if you remember, you can reveal it to me. And surely I have no way of knowing what you dreamed last night – unless you reveal it to me or someone else. Indeed, there are many truths that we cannot know about a person's inner life unless she reveals them to us.

When I mentioned "revelation" as a source of truth, you may have thought I meant "divine revelation," so let me say something about that. If there is a God, then surely God can reveal to us truths that we could not find out otherwise. Hence, if there is a God, then divine revelation is a possible source of truth.

It seems, then, we are rich with sources of truth. Consequently, our minds should be overflowing with truths! And perhaps they are. But an obvious problem is that all of these sources produce false beliefs as well as true ones. Illusions of sensation are common: the partially submerged stick that looks bent but isn't; the speeding locomotive whistle that seems to change pitch but doesn't; the steel that feels hot but is cold. There is also the answer that we reason out and are confident of but mistaken about. Even introspection can mislead us. I had a friend who suddenly experienced sharp pains in his chest. He believed he was having a heart attack, so he had someone rush him to an emergency room. An examination revealed that his heart was fine; the problem was his back. He had a back problem that caused radiating pains which felt as though they originated in his chest.

Then, too, many a dream has not come true. I am especially fond of the following story. An NBC sports producer, Peter Rolfe, said that before the 1984 boxing match between heavyweight champion Larry Holmes and Marvis Frazier (son of another great heavyweight champion, Joe Frazier), a friend who was an editor of a major boxing magazine phoned Rolfe and said that he had dreamt that Marvis beat Holmes by a decision. Because of the dream, Rolfe's friend was emphatic: "I *know* that Marvis is going to win," he said. What happened? Marvis was crushed in the first round. Let's hope Rolfe's friend didn't bet his home on his dream.[3]

Finally, there are problems with revelation. First, when a person is mistaken about his or her own past or inner life (as, for example, my friend was when he mistook his back problem for a heart problem), then when we accept what he or she reveals to us, we, too, will be mistaken. Further, when people *do* know the truth about these kinds of things, they can easily deceive us when they want to – as every good liar knows.

As to divine revelation, assuming that we are speaking of a God who knows all truths and is perfectly good, if God reveals something to us, then it will be the truth and we can trust it. God won't be wrong and won't deceive us. The problem here is knowing whether something that *seems* to be a revelation from God *is* a revelation from God, rather than a trick by the devil or a figment of our imagination or a trick that someone is trying to play on us. Recall from chapter 4 a statement by Thomas Hobbes: "If someone says that God spoke to him in a dream, he has said nothing more than that he dreamt that God spoke to him." That is a delicious turn of phrase. Ponder it carefully. If there is no God, then certainly Hobbes is correct. If there is a God, then Hobbes may be correct or he may be incorrect, for if there is a God, then surely God can communicate truths to us through our dreams. However, that fact doesn't relieve us of the responsibility and difficulty of deciding whether the message in a dream is truly from God – and whatever we decide, we might be mistaken.

It seems, then, that all six of these means can lead us to truth, but each can also lead us into error. Because of that unhappy fact, we want more than just to *have* the truth; we want to have it in a special way. We want to have it in such a way that we *know* that we have it. We want *knowledge* of the truth – and that leads us to ask next: "What is knowledge?"

Notes

1 Bertrand Russell, *The Problems of Philosophy* (NY: Oxford University Press, 1959), p. 129. Originally published 1912.
2 *Parade Magazine*, October 18, 1987, p. 9.
3 *The Ithaca College News*, 6, 15 (April 25, 1984), p. 2.

Reading Further

Brand Blanshard, Chapters XXVI and XXVII, *The Nature of Thought*, volume 2.
William James, "Pragmatism's Conception of Truth," in *Pragmatism and Other Essays*. "Preface," *The Meaning of Truth*.
Bertrand Russell, *The Problems of Philosophy*, Chapter XII, "Truth and Falsehood."

Chapter 11: What is Knowledge?

- Hope
- Faith
    Why it is easily confused with knowledge
- Belief based on evidence
- True belief based on evidence
- Justified belief
- Justified true belief
- The justification theory of knowledge
    Conclusively justified true belief
- The causal theory of knowledge
    Knowledge without justification
    Knowing that one knows

One of the best ways to proceed toward understanding what something *is* is to examine things that it is not and to understand why it is not the same as those things.

Hope

For example, is knowledge the same thing as hope? It would seem not. In fact, hope and knowledge about one and the same thing are not even compatible. It makes sense to hope that something is true only when we think we do not know it is true. It makes sense to hope that your answer to a test question is correct only if you think you do not know it is correct. More generally, to hope that something is true is to be intellectually uncertain as to whether it is true but to *want* it to be true because we think it would be good if it were true. For example, we hope that our severely injured leg will be saved from amputation or that we will get the job for which we have applied.

Furthermore, what we hope is true might be false, whereas what we know cannot be false. That is a difficult but very important point. It follows from the fact that we can think we know something without actually knowing it. If something that you think you know turns out to be false, then you *thought* you knew it, but the fact that it turned out to be false means that you did not know it – you only thought you knew it. When people 'knew' that the sun went around the earth, they didn't know it; they just thought they knew it.

Faith

Okay, knowledge is not hope. Is it perhaps faith? If you have faith that something is true, then don't you know that it is true? The answer is "not necessarily." To have faith that something is true is to feel confident beyond any evidence that we have that it is true, but we can feel confident that something is true, yet be wrong. I once loaned money to a stranger who gave me a hard luck story; he even took me to his apartment to visit his wife who was sick in bed. I felt that I could trust him. A few days after I was supposed to be repaid (and I needed the money!), I checked back, only to be told by the landlord that the couple was gone, the wife hadn't been sick, and I had been the victim of a scam. I had faith that the couple's story was true. Obviously I did not *know* that it was true. Here is another example. When my mother was in a post-operative coma caused by a stroke, a relative whom I love dearly visited with my mother and me in the hospital; afterwards she assured me emphatically that my mother was "going to be okay." My relative had faith that my mother would come out of her coma. Alas, once again the fact that faith is not the same thing as knowledge was made clear. My mother died without coming out of her coma.

The preceding are examples of faith being false, but what if the man had paid me back and my mother had recovered? Would it then have been the case that my relative and I knew our beliefs were true? When faith turns out to be true, does that mean it was knowledge? The answer seems to be "no"; the reason is that when one has faith that something is the case, that means, by definition, that he or she does not know it is the case but only has a strong conviction that it is the case. When knowledge enters, faith, like hope, fades away.

Yet if faith and knowledge are not the same thing, why are they so often mistaken for one another? Why do people who only have faith that something is the case often think they know it is the case? The answer is that faith and knowledge generally *feel* the same way – quite confident and without doubt. To have *faith* that something is the case is to think it is the case and to have no serious doubt that it is the case. To *know* that something is the case is to think it is the case and to have no serious doubt that it is the case. So how do faith and knowledge differ? Faith is confidence that is not based primarily on reasons or evidence whereas knowledge typically is based on reasons or evidence. To be sure, people who have faith that something is true may, and often do, have *some* reasons or evidence which support what they have faith in, but what makes their belief a matter of faith, rather than of knowledge or belief based on probability, is that its confidence arises from some source other than reasoning, and so it usually remains strong and stable in spite of fluctuations in the evidence.

This difference between faith and knowledge comes out wonderfully in how Grace Vaughan corrected herself in an interview. Vaughan was an 80-year-old woman who got lost in the rugged terrain of rural Virginia while hunting for mushrooms with her 85-year-old husband and a neighbor. She was lost for a full week with little water and no food. One more day, said the head of the rescue squad, and she would have died. Vaughan said that her first words when she saw the rescue team were "Praise the Lord!" Later that day she told someone, "I had a regular prayer session this morning and I just knowed, or I felt like I knowed, that I was gonna be taken out of there that day." Notice how she qualified her statement: first she said that she *knew* she was going to be rescued that day; then she sensed that her statement was too strong, so she changed it to say that she *felt* like she knew she would be rescued that day. In other words, after her prayer session she had *faith* that she would be rescued that day. Her grammar wasn't perfect, but her mind was working fine. She recognized that her confidence was a matter of faith, not knowledge.

Belief based on evidence

Knowledge, then, must be based primarily on reasons or evidence, not feeling or intuition. If, then, I believe something because of evidence, does that mean I know it? Not necessarily. Evidence can be misleading, misunderstood, or inadequate, so even though my belief is based on evidence, it might be false. Hence, because knowledge cannot be false, belief based on evidence cannot be the same thing as knowledge.

To be sure, a person always *thinks* that his or her beliefs are true; that is part of what it means, by definition, to believe a proposition, namely, to think it is true. In popular terminology: a belief is always true to the person who has it. That, however, does not mean that the belief *is* true. Believing for a reason that something is the case doesn't necessarily mean that it is the case, nor does it make it the case; it means only that for some reason, perhaps a very good reason, a person *thinks* it is the case, but they might be mistaken. Remember, it is reality that makes a belief true or false. Believing doesn't make it so.

Another tricky but important distinction is the following: knowledge cannot be false, but we can have knowledge of that which is false. For example, given what is typically meant in geometry by the words "triangle" and "square," I know it is false that a triangle has more sides than a square. My knowledge is true, however, because what I know is that *it is true that it is false that a triangle has more sides than a square*. Hence, my knowledge is not false; rather, my knowledge is knowledge that it is true that a certain proposition is false. Knowledge itself is always true. Belief is not always true, so they cannot be the same thing.

True belief based on evidence

Is knowledge, then, *true belief based on evidence*? If you think for a reason that something is true *and it is true*, then do you have knowledge? Let's think about it. First, a true belief can be based on very weak evidence. As a novice at a race track, you might read the racing forms and tip sheets, place your bet and win, so you had a true belief, and it was based on information, but if it was a fair race, you didn't *know* which horse would win. Right? Second, consider the fact that a few people in ancient Greece were heliocentrists; they believed that the earth revolves around the sun (see Aristotle's *On the Heavens*, II.13–14). Hence, ancient heliocentrists had a true belief, and they had some reasons for their belief, but they did not have enough evidence to *know* that the sun went around the earth. Hence, a true belief based on evidence is not necessarily knowledge. Something more than truth and evidence is needed to turn belief into knowledge.

Justified belief

Is justification the "more" that is needed? Is knowledge the same as *justified belief*? If you *think* that something is the case, and you are genuinely *justified* in thinking it is the case, then do you *know* that it is the case? Let's think about it. To be justified in thinking that something is the case means to have sufficiently good reasons for thinking it is the case that no rational person would criticize you for thinking as you do. However, could a person be *justified* in believing something that is false? Unfortunately, the answer is "yes." Justified belief is not necessarily true. For example, most people in ancient Greece were geocentrists; they thought the sun went around the earth. Were they justified in thinking that way? Yes. As we say, "Seeing is believing." It looked to them (as it does to us) like the sun circles the earth, and they had no good reason to think otherwise. Hence, they had a justified *false* belief.

Consider also the honest, hard-working jury that convicts an innocent person.

The members of the jury listen carefully to all of the testimony, discuss it vigorously, and it all points in the direction of the guilt of the defendant, so they are justified in their conclusion that the defendant is guilty. However, unknown to the jury, the defendant and some of the witnesses colluded in lies to deceive the jury and protect the person who actually committed the crime. Hence, the jury's decision was justified but false. Knowledge, as we have seen, cannot be false, so justified belief cannot be the same thing as knowledge.

Justified true belief

Knowledge, then, is not simply true belief, and it is not simply justified belief. Knowledge needs to be both true and justified, so now the answer to our question seems clear: knowledge is *justified true belief*. If you think a proposition is true, and it is true, and you are justified in thinking that it is true, then you *know* that it is true!

We've come a long way from asking whether knowledge is hope or faith or mere belief. We've made definite and important progress, but we're still not home. The problem is that a justified true belief might not be justified *enough* to constitute knowledge. For example, weather forecasting is an applied science; it presupposes a lot of knowledge and training. More often than not, short term forecasts by trained meteorologists are correct – but sometimes they are not. When a trained, experienced meteorologist makes a responsible weather forecast that turns out to be true, then her belief about the weather was true and justified, but that does not mean that she *knew* what the weather would do; in fact, she would probably be the last person to say that she *knew* her forecast was correct. Prior to the forecast becoming true she would say, "Because of the evidence, I think I am probably correct – but I could be wrong." The same is true of a jury which correctly convicts a defendant on the basis of evidence that is strong enough to indicate guilt "beyond a reasonable doubt" but which is not conclusive.

The justification theory of knowledge

That word "conclusive" might hold the key we are looking for. Consider the following: knowledge is *conclusively justified true belief*. If you think a proposition is true, and it is true, and you think it is true because of reasons which show *conclusively* that it is true, that is, which *prove* that it is true, then you have knowledge. Note, this means that you cannot just *think* that your reasons prove the belief to be true; in order for you to have knowledge, they really must prove it, and you must understand how they prove it. When you do understand this, then you know that what you think is true is true.

To summarize, according to the justification theory of knowledge there are four conditions that are necessary for you to have knowledge: (1) you must think that a specific proposition is true; (2) that proposition must be true; (3) you must have reasons which prove that that proposition is true; (4) you must understand how those reasons prove that that proposition is true. Why are these conditions necessary for knowledge? Without (1) you wouldn't have a belief about A (the proposition in question), and it is not possible to know that A is true if you think A is false or if you don't have a belief about whether A is true or false. Without (2) A would not be true, and you cannot know that a proposition is true if it is not true. Without (3) you would not have a strong enough justification for it rightly to be said that you *know* that A (though you might be justified in thinking that A is *probably* true or that it is *plausible* that A is true). Without (4) *you* wouldn't know that A is true because you wouldn't understand how the reasons in (3) prove that A is true. For example, if in a mathematics book I find what claims to be and really is the only proof that A is true, but I cannot understand the proof, then I do not know that A is true – even though on the basis of someone else's authority I may be justified in accepting that the solution works and A is true. Hence, all four of these conditions seem necessary for knowledge, and on the version of the justification theory of knowledge being presented here, together they are sufficient for knowledge. That is, if you think that A is true, and A is true, and you have and understand a proof that A is true, then you know that A is true.

The causal theory of knowledge

Some people are dissatisfied with the justification theory of knowledge because, they say, "If that's what knowledge is, then we have very little of it – if any!" It does seem true that we have few beliefs that are conclusively justified or that we can prove, and therefore we have little knowledge in the preceding sense, but perhaps that is just an unhappy truth about the human situation. We shouldn't arbitrarily define knowledge so that we can have as much of it as we want. That would be like defining "wealth" in terms of however much money we have. Such a move might make us 'wealthy,' but if we were poor (barely had enough money to feed ourselves) before we became wealthy by redefining "wealth," we would still be poor afterwards. Similarly, defining "knowledge" in a broad, permissive way would not give us any more knowledge than we had before we introduced the new definition.

In spite of that fact, many people are convinced that we know much more than we can prove, so they reject the "conclusively justified true belief" conception of knowledge. Perhaps you are not satisfied with it, either, so let's look at an influential alternative which is called "the causal theory of knowledge." (Please note: the word is "causal" not "casual". Big difference! If you are not sure of the difference, look up both in a dictionary.)

According to the causal theory of knowledge, a person can know that something is the case without personally having a proof of it or even a justification for it. A person has knowledge that something is the case whenever his or her belief that it is the case is *caused in the right way*. For example, if you think the sun is shining, and the sun is shining, and you think the sun is shining because you are looking at the sun, your eyes are in good condition, and the sun is *causing* you to see it shining, then you know the sun is shining. Similarly, if you think the telephone is ringing, and the telephone is ringing, and you think the telephone is ringing because the ringing of the telephone is causing you to hear its ringing, then you don't merely *think* the telephone is ringing, you *know* that it is ringing. Notice that in these examples you do not need to be able to prove that the sun is shining or the telephone is ringing in order to know that it is. You just need to think it is shining or ringing because, in the one case, the shining of the sun is causing you to think it is shining, and, in the other case, the ringing of the telephone is causing you to think it is ringing.

This approach to knowledge has the virtue of accounting for our commonsense conviction that we *know* many things that we cannot prove. When you say, "It has stopped raining," and someone says, "Prove it," and you take them outside where the two of you agree that you neither see nor feel raindrops, and the other person says, "But maybe we are hallucinating that the rain has stopped. Or maybe you are dreaming that it has stopped. How do you *know* it has stopped? Prove it!", you dismiss that person as silly, boring, or insane. The causal theorist agrees and says, "If you think the rain has stopped, and you think the rain has stopped because the stopping of the rain has caused you to think it has stopped, then you *know* it has stopped – even if you cannot prove conclusively that it has stopped."

Still, it does seem odd to say that we know something about which we are not certain – about which we must acknowledge that we might be mistaken. After all, surely it is possible to hallucinate or dream that rain has stopped when it has not. Therefore, perhaps you *are* hallucinating or dreaming that the rain has stopped when it has not! So how can you know for certain that the rain has stopped?

The causal theorist might respond to this question by making a distinction between *knowing*, on the one hand, and *knowing that one knows*, on the other hand. According to the causal theory of knowledge we can know that something is the case without knowing that we know that it is the case. In order to know that we know that something is the case, our belief that we know that we know must be caused in the right way. One appropriate cause of such a belief would be having and understanding a proof that our belief is true. If I cannot rule out the possibility that I am hallucinating or dreaming or being deceived by someone who is manipulating my brain, then perhaps I cannot know that I know that it has stopped raining. But it seems that I can and do know that I know that the cube root of 8 equals 2 because of what I mean by "cube root", "8", "2", and "equals", and because I understand how the relations between those four concepts prove that the cube root of 8 is 2. In brief, my comprehension of these concepts and their relations has caused me to know that I know that the cube root of 8 is 2. More generally, having and understanding a proof that one's belief is true causes one not only to know but also to know that one knows, and perhaps there are other causes of knowing that one knows (some people argue that a revelation from God could have that effect).

Whether we should think of knowledge as proven true belief or as belief that has been caused in the right way is still quite controversial. Moreover, there are additional, important theories of knowledge, such as the coherence theory and the pragmatic theory, which, alas, you will have to learn about elsewhere. (See a good dictionary of philosophy or encyclopedia of philosophy.) Meanwhile, you must decide whether you think the word "knowledge" should be restricted to beliefs that we can prove, as in the justification theory of knowledge, or whether you think knowledge is something we can have without being able to prove it, as in the causal theory of knowledge.

Reading Further

Plato, *Theatetus*.
Robert Audi, *Epistemology: A Contemporary Introduction to the Theory of Knowledge*.

Chapter 12

Logic – Understanding and Evaluating Arguments

- What "argument" means in logic
    - Not a disagreement, debate, or dispute
    - The nature of an argument
    - The nature of a debate
- The deductive argument
    - Two parts
    - Mode of necessity
    - Fallacious reasoning
    - Valid or Invalid
        - Necessary and sufficient conditions
    - Sound or unsound
    - Not true or false
- The inductive argument
    - Two parts
    - Mode of probability
    - Warranted
    - Unwarranted
        - Irrelevant (superstition)
        - Too weak

We do not merely want to *have* the truth; we want to *know* that we have it. We usually do not think we have the truth unless we can give good reasons for thinking that we have it. Similarly, we do not think that other people know what they claim to know unless they can give us convincing reasons for their claims. Indeed, when we are unsure of what someone claims to be the case, our almost automatic response is, "How do you know that?" Or, "What reasons can you give to convince me that you know that what you are claiming to be true is true?"

John Wilson pointed out the following reason for the importance of giving reasons: "There are a great many occasions on which people say things like, 'Well, you may argue as much as you like, but I just KNOW that [so and so is true]', or 'I just FEEL that [so and so is true].' What is missing here," Wilson continues, "is the notion of GIVING REASONS for beliefs. If we abandon this notion, there is really nothing to distinguish sane human beings from lunatics."[1] Why? Because lunatics are notorious for being absolutely convinced of what they see or think or feel. So if it is merely the degree of confidence one has in one's beliefs that makes those beliefs true or worthy of serious consideration, then the lunatic's beliefs are at least as true or worthy of serious consideration as are anyone else's. But surely that is not true – unless the lunatic has good reasons for his or her beliefs – in which case perhaps he or she is not a lunatic!

We want, then, to hear the reasons for a person's belief. *Argumentation* consists of giving a reason or reasons for one's belief, for thinking that a claim is true, so we want people to give arguments in support of their beliefs. Logic is the study of argumentation. More fully, *logic* is that part of philosophy which helps us understand, recognize, and evaluate arguments – and arguments are the tools by means of which we justify our beliefs to ourselves and others. A knowledge of logic helps us do a better job of constructing arguments of our own and evaluating arguments of other people, so let's examine some of the basic concepts of logic, beginning with the question, "What is an argument?"

What "argument" means in logic

Like so many words in English, "argument" has many meanings. Its meaning in logic is related to several of the more common meanings, but it is also different, so let's begin by noting what an argument in logic is *not*. First, by "argument" in logic we do not mean *a disagreement*. When one child says, "There is a Santa Claus," and another child says, "There is not," and the first child says, "There is!," and the other says, "There is not!," they are having a disagreement, not an argument. Why? Because they are simply taking different positions on the question of the existence of Santa Claus. They are not giving reasons for their positions; they are simply disagreeing, not arguing. Arguing, as it is defined in philosophy, requires the giving of reasons.

Second, "argument" in logic does not mean *a debate*. Let's assume that ten years later those same children, now 16-year-old friends, disagree over whether God exists. Now, however, they do not merely disagree; rather, each gives reasons for her position. It is common to call such a debate an "argument," but a debate requires two people plus a disagreement between them, whereas what is meant by "argument" in logic does not require two people or a disagreement.

Third, "argument" in logic does not mean *a dispute*. By "dispute" I mean a quarrel, a debate that is nasty or unpleasant in tone, usually involving explicit or implicit insults. Because a dispute is a kind of debate, and we have seen that a debate is not an argument, therefore a dispute is not an argument. Of special importance here is that an argument in logic does not have to be insulting in tone. Indeed, as we have seen, insult is antagonistic to philosophy, of which logic is a part. Moreover, in the logic sense of "argument," an argument doesn't have an emotional tone at all!

So what is an argument? To answer that question, I will start with a technical definition and then explain it in detail. Here is the technical definition: *An argument is a set of two or more assertions, one of which is claimed to follow logically from the other or others.* (A person can, of course, construct an argument in his or her mind without stating it publicly. Indeed, we usually do go over arguments in our minds before we state them publicly.)

Notice from this definition that an argument must consist of at least two assertions. Why? Because a single, simple assertion, such as, "There is life after death," is not an argument; it is simply a bare, 'flat' claim. Someone who makes such a claim has not presented an argument. The reason that at least two assertions are needed for an argument is that an argument *consists* of using one or more assertions *to justify another assertion* – or, to put it the other way around: an argument consists of one assertion that is being justified by one or more other assertions.

For example, "People should be judged on their individual merits. Therefore, racism is wrong," consists of two assertions which constitute an argument because one assertion, "People should be judged on their individual merits," is being used to justify the other assertion, "Racism is wrong" (or to put it the other way around, the assertion "Racism is wrong" is being justified by the assertion "People should be judged on their individual merits").

What about the following two assertions: "The sky is sunny. I am listening to Mozart." Do they constitute an argument? Because we read from left to right, top to bottom, the one statement does *follow* the other statement, but it does not follow from it *logically*. Moreover, I did not claim that either of those statements follows logically from the other. I did not mean to present an argument, and I did not. I was just making two independent observations.

Notice also that just because two statements *can* be made into an argument does not mean that they *are* an argument. Two statements do not become an argument until someone *claims* that one of the statements supports the other. Consider these two statements: "An equilateral triangle has three sides equal in length. An equilateral triangle has three angles equal in degrees." Many children know those two facts and can get them right on a test without realizing that those facts can be turned into an argument. That is, the children do not realize that the one truth can be deduced by logic from the other truth: "An equilateral triangle has three sides equal in length. *Therefore*, an equilateral triangle has three angles equal in degree" (of course some intervening steps need to be added!). That illustrates one reason why it is important to say that an argument consists of two or more assertions one of which someone *claims or thinks* follows logically from the other or others.

Now let's note how an argument in the logic sense relates to an argument in the debate sense. A debate between two people consists of *two logic arguments* pitted against one another. A debate requires that the debators disagree about a single topic, and each debator must give reasons for his claim; therefore, each debator gives an argument for his position. This means that a debate type argument consists of two logic type arguments. For example, the atheist and the theist disagree about the topic of the existence of God. In a debate between them the atheist presents an argument that God does not exist, and the theist presents an argument that God does exist.

Whenever you hear or read someone giving a reason in support of a claim, you are encountering a logic type argument. Whenever you hear or read a debate between two people, you are encountering *two* logic type arguments – one argument being developed by each debator and directed against the conclusion of the other debator. If the relation between a debate (in the ordinary sense) and an argument (in the logic sense) is not already clear, perhaps the following will help. A geometrical intersection (such as that which creates an X) is not the same thing as two lines(two parallel lines do not create an intersection), even though an intersection cannot exist apart from two lines. Similarly, even though a debate cannot exist apart from two arguments (in the logic sense of "argument"), a debate is not simply the same thing as two arguments (in the logic sense of "argument"). For two arguments to constitute a debate they must be on the same topic and arrive at opposite conclusions. If at a party you argued that space is infinite and I argued that the arts are as important as technology, then two arguments were made, but we did not engage in a debate.

Now let's distinguish the *two most basic types of argument*: deductive and inductive. Nearly every argument that you encounter will be of the one type or the other, so let's examine each, beginning with the deductive argument.

The deductive argument

A deductive argument has two parts: a *conclusion* and a *premise* or premises. The premise is (or the premises are) used to justify that part of the argument which is called "the conclusion." The conclusion is that part of the argument that is claimed to be justified by the other part of the argument (the premise or premises).

We usually think of the conclusion of an argument as being stated at the end of the argument, but a conclusion can be stated at the end of an argument or at the beginning or (when there are two or more premises in addition to the conclusion) in the middle. Consider the following argument (to be understood along biological lines): "All mothers are women, therefore no man is a mother because no man is a woman." "All mothers are women" and "no man is a woman" are being used to justify "no man is a mother"; therefore, "no man is a mother" is the conclusion of the argument even though it is stated in the middle of the argument.

What distinguishes a deductive argument is that its conclusion is claimed to follow from its premise(s) with *necessity*. That is, a deductive argument is a set of two or more assertions one of which is claimed to follow *necessarily* from the other(s). The arguer is saying: "Given this premise, that conclusion *has* to follow (that is, follows necessarily; follows with certainty)." For example, "All mothers are women, and no man is a woman. Necessarily, therefore, no man is a mother."

But note: all arguers are human, and all humans are fallible; therefore it is possible for an arguer to *claim* that a conclusion follows logically from a premise when in fact it does not. That is a second reason why it is important to define an argument as something someone *claims* to be the case. Just because an argument is not a good argument, that is, does not succeed in showing what it claims to show, does not mean that it is not an argument; it means only that it is not a good argument. In general, when an argument does not succeed, we say it is "fallacious." That means the argument contains a fallacy. A *fallacy* is a mistake or flaw in one's reasoning from one's supporting ideas to one's conclusion.

Now let's look at the special terminology used to indicate that a deductive argument is good or is fallacious. When the conclusion of a deductive argument follows necessarily from its premise or premises, the argument is said to be *valid*. To say that the conclusion "follows necessarily" is to say that *given* the premise(s), the conclusion *has* to be the case. For example:

> All humans are mammals.
> I am a human.
> Therefore, I am a mammal.

From those two premises, it follows necessarily that I am a mammal, so the preceding argument is valid.

When the conclusion of a deductive argument does *not* follow necessarily from the premise(s), it is said to be *invalid*. Consider the following argument: "All bachelors are men. Therefore, all men are bachelors." Its premise ("All bachelors are men") is true because by definition a person cannot be a bachelor without also being a man; that is, being a man is a *necessary condition* of being a bachelor.

However, the premise doesn't say or imply that being a man is a *sufficient condition* of being a bachelor; it leaves open the possibility that a person can be a man without also being a bachelor. Therefore, it is not necessarily the case that all men are bachelors. Hence, this argument is invalid because its conclusion does not follow necessarily from its premise. For the same reason the following argument is invalid: "All mothers are women. Therefore, all women are mothers."

The preceding distinction between *necessary and sufficient conditions* is an extremely valuable tool for the analysis of concepts, so let's examine it more closely before continuing our study of deductive arguments. A necessary condition is a condition that is necessary for something else to exist or take place. For example, oxygen is a necessary condition for fire: if no oxygen, then no fire. By contrast, a *sufficient* condition of something will *make* that something exist or occur. Obviously oxygen is not a sufficient condition of fire. If it were, then the atmosphere and the contents of every container of oxygen would be on fire! It is rare for a single thing to be a sufficient cause of something else, so usually we speak of "sufficient conditions," in the plural. For example, being a male is not a sufficient condition of being a bachelor, but being a male, being an adult, being human, and never having been married *are* sufficient conditions of being a bachelor.

Sufficient conditions are notoriously hard to pin down because they have to be exhaustive. For example, do I need to add to my definition of bachelor that the adult, human, never married male in question is alive? Can a corpse be a bachelor? Also, part of a sufficient condition of something need not be a necessary condition of that something. For example, wood is part of one set of conditions sufficient for fire, but wood (unlike oxygen) is not a necessary condition of fire. Fire can also be generated from oil, coal, or gas.

Now let's return to our study of deductive arguments. The primary objective of an argument is to justify confidence in its conclusion; therefore, an invalid argument fails to achieve its objective. However, that does not mean that the conclusion of an invalid argument is false. The conclusion of an invalid argument can be true! The conclusion of the preceding invalid argument, "All mothers are women. Therefore, all women are mothers," is false. Consider, however, the conclusion of the following argument:

>All frogs are amphibians.
>Some amphibians are red.
>Therefore, some frogs are red.

Both of the premises of this argument are true, but all that follows necessarily from those premises is that some frogs *may* be red. It does not follow *necessarily* that even a single frog is red. Perhaps all the amphibians that are red are salamanders or newts or turtles.

The lesson here is that once you see that an argument is invalid, then you know you cannot tell from the reasoning of the argument whether the conclusion is true or false (even if all its premises are true!). You'll have to find some way other than that argument to determine whether its conclusion is true or false. For example, we know by exploration that there are frogs of amazingly diverse colors in South America, including yellow, blue, and red (the skin of the red ones is poisonous for humans even to touch!), so the conclusion of the preceding argument ("Some frogs are red") is true even though we can't know that from the argument, since the argument is invalid.

Similarly, the conclusion of a *valid* argument with *even one false premise* can be true or false, so a false premise renders a valid argument useless and sometimes dangerous because if you do not know that a premise is false, the *validity* of the argument may mislead you into assuming that the conclusion is true. Consider the following:

> All creatures which have fins and live exclusively in the water are fish.
> Porpoises have fins and live exclusively in the water.
> Therefore, porpoises are fish.

This is a valid argument because *necessarily* if all creatures which have fins and live exclusively in the water are fish, then porpoises are fish. However, the first premise is false. Having fins and living exclusively in the water are *necessary* conditions of being a fish, but they are not *sufficient* conditions of being a fish. To be a fish a creature must also be cold-blooded, but porpoises are not. Hence, because one of its premises is false, we can't tell from the preceding argument alone (even though it is valid!) whether porpoises are fish. To identify this particular problem, a valid argument with an untrue premise is called "unsound."

To summarize, we cannot trust the conclusion of an invalid argument or of a valid argument with a false premise because in neither case can we tell from the argument alone whether the conclusion is true or false.

Is there, then, any type of argument the conclusion of which we can be certain about? Yes. It is called a "sound" argument, as distinguished from an "unsound" argument. A *sound* argument is *a valid argument with all true premises*. When an argument is *sound*, the conclusion has to be true! Here we see why validity is so important even though it cannot establish truth by itself alone: when all of the premises of an argument are true, valid reasoning *preserves* the truth of the premises and *transmits* that truth to the conclusion of the argument, so when all of the premises of an argument are true, valid reasoning from them to the conclusion ensures that the conclusion will be true. (Trust me on that. To *understand* how validity ensures a true conclusion from true premises most people need to take a logic course.) Hence, in our pursuit of truth by means of deductive logic our objective is to discover or construct *sound* deductive arguments.

Now let's pause for a technical point that is important but difficult to get used to: *arguments are not true or false.* Deductive arguments are valid or invalid, sound or unsound. Assertions are true or false. Because each premise of an argument is an assertion, and because the conclusion of an argument is an assertion, therefore each premise of an argument is true or false, and its conclusion is true or false. But an argument, unlike its parts, is not an assertion. So we evaluate each *part* of an argument for truth or falsity, but we evaluate the *whole* argument for validity or invalidity, soundness or unsoundness. When doing logic, then, we need to stifle our tendency to say that an argument is true or false. Strictly speaking, it is not.

Note that in the preceding paragraph I was doing prescriptive philosophy. In line with common philosophical usage I was prescribing how to use certain words in order to avoid confusion and make ideas clearer. People do in fact say that arguments are true or false, and they do say that an assertion is valid or invalid, sound or unsound. In logic it is recommended that we not speak in those ways in order to keep important differences clear between a simple assertion, for example, "All dogs are mammals," and an argument (which could be considered to be a complex assertion), for example, "All dogs are mammals, and all beagles are dogs, therefore all beagles are mammals."

It is okay to speak in the old ways in appropriate circumstances when we know better, just as it is okay to say, "The sun is rising," when we know better. But when we need to be careful and accurate in what we are thinking and saying, it is just as important to know the difference between truth, on the one hand, and validity and soundness, on the other hand, as it is to know the difference between the sun rising and the sun appearing on the horizon because the earth is rotating.

The inductive argument

Now let's turn our attention to the other basic type of argument: the inductive argument. Every argument, whether deductive or inductive, has two basic parts: (1) the conclusion and (2) that which is used to support or justify the conclusion. In deductive arguments that which is used to support the conclusion is called a premise. In an inductive argument that which is used to support the conclusion is *not* called a premise – nor is there a single word for what it is called. Two of the most common words for what supports the conclusion in an inductive argument are "data" and "evidence." The conclusion is "based on" certain data or evidence. Sometimes the conclusion is called an "hypothesis" or "theory" rather than "a conclusion."

DATA or EVIDENCE – *supports* – →HYPOTHESIS or THEORY or CONCLUSION

The mode of an inductive argument is that of *probability* rather than certainty. This is the critical difference between a deductive argument and an inductive argument. A deductive argument claims that its conclusion follows with necessity from its premise(s), whereas an inductive argument makes a weaker claim, namely, that the conclusion follows from the evidence not with necessity but with some degree of probability.

Let's assume that degrees of probability range from anything just above zero percent probability to anything just short of 100 percent probability. Anything that is 100 percent probable is not probable in the ordinary sense; it is definitely, perhaps necessarily, the case. Anything with zero probability is not improbable in the ordinary sense; it is definitely not the case, and perhaps is impossible. That which is impossible lies outside the category of probability, and so does that which is necessary.[2]

1% → increasing probability → 50/50 ← decreasing probability ← 99%
⇑ Very weak   Neither probable ⇑ Nor improbable   Very strong ⇑

Inductive reasoning, that is, probabilistic reasoning, is the kind of reasoning that is most common in science and everyday life, where few things are certain. As Joseph Butler stated so well: "Probability is the very guide of life." However, because we are fallible, inductive arguments, like deductive arguments, can be good or bad. A good inductive argument is called "warranted" because the conclusion is sufficiently supported by the evidence submitted to support it. A bad inductive argument is called "unwarranted" because its conclusion is not adequately supported by the evidence cited on its behalf.

Because an inductive argument claims some more or less specific degree of probability for its conclusion, if the evidence on which the conclusion is based *supports that degree of probability* for the conclusion, then the conclusion is *warranted*; if the evidence *does not support that degree of probability* for the conclusion, then the argument is *unwarranted*. If, for example, someone asks me, "Is it going to rain here tomorrow?," and I say, "Probably" (meaning that there is more than a 50% chance that it will rain), and if I believe that claim because today I both heard it on the weather report of a local radio station and read it in the weather section of a local newspaper, then my inductive reasoning that it will rain is warranted.

By contrast, if I say "Probably" only because I remember hearing "Expect rain tomorrow" on a national television weather station to which I was not paying close attention, and because I glanced at the floor where a month's worth of newspapers were strewn and noted that the one on top said, "Rain tomorrow," then my belief would not be warranted. The television prediction may not have been about my locale, and the newspaper I saw may have been from earlier in the month. (I have *got* to stop working on this book and clean up that mess!)

What makes inductive arguments unwarranted? There are two major reasons. The first reason is *irrelevance* of evidence. Sometimes people think that one thing is good evidence for another thing when it is not evidence at all for that thing. Here is a silly example. On "What's Happening?," a kids' television comedy series, there was an episode in which one of the characters was having a fabulous string of successes at predicting which professional football team would win its game each week. A gambler got wind of this and started making bets based on the boy's choices. Eventually he asked the boy what his method was for picking the winner. The boy replied, "I pick the team with the prettiest helmets." Soon, of course, his string of luck snapped, and the gambler, who had already placed big bets, was now in big trouble. Why? Because the comparative attractiveness of team helmets has nothing to do with whether a team is likely to win.

On a more serious level, drawing conclusions from irrelevant evidence is a source of superstition. Imagine the following: A black cat crosses a man's path; almost immediately thereafter he trips and falls. Early the next day the same man's path is crossed by another black cat; by the end of the day he catches a cold. The next day he discovers that a friend who has also caught cold was crossed by a black cat. Soon he is convinced that black cats are bad luck.

It is fascinating and understandable that such mistakes are made. It is important to realize the extent to which inexperienced people, and especially children, make such mistakes and need the help of experienced people to help them avoid or correct such mistakes. It is equally important to realize that we, too, can make such mistakes. When we do, we commit the fallacy called "*post hoc, ergo propter hoc*," which literally means, "after this, therefore because of this." As we have seen, just because one thing comes after another does not mean that it was caused by the thing after which it came. Consequently, we need to ask whether our evidence is genuinely relevant to the conclusion we are drawing from it. Is the pattern we have observed merely an *accidental correlation* (the black cat and bad luck), or does it involve a genuine *causal connection* (drinking alcohol and becoming motor impaired)?

A second major reason for unwarranted inductive arguments is evidence that is relevant to the conclusion but *too weak* to support the strength of the conclusion. If someone claims that something is highly probable, whereas her evidence indicates it is only weakly probable, then her claim is too strong and is, therefore, unwarranted. Similarly, if she claims that something is probable whereas the evidence indicates that it is less than probable, then, again, her argument and its conclusion are unwarranted. For example, imagine a novice economist who makes a stockmarket prediction which is unwarranted because although he has taken all the proper factors into account, he has forgotten that a change in the prime lending rate is twice as important as he factored it in to be. Or consider a meteorologist who says there is a high probability of clear skies tomorrow, but who holds that belief because, in a rush, she assumed the barometric pressure was still going up when in fact it had begun going down. Her conclusion was unwarranted because it was based on a mistaken belief about the evidence.

Here are two real life examples of unwarranted inductive reasoning. In 1936 Alf Landon ran against Franklin Roosevelt for president of the USA. *The Literary Digest*, a highly regarded magazine, decided to conduct a poll to figure out who would be elected. *Digest* pollsters chose names randomly from telephone books and lists of automobile owners around the nation. Then they sent questionnaires to ten million of these people. More than two million responded! Such a large percentage of the respondents said they would vote for Landon that the *Digest* published a prediction that Landon would win by a landslide.

What actually happened? Not only did Landon lose – Roosevelt won by a landslide! The *Digest* became a laughingstock and soon went out of business. What went wrong? In 1936 a large percentage of the nation's population could not afford a telephone or a car. Consequently, what the *Digest* pollsters were doing was polling relatively wealthy people. Yes, *those* people voted for Landon by a landslide, but they did not constitute a large majority of the nation's voters. Hence, the information that the *Digest* gathered was *relevant* to the question of who would be elected, but it was not strong enough to justify the conclusion that Landon would win – and much less that he would win by a landslide.

Here's another example. I once saw a television documentary about the federal government's protective witness program. This program releases and relocates convicted, jailed criminals in exchange for valuable information that these criminals can provide about more important criminals who have not yet been convicted. The focus of the documentary was on the extent to which convicted criminals, after they were released in exchange for information, went on to commit further crimes, including murder of innocent victims.

If those criminals had been kept in jail, obviously they would not have robbed or murdered people on the outside again. The journalist who reported how often they had done so was asked toward the end of the program, "How do the protective witness program people respond?" He replied, "They point out that only 17%, or about 1 out of 5, return to crime."

In addition to being a report of how protective witness criminals had behaved in the past, the reporter's statement was an implicit prediction of how they would behave in the future. Something about the numbers made me uneasy. At first I did not know why. Then I realized that 17% is *much* closer to 1 out of 6 (which is 16.7% – almost exactly 17%) than it is to 1 out of 5 (which is 20%). Hence, the journalist's 'rounding off' from 17% to 1 out of 5 instead of 1 out of 6 was unwarranted. One has to wonder whether the journalist simply made a careless mistake or whether he knowingly bent the truth to continue doing what the documentary had been doing all along, namely, arousing viewer anxiety about the protective witness program.

The moral of these examples is that we should pay very careful attention to the arguments, written and oral, that people use to try to influence us. To be sure, an argument can't bend the truth too outrageously or it will be useless because it will be obviously false, and the arguer will lose credibility. For example, if the reporter had said "17% or 9 out of 10 people in the protective witness program return to crime," everyone would have picked up on the error. By contrast, "17%, or 1 out of 5" is not such an obvious error.

People who would deceive us have to work the margins within which we can be misled. We have to work to prevent them from succeeding – people rarely try to mislead us for our own good! Hence, ask of a deductive argument: Are all of its premises true, and does the conclusion follow validly from the premises? Of an inductive argument ask: Is the evidence relevant, and is it strong enough to warrant the strength of the conclusion?[3]

Parenthetically, it is important to understand correctly how strong a claim is being made by an arguer. If someone claims that something is *probably* the case, it is not fair to respond, "That's not necessarily so." She hasn't claimed that it is necessarily so. Other times it is not clear how strong a claim an arguer means to make. In that case it is important to ask for clarification before proceeding to agree or disagree. For example, someone who says, "I believe God exists," might mean any of several very different things: "I have a feeling that God exists" or "I think the evidence for the existence of God slightly favors the probability of the existence of God" or "I think the evidence for the existence of God is so highly probable as to be almost certain" or "I can prove that God exists." To respond as fairly as possible, we need to know exactly what someone means – and if they aren't sure, then we need to use the Socratic method to help them clarify their belief for both of us. Conversely, we need to keep our ears tuned to how other people are understanding us so we can make sure they are understanding us correctly.

Finally, keep in mind that reasoning concepts and skills are very much like athletic and musical concepts and skills. To play golf well you need to know which club to use in different situations and what kind of a swing to take in each type of situation, but you also need to be able to *execute* that swing smoothly and effectively. Similarly, in order to reason well you need to know numerous concepts, including the ones we have just studied, but you also need to be able to *apply* those concepts in real life situations. As in sports and music, it takes a great deal of *practice* to ensure that when the time for action comes, you can do what needs to be done philosophically. Mere book knowledge is not adequate in music, sports, or reasoning. Consequently, I urge you to take a course or read a book that will require you to practice with the concepts of logic until you can use them smoothly and effectively. One thing is for sure: you will have plenty of opportunities throughout your life to use those skills.

Notes

1 John Wilson, *Philosophy* (London: Heinemann, 1968), p. 4.
2 You may prefer to speak of "100% probable" as meaning "definitely true," and "0% probable" as meaning "definitely false". However, the use of "probable" in ordinary language is so closely tied to the notion of a thing *not being certain* (whether by virtue of not being necessary or not being impossible) that I prefer to restrict degrees of probability to less than 100% and more than 0%. Whichever approach you take, the important thing to keep clear in your thinking and communication is the radical difference between, on the one hand, evidence or reasons which are claimed to *prove* that a proposition is true, and, on the other hand, evidence which is claimed to support to some degree of *probability* (whether weak, moderate, or strong) a claim that a proposition is true but also leaves open the possibility that it is false.
3 The concept of evidence is tricky. Is evidence something that is objective, subjective, or both? Is the evidence for a claim what someone *thinks* it is (even if mistakenly, as with the meteorologist and the barometer), or is it what is *actually* the case and relevant to the hypothesis? Here are some suggestions.

  If I mean to submit something as evidence which, it turns out, does not exist (I thought there were fingerprints but there were not), then I simply did not succeed in submitting evidence – just as I cannot make a withdrawal from my checking account if the balance is $0.00. Yes, I can write a check on my account and give it to someone, but I cannot make a withdrawal; there is nothing to withdraw. Similarly, I can *say* (and believe) that something is evidence, but if it doesn't exist, it cannot serve as evidence – just as nonexistent dollars in my checking account cannot serve to cover a check that I write.

  What if I write a check for $200 but have only $50 in my account? That would be analogous to making an unwarranted inductive argument. It is like making a strong claim, "Landon will win by a landslide," on the basis of weak evidence that does not justify it (how a wealthy minority says it will vote). As we sometimes say in philosophy, the conclusion is "underdetermined" by the evidence. Evidence for that claim about Landon existed and was relevant but was too weak to support the conclusion – just as my $50 checking account was too small to cover my $200 check.

Next consider a claim which is supported by the evidence that I submit but just barely – which would be like writing a $100 check on a checking account with exactly $100 in it. Finally there is the case of a claim which is very strongly supported by the evidence submitted – as when I write a $100 check against an account with $5,000 in it.

Reading Further

Irving Copi and Carl Cohen, *Introduction to Logic*, 8th edition or later.
Patrick Hurley, *A Concise Introduction to Logic*, 5th edition or later.

Part III Theory of Value

Chapter 13: Axiology and Happiness

- Our innate craving for happiness
    - Aristotle, Aquinas, and Pascal
    - Why we seek happiness
- Aristotle's definition of happiness
- Critiques of happiness
    - Friedrich Nietzsche, Albert Einstein, John Dewey
    - Immanuel Kant
    - Arthur Schopenhauer
        - Our one innate error
        - The hedonistic paradox
            - *The fallacy of hasty generalization*
        - The decoy effect
        - Euthansia of the will
- Axiology: Its nature and purposes
    - Alfred North Whitehead
- The Good of enjoyment vs. Enjoyment of the good
- Three conceptions of the good
    - Counterexamples
    - Argumentum ad ignorantiam
- The interaction theory of experience
- The package theory of alternatives
    - Cost, felt-qualities, consequences
    - George Santayana
- Ignorance as cause / Intelligence as cure
    - Aristotle; Diotima and Socrates
    - The Upanishads
    - The hedonic treadmill
- Jeremy Bentham's Hedonic Calculus
    - Intensity, duration, purity, fruitfulness, number, probability, propinquity
    - John Dewey: cost
- John Stuart Mill's criticism of Bentham
    - Quality vs. Quantity

- Intrinsic Values and Instrumental Values
- A summary of concepts in value theory
  Definitions: value, valuable, to value, to evaluate
  Intrinsic value: positive, neutral, negative
  Instrumental value: positive, neutral, negative

Many people turn to the writings of philosophers for the first time because they are seeking wisdom regarding happiness and morality. St. Augustine, a brilliant pivotal figure between classical Rome and medieval Europe, went so far as to say that there is no other reason for a person to philosophize than that he or she might be happy.[1] Whether that is true or not, questions concerning value are rarely far from the forefront of any person's mind, so let's take up the topic of happiness in this chapter and the topic of morality in the next chapter.

Our innate craving for happiness

Questions regarding value frequently arise in relation to action. In response to what we have already done, we ask ourselves and others whether it was a good thing to do. In response to what we are doing, we ask whether it would be good to continue doing it. In response to what we can do, we try to determine what would be good, what would be even better, and what would be best. In order to understand values, then, it is helpful to think of them in the context of action. The reverse is also true. To understand human action we must comprehend it in relation to the values that motivate it.

Aristotle examined the relationship between action and value in his *Nicomachean Ethics*, especially Books I and X. According to Aristotle, action is the substance of human existence. We humans are almost always doing something; our waking lives consist of activity. Even in the womb we are restless. But our activity is not random or aimless. It always has some aim. Indeed, according to Aristotle we are by nature goal-oriented creatures. By nature we aim at ends and act to achieve them. Those ends at which we aim are the values that motivate us.[2]

Aristotle's claim that we are goal-oriented creatures seems non-controversial. Every normal, healthy human is busy with something most of the time. Illness, by contrast, is generally accompanied by a decrease of activity and a loss of interest in one's usual sources of pleasure. When a child becomes abnormally inactive, a parent becomes worried.

The things with which humans busy themselves are, of course, extremely diverse: I am writing, you are reading, someone else is making lasagna, someone else is speeding on a surfboard. Yet in spite of the myriad diversity of human activities, Aristotle claimed that they all have something in common. In all of our activities we are each aiming at something that we perceive to be good. If we didn't perceive a thing to be good in some respect, we wouldn't pursue it. Things perceived to be good are the beginning and end of all human action; human action *begins* because of desire for something good, and it *ends*, when successful, in enjoyment of that good. Hence, action is always for an end (a goal or objective), and a rational being always believes that end to be a good of some kind.[3]

Again we may be willing to go along with Aristotle: yes, people are goal-oriented; yes, people always pursue a goal because they perceive it as good in some respect. But Aristotle went even further – and here is where the controversiality of his position comes out most sharply. He claimed not only that we are each in pursuit of good things according to our individual judgments; he also claimed that ultimately we are all in pursuit of the same good. Yes, individual actions aim at specific goods that are quite diverse from person to person, but action in general is for the sake of a single good. That single, unique good, according to Aristotle, is *eudaemonia*. "*Eudaemonia*" is a Greek word that in English literally means "good spirit" or "well spirit"; it is generally translated into English as "happiness" or "well-being." Note that we use "good spirits" as a synonym for happiness: "How are you today?" "I'm in good spirits." Happiness, then, is the one thing that all humans are seeking in all that they do.

According to Aristotle, the pursuit of happiness is not something we just happen to do, as though we could do otherwise; rather, it is our nature to seek happiness above and through all other things. Thomas Aquinas, a brilliant medieval philosopher who was strongly influence by Aristotle, reasoned that since it is our nature to seek happiness, therefore *we cannot not seek happiness*. Here are several closely related statements by Aquinas: "Man cannot not will to be happy"; "Every man necessarily desires happiness"; "The will tends to happiness naturally and necessarily"; "Man wills happiness necessarily, and he cannot will not to be happy, that is, to be miserable."[4]

Blaise Pascal, a seventeenth-century French philosopher, scientist, and mathematician, echoed in the following words the sentiments of Aristotle and Aquinas:

> All men seek happiness. There are no exceptions. However different the means they may employ, they all strive towards this goal. The reason why some go to war and some do not is the same desire in both, but interpreted in two different ways. The will never takes the least step except to that end. This is the motive of every act of every man, including those who go and hang themselves.[5]

If Aristotle, Aquinas, and Pascal are correct, then it is our nature to seek happiness, and we can no more act contrary to our nature than a lion or a lizard can act contrary to its nature.[6]

What is it about us that causes us to seek happiness above all other things? It is our nature to seek happiness because we are by nature *sentient*, *social*, and *rational*. To say that we are "sentient" is to say that we are capable of feeling physical pleasures and pains – such as the pleasures of eating and swimming, and the pains of cramps and suffocation. Because pleasure is a quality of experience that we like, and pain is a quality of experience that we dislike, we naturally seek to secure physical pleasures and avoid physical pains.

To say that we are "social" is to say not only that we depend on other people for our physical survival and comfort, but also that we are capable of feeling social pleasures and pains – for example, the pleasures of friendship and being admired, and the pains of loneliness and ridicule. Hence, it is just as natural that we seek to achieve social pleasures and avoid social pains as it is that we seek to achieve physical pleasures and avoid physical pains.

However, we are not simply metal filings irresistibly attracted or repelled by every pleasure or pain that catches our attention. Because we are *rational*, we are able to *imagine* the possibility of an ideal state of affairs (happiness), and we are able to *evaluate* particular actions in relation to achievement of that ideal. George Santayana, an early twentieth-century philosopher who taught at Harvard University, expressed in the following couplet the important bearing of rationality on happiness: "Pleasure is the aim of impulse. Happiness is the aim of reason."[7] It takes reason to understand the concept of happiness (as distinguished from the concept of pleasure), and it takes reason to guide our actions so that we can achieve, or at least approach more closely, the ideal of happiness. Once the concept of happiness looms clearly in our minds, it becomes the ultimate criterion by which we evaluate the promises of possible actions and the outcomes of actual actions.

Aristotle's definition of happiness

What does reason understand happiness to be? According to Aristotle, reason reveals happiness to be *the supreme, complete, final human good*. It is the *supreme* human good because there is no good better than it or equal to it. It doesn't make sense to ask "What good is better than happiness?" because on reflection we realize that happiness is the good par excellence.

Happiness is the *complete* human good in the sense that it lacks nothing that is important to us. When we are happy, we want nothing more. To put the point another way, happiness is a self-sufficient state of affairs in which we desire nothing more because we have everything that is important to us.

Finally, happiness is the *final* human good because it is always sought simply for its own sake rather than as a means to something else. Happiness is that which we pursue above all other things and through all other things. It is that for the sake of which all other things are pursued. Some things (such as money), we pursue for the sake of other things (such as food and recreation). But we never pursue happiness for the sake of anything else. After all, because happiness is the supreme human good, there is no good beyond it (in the sense of being superior to it), and because it is the complete human good, there is no good in addition to it. Hence, there is no other good to which happiness could be a means. All goods other than happiness are good by virtue of being a part of happiness or a means to it. Hence, happiness is the final human good.[8]

Critiques of happiness

However, not everyone agrees with Aristotle that happiness is the supreme human good. Friedrich Nietzsche, a nineteenth-century German philosopher, once quipped that happiness is something for British shopkeepers. Albert Einstein, the brilliant physicist, once stated that happiness is something for swine. In John Dewey's opinion, however, Nietzsche, Einstein, and others who make statements that appear to reject the supremacy of happiness are usually not rejecting the supremacy of happiness at all; what they are rejecting is a specific conception of the means to happiness. Nietzsche, for example, was contemptuous of what he perceived to be the "comfort and security above all" mentality of the British middle-class of his time; Einstein seems to have been attacking the "happiness through sensuality" lifestyle. But each man had a vision of a lifestyle that he considered worthy of, and fulfilling to, human beings.[9]

In brief, it seems that people who begin by rejecting the supremacy of happiness usually end by restoring it under some other name, such as "the will to power" (Nietzsche) or "a sense of the mysterious" (Einstein). As John Dewey noted in his *Reconstruction in Philosophy*, "Happiness has often been made the object of the moralists' contempt. Yet the most ascetic moralist has usually restored the idea of happiness under some other name, such as bliss."[10]

If Aristotle and Aquinas are correct about the nature of humans, then Dewey's conclusion is to be expected – that is, we cannot really differ about our supreme goal (though due to confusion, we might think we are differing about it); we can differ only about the means to that goal. Hence, people who believe they are rejecting happiness as the supreme human value are mistaken. What they are really rejecting is a specific conception of the means to happiness – such as security, wealth, fame, power, or physical pleasure. In place of that conception, they substitute their own conception of the best means and thereby affirm unwittingly the supremacy of happiness.

Dewey's point is plausible. I incline toward it myself. Yet there are arguments that give me pause. Two of those arguments were stated well by Immanuel Kant and Arthur Schopenhauer. Kant argued in his *Groundwork of the Metaphysics of Morals* that in this life there is an inevitable and frequent conflict between what we *desire* to do and what we *ought* to do, that is, between our desire for happiness and our obligation to be moral. Clearly, Kant said, whenever such a conflict arises, it is happiness that should be sacrificed to morality, not vice versa. When what we want to do is not what we ought to do, what we ought to do is what we ought to do.

Kant further supported his position that happiness is not the supreme human good by reasoning that if we were meant to seek happiness above everything else, Nature would have gifted us differently. The eagle, for example, is obviously meant to be a bird of prey, so it has telescopic vision with which to spot game, needlelike talons with which to catch it, and a strong sharp beak with which to rend it. The mole, by contrast, is meant to grub underground for its fare. Consequently, it has poor eyesight but keen hearing and thick front claws.

Nature, then, is efficient and accurate at gifting creatures with the tools they need for what they are meant to do. By inference from their gifts, we can determine what a creature is meant to do and not do. An examination of human abilities makes it clear that we are not meant to seek happiness. If we had been, we wouldn't have been given a conscience, since it interferes so often with the pursuit of pleasure. If we were meant to pursue happiness above all other things, we would have been more like cats, which seem not to be morally troubled at all when they catch, cripple, and then play with their still living prey – sometimes finally killing and eating it; sometimes simply walking away from their living but crippled toy. (I once caught my cat 'toying' with a chipmunk in my backyard.)

Furthermore, reason is not a very effective tool for achieving happiness; if it were, we would all be a lot happier. If we were meant to seek happiness in this life, then in place of reason we would have been given instincts as unerring as those of the beaver and the bee – only in our case they would have been unthwartable, unerring instincts for happiness. Instead we have been given (1) a *conscience* (because of which we have a sense of moral right and wrong, a feeling of obligation to do what is morally right even at the expense of happiness, and feelings of guilt when we do not) and (2) a *free will* (because of which we are *able* to do what is wrong, yet are blameworthy if we do!). On two counts, then – (a) what is right, and (b) what we are best built for – the obvious purpose of our lives on earth is not happiness; it is morality. Clearly, Kant's reasons for placing morality above happiness are worthy of serious consideration.

In fairness to Kant it should be noted that he did place a very high value on happiness. Indeed, he said we can *hope* that there is a God and an afterlife during which God rewards the virtuous with happiness – virtuous people being those who place morality above happiness in this life and who therefore deserve to be happy.[11] However, we can only *hope* for a conjunction of virtue and happiness in a life after death because in this life, according to Kant, we cannot know whether there is a God or a life after death. But we do know, he thought, that when we must choose between happiness and morality (as happens all too often in this life), we should choose morality – and that makes it clear that morality is a higher value than happiness.

A half-century after Kant's death in 1804, Arthur Schopenhauer – who greatly admired Kant – carried to an extreme Kant's point about our unfitness for the pursuit of happiness. Kant, as we saw, was sympathetic to the *desire* for happiness. Though he placed morality above happiness in this life, he hoped that happiness would be granted to the virtuous in a life after death. Schopenhauer, by contrast, saw the very desire for happiness as a vicious disease to be overcome. "There is only one inborn error," he wrote, "and that is [the belief] that we exist in order to be happy. It is inborn in us because it is one with our existence itself, and our whole being is only a paraphrase of it, nay, our body is its monogram."[12]

Schopenhauer agreed, then, with Aristotle and Kant that the craving for happiness is deep and pervasive in all human beings. However, because of his understanding of reality, Schopenhauer rejected belief in life after death. Furthermore, he was convinced that in this life – the only life he thought there is – the desire for happiness is like a maddening itch that only gets worse the more we scratch it. Consequently, he reasoned, the most miserable of all human beings are those who, convinced that the itch can be got rid of only by scratching all the more vigorously, abandon themselves to scratching without reserve. Giving full rein to their craving for happiness, believing that happiness can be achieved by their efforts, they drive themselves into deeper and deeper frustration as they fail ever more dramatically.

This phenomenon of making things the opposite of what we mean to make them is called "the hedonistic paradox." A paradox occurs when something leads to the opposite of what we expect, and we don't understand why – such as when we mean to heal a damaged friendship by making a sincere apology, but the apology only makes the other person angrier. According to the hedonistic paradox, the more intensely people seek happiness, the more unhappy they become.[13] Hence, one of the biggest causes of unhappiness is the pursuit of happiness! The self-conscious pursuit of happiness is counterproductive and self-defeating.

People are victimized so readily by the hedonistic paradox because of two mistaken assumptions, said Schopenhauer. First they mistakenly infer that every natural desire can be satisfied by taking appropriate steps. To be sure, most natural desires, for example, our desires for water, food, sleep, and sex, can be satisfied by taking action, but the desire for happiness cannot be satisfied by taking action. Those who insist that it, too, must be satisfiable because our other natural desires are satisfiable are guilty of the fallacy of hasty generalization. According to the *fallacy of hasty generalization* (sometimes known as "jumping to conclusions"), it is invalid to infer that because some X's are Y (for example, some natural desires can be satisfied by taking action), therefore all X's are Y (all natural desires can be satisfied by taking action).

Second, it is also fallacious to reason (as people commonly do) that because there are some happy human beings, therefore happiness can be achieved by anyone who makes an intelligent, vigorous, sustained effort to achieve it. Schopenhauer called this "the decoy effect." A few people here and there who appear to be happy delude us into thinking that we can be happy, too – just like decoys delude ducks into thinking it is safe to land. What people who reason this way fail to appreciate is that the few people who are happy are not happy by virtue of their own efforts – though they usually think they are. Rather, they are happy as a result of extraordinary luck: they inherited the right genes, had the right parents or guardians, were born into and reared in the right circumstances – none of which they had anything to do with!

Yes, happiness does occur; but it is an occurrence as rare as the music of a Beethoven or the insights of an Einstein. The masses who believe they can achieve happiness by their own efforts are just as pathetically mistaken as people who believe that by means of sheer willpower they can become as great as Beethoven or Einstein. If such people have any genuine appreciation of Beethoven and Einstein, then the harder they try to reach their level of achievement, the greater will appear the gap between their accomplishments and the accomplishments of genius – not because they will not make genuine progress, but because their efforts to close the gap will cause them to become all the more conscious of how vast the gap really is. As with the occurrence of genius, the occurrence of happiness is dependent on a multitude of complex, subtle, fragile variables that are beyond our control. To try to control them is to set oneself up for failure.

According to Schopenhauer, the only cure for our natural craving for happiness and our illusion that we can achieve it is "euthanasia of the will."[14] "Euthanasia" in Greek literally means "good death" ("*eu-thanatos*"). To "euthanize" an animal is to kill it while causing it little or no discomfort. How does the concept of euthanasia apply to the problem of happiness? We must kill our will for (our craving and striving for) happiness.

Individual desires are like the heads of the mythical hydra: destroy one by satisfying it and two more appear in its place, clamoring for their own satisfaction; destroy a weak desire by satisfying it, and soon it is reborn as a desire of moderate rather than weak intensity; make a desire of moderate intensity go away by satisfying it, and soon it returns as a desire of strong intensity. Obviously, then, indulging our desires will never bring happiness; it will only breed more desires with ever greater clamor. What must be done is to go to the heart of the hydra – to kill desire itself. Even then we will not have happiness, but at least we will not be increasingly miserable.[15]

Whether Schopenhauer was right that happiness is a will-o'-the-wisp is an important question. If he was correct, his position has profound significance for how we should understand ourselves and live our lives. Unfortunately, I know of no way to determine whether he was correct. His claim is an item that will remain on the agenda of philosophers and social scientists for some time to come. Meanwhile, life goes on; we must make choices and act upon them. Schopenhauer chose one direction; most people seem to prefer the eudaemonistic direction of Aristotle and Aquinas, so let's sail again with the winds of eudaemonism, leaving behind but keeping in mind the thoughts of Kant and Schopenhauer.

Axiology: Its nature and purposes

Earlier I claimed in the spirit of Aristotle and Aquinas that since happiness, our ultimate, overriding goal, is fixed for us (whether by God or evolution), the only important questions are questions about the *nature* of that goal and the *means* to that goal. Axiology is the area of philosophy which is most concerned with helping us formulate and answer those questions. To borrow the language of Alfred North Whitehead, the purposes of axiology are to help us *live*, to help us *live well*, and to help us *live better*.[16] Let's look at the meaning of each of these points.

First, if we do not live, then we certainly cannot enjoy this life. Hence, our first goal must be to survive (though perhaps not at any cost – a point we will discuss later). If we do survive, we don't want just to live and nothing more. We want to live well! And that is Whitehead's second point. The fact of suicide is a clear indication that the human drive is not just for survival but for happiness. Yet even living well is not enough, for as Whitehead goes on to say, we want to live better! Better than what? Better than we are living now!

By "better" Whitehead does not necessarily mean better in a material sense – though it may include that. He is thinking about the *quality* of life. We want to live better (and better) qualitatively – the quantitative aspect of life being a servant to the qualitative. We want an ever-flowing supply of novelty and freshness in our lives in order that "living well" might never turn boring or stale.

Axiology, the philosophical study of non-moral values, is the theoretical part of our endeavor to reach these goals that Whitehead articulated so nicely. It is the art and science of maximizing good and minimizing evil in our lives. It exists because, as John Dewey put it, "Nothing but the best, the richest and fullest experience possible, is good enough for man." Dewey added that "the attainment of such an experience is not to be conceived as the specific problem of 'reformers' but as the common purpose of men."[17] It is to that common purpose that this section on axiology is devoted.

Aristotle believed that happiness is difficult, if not impossible, to achieve without good luck. By definition, though, luck is something we cannot control, so let's focus our attention on what we can do toward the achievement of happiness.

The Good of enjoyment vs. Enjoyment of the good

Obviously, we should seek happiness *intelligently*. What does that mean? In part it means we need to learn to distinguish *the good of enjoyment* from *enjoyment of the good*. Enjoyment is always good in the sense of being pleasurable. But *what* we enjoy may or may not be good. It may, for example, cost too much or have adverse consequences. The "cost" of enjoying something comes *before* the enjoyment and is what we must "give up" in order to enjoy it. What we must give up may take the form of money, time, energy, personal relationships, or something else, small or great.[18] Enjoyment of a new car may cost thousands of dollars; enjoyment of a sunset may cost nothing more than the time and energy it takes to look out a nearby window.

Every enjoyment has its cost. There are no exceptions! This is not a pessimistic point but a realistic point. Likewise, every enjoyment has its consequences. The "consequences" of enjoying a thing are the effects which follow from enjoyment of it. Such effects may be positive (such as a feeling of relaxation), negative (such as a stomach-ache), or negligible (when the consequences are indifferent). If, for example, you thought you had enjoyed a delicious dish of mushrooms only to be informed later by your stomach that they were toadstools, then although eating the toadstools was a very real enjoyment, you would hardly say that eating the toadstools was a good thing to do. As Boethius, another pivotal figure between classical Rome and medieval Europe (like Augustine), said in his treatise *The Consolation of Philosophy*, "I deny that anything can be considered good, which harms the one who has it."[19]

In conclusion, *every enjoyment has both costs and consequences* which must be taken into account if we are to make wise decisions. Something may cost so much that afterwards we realize it wasn't worth it, or it may cost so much that it plunges us into painful debt. Other things may cost very little but have disastrous consequences (such as playing Russian roulette – literally or, as in unprotected sex, figuratively).

Usually the more intense our immediate pleasure is (or the more intense our craving), the more oblivious we become to whether what we are enjoying is good rather than just pleasing. Something is *good*, as I am using the term here, if it gives us pleasure now and does not later cause us regret that we enjoyed it. (We don't want to get tangled up with morality at this point, so let's assume for now that we are speaking of things which give us pleasure and are morally okay.) The toadstools were a pleasure to eat (the good of enjoyment), but they were not a good thing to eat (enjoyment of the good). The good, then, is not necessarily that which I desire or enjoy – no matter how much I crave it or how much I enjoy it – for that which I desire may disappoint me and that which I enjoy may later cause me regret. Hence, the good is that which will give me pleasure without exacting a cost or producing consequences that do make me or would make me regret that I enjoyed it.

Three conceptions of the good

The preceding analysis involves three conceptions of what is good. Let's examine each. First there is "the desire theory of the good." According to the desire theory of the good, what makes a thing good is that someone desires it. After all, they wouldn't desire it if they didn't think it was good, and if it's good to them, then that settles the issue. Who's to say that they're wrong? Hence, it is desire that makes a thing good. If someone desires something, then to them that *makes* it good.

Is this position defensible? There are reasons for thinking it is not. Sometimes the very person who thinks that something is good gets it and then, as a result of experiencing it, says, "I was wrong. It wasn't good after all." But that means their desiring it did *not* make it good – it only made them *think* it was good. I can give an example of this point from my childhood. Because this example will be a *counterexample*, and because counterexamples are powerful philosophical tools, let's pause to examine what a counterexample is and then turn to my example.

A counterexample is an example which proves that a claim is false. Consider, for example, a claim which says that something cannot be the case; a counterexample would show that it *can* be the case. If I claim that all mammals give birth to live young, and you point out that the duckbill platypus is a mammal which lays eggs, then you have given a counterexample to my claim. You have countered (disproven) my claim with an example. When anyone makes a claim in philosophy, it is standard procedure to try to think of a counterexample to it. If we think of a counterexample, then we know the claim is false – though we must, of course, make a distinction between examples that really are counterexamples and examples that seem to be counterexamples but are not.

If we do not succeed in thinking of a counterexample to a claim, then the claim *may* be true. Just because we cannot think of a counterexample to a claim does not necessarily mean that the claim is true; perhaps there is a counterexample that we just haven't thought of yet. That is why the following reasoning is fallacious and is called the "*argumentum ad ignorantiam*" (the argument from ignorance): "If you cannot show that my claim is false, then you should accept it as true." Our ignorance (lack of knowledge) of how to show that a claim is false does not mean that it is true. Still, the more frequently a claim survives serious efforts to prove it is false, the more it becomes plausible to think that it is true. (The last point should remind you of the section on skepticism in which it was said that a good theory is "hardy," that is, can survive rational scrutiny.)

Now for my counterexample to the desire theory of the good. When I was a child I thought that because perfume smelled so wonderful, it should taste great! So I sipped some. Guess what? It was the nastiest stuff I had ever tasted! Hence, at a tender age I learned that the desire theory of the good is false. My desire to taste the perfume, my belief that it would taste good, did not make the taste of the perfume good.

But what if the perfume *had* tasted as wonderful as it smelled? That question leads to *the pleasure theory of the good*. According to the pleasure theory of the good, it is pleasure that makes a thing good. If a thing gives a person pleasure, then for that person it is good. This means that the goodness is in the thing – whether you desire it or not. Sometimes a thing surprises us with pleasure. After that we desire it because it gave us pleasure – pleasure because of which we discovered that it was good.

Now let's shift gears from justification to evaluation. Is the pleasure theory of the good an adequate theory of the good? The answer, I think, is "no." What if the perfume had tasted as wonderful as it smelled but ten minutes later gave me a terrible headache? From that vantage point I would not have thought that the perfume was a good thing to drink. Also, think back to the example of the delicious toadstools that later caused a ferocious stomachache. Clearly, the fact that something gives us pleasure does not make it good, so next let's consider "the worth theory of the good."

According to *the worth theory of the good*, what makes something good is that it gives us pleasure or some other benefit and does not cause us regret. A truly good thing is, then, one that is worthy of being chosen because it will give us pleasure or some other benefit and not cause us later to regret that we chose it. Hence, a truly good thing may not be desired, but it is "desirable" in the sense of being worthy of being desired.

Is the worth theory of the good an adequate theory? I think it is definitely better than either the desire theory or the pleasure theory. However, it, too, needs important qualifications. Presumably we don't want to allow immoral things to be good, or things which we *would* regret but don't regret only because we don't know some of its costs or consequences. Some of these qualifications will be made later in this section; others will be made in the section on morality. For now let's examine more of the insights that reason can give us about the pursuit of happiness within the boundaries of morality.

The interaction theory of experience

Experience seems for the most part to be the result of interaction between an organism and something with which it interacts – or between an organism and something that it literally or metaphorically rubs against. Think of a cat rubbing against a tree, a dog "rubbing" its teeth and gums against a bone (by chewing it), or a person "rubbing" her mind and imagination against a poem (by reading it).

Enjoyment is a quality of experience, so enjoyment, too, is the result of interaction between an individual and something that causes pleasure to the individual. For a human being the something that causes pleasure could be another person, a book, a scene of natural beauty, an idea, a piece of cloth, a motorcycle, a basketball, or any of an endless list of things. Hence, in order to make our lives more enjoyable, we need to interact with good things, that is, with things that will give us pleasure and not cause us regret.

The package theory of alternatives

As we have seen, whether the something with which we interact is *good* cannot be determined by the mere fact that we desire it or that it causes enjoyment. To be able to ascertain whether an alternative would be a good thing to choose, we need to recognize that each alternative is a *package* of at least three components: *costs*, *immediately felt qualities*, and *consequences*.

The less mature we are, the more we tend to think that there is nothing more to an alternative than the immediate feelings that it gives or promises to give.

However, the thing that *causes* those positive qualities of feeling will have costs and consequences. Consequently, an adequate evaluation of an alternative must take into account its full, three-fold impact on our lives. Only then will we have adequate information for deciding whether an alternative would be a good choice, for only then will we have an adequate understanding of its total impact.[20] Only then can we decide wisely whether we want *the whole package* (the costs, the immediately felt qualities, and the consequences) – for the whole package is what we always get, whether we like it or not.

We also need to *use our intelligence* to examine the whole package which each alternative is because it is happiness that we want and not merely pleasure. Recall George Santayana's point: *Whereas pleasure is the aim of impulse, happiness is the aim of reason.* Pleasure is of the moment and is caused by a specific stimulus, such as scratching an itch. Impulse cares only about immediate enjoyment or relief from discomfort; it does not care about costs and consequences.

The desire for happiness, by contrast, is the desire for pleasures that we will not regret. We want a happy life – not just moments of pleasure. A happy life is achieved by choices that cause pleasure but not regret. Consequently, whereas impulse says, "I want what I want!," reason says, "I want what is good!" Apart from reason we serve our passions. By means of reason our passions serve us. Blind emotion says, "I don't care about costs and consequences. I just want pleasure – now!" Reason says, "You may not care about costs and consequences now, but you will later!" The inability to take costs and consequences into account is why children so often get themselves into trouble – they just want what will give them pleasure now. The more mature a person becomes the more he or she takes costs and consequences into account.

If we allow a specific impulse of desire to dominate us, it causes a kind of tunnel vision that fixates our attention on the object desired and is oblivious to wider considerations regarding costs and consequences. We just want that thing (or person) now! Clearly that is a path to misery rather than happiness, for though our pleasure may be intense for awhile, sooner or later acting on impulse has dire consequences – and then we're sorry we didn't pay attention to what we were getting ourselves into.[21]

Ignorance as cause/Intelligence as cure

What is the cause of our downfall in such situations? Many great thinkers, eastern and western, believe that the main cause of unhappiness is *ignorance*. Why ignorance? Because if we *knew* the costs and consequences of each of the alternatives before us, we would naturally choose the alternative yielding the most pleasure and the least pain. When we choose a lesser alternative, it is, they say, only out of ignorance. Hence, the importance of intelligence in living a good life cannot be overestimated, for it is by means of intelligence that we estimate the costs, felt qualities, and consequences of the alternatives available to us.

Indeed, we might say with Socrates that intelligence is humankind's greatest instrumental good, that is, humankind's most important instrument for achieving a good life. Why? Because it is by means of intelligence that we can most realistically evaluate alternative ways of life. (Socrates argues this point brilliantly in Plato's dialogue *Philebus*.) Consequently, we should, as Aristotle stated in his *Nichomachean Ethics*, "Strain every nerve to live in accordance with the best thing in us," namely, intelligence or reason, "for even if it be small in bulk, much more does it in power and worth surpass everything." Indeed, Aristotle concluded, "the life according to reason is best and pleasantest, since reason more than anything else is man. This life therefore is also the happiest."[22]

When I speak here of intelligence, I am not talking about an inherited capacity. I'm talking about *acting* intelligently as distinguished from *being* intelligent. The latter is inherited, but the former is learned. You have inherited enough intelligence (intellectual capacity) to read this book, and that is also enough for you to be able to benefit from the points I am presenting. But whether you or I *act* intelligently is another matter. To do so means, in part, to choose only those pleasures that are approved by intelligent reflection on the cost/felt-quality/consequence packages in which they are contained.

The distinction between pleasure, good, and happiness was noted in ancient as well as contemporary times. In ancient Greece Socrates engaged in the following exchange with a priestess named Diotima, whose wisdom he greatly respected. After a discussion of the nature of beauty, Diotima changed the subject:

> "Well then," she went on, "suppose that, instead of the beautiful, you were being asked about the good. I put it to you, Socrates. What is it that the lover of the good is longing for?"
> "To make the good his own."
> "Then what will he gain by making it his own?"
> "I can make a better shot at answering that . . . He'll gain happiness."
> "Right," said she, "for the happy are happy inasmuch as they possess the good, and since there's no need for us to ask why [people] should want to be happy, I think your answer is conclusive."[23]

Note first that Diotima expressed the same sentiment that Aristotle and Aquinas expressed later: the pursuit of happiness is ultimate. Other things are pursued for its sake. What do we love? That which appears to bring us happiness or which we believe will bring us happiness. But we can be wrong as to *what* will bring us happiness. The *truly* good is that which *will* bring us happiness, but what we *think* is good and what *is* good may be different. Hence, there is need for intelligent analysis of that which we think is good – for as we have seen, pleasures always come to us in packages containing costs and consequences in addition to felt qualities.

Since whenever we choose something, we are choosing a package willy nilly, we should be aware of this and act accordingly. When looking for a marriage partner, for example, if I put physical attractiveness above all else, I will eventually discover that physical attractiveness is always accompanied by other characteristics. My good-looking mate will be selfish or unselfish, reliable or unreliable, interesting or uninteresting, faithful or unfaithful, and so on. The same thing is true of choosing a job, a vacation, or anything else. We never get just one aspect of it. We get the whole package – whether we like it or not. Hence, we should consider in advance whether we want the whole package, for that – and not just its immediately felt aspects – is what we will get.

The distinction between pleasure and happiness can be found in the ancient Orient, as well as the ancient Occident. Consider, for example, the *Upanishads*, ancient Hindu literature which makes the distinction between pleasure and happiness by speaking of pleasure and joy. "There is the path of joy," according to the *Upanishads*, "and there is the path of pleasure. Both attract the soul. Who follows the first comes to good; who follows pleasure reaches not the End. The two paths lie in front of man. Pondering on them, the wise man chooses the path of joy; the fool takes the path of pleasure."[24] We've all chosen pleasures that have led to regret, but not all pleasures end regrettably. Those pleasures that do not end regrettably, I think the *Upanishads* are saying, are the pleasures that constitute the path of joy.

Opposite to the path of joy is what psychologists Philip Brickman and Donald Campbell call the "hedonic treadmill." "Hedonic" means "having to do with pleasure" (recall the Greek word "*hedone*"), so Brickman and Campbell are speaking of a "pleasure treadmill." One of their points is that certain kinds of pleasure promise satisfaction but don't deliver it. The more we get, the more we crave. Hence, to pursue such pleasures is self-defeating. Instead of enjoying ourselves, we become more and more frustrated. We keep thinking we are going to reach a state of enduring satisfaction, but we never do. So either we wake up to the fact that pursuing happiness in that way does not work, and we kick the habit, or we keep running the treadmill ever faster like a rat in a cage.[25]

Maybe you've heard the following "good news/bad news joke" about the people who took the first flight on a super-fast, trans-pacific aircraft. After a couple of hours in flight, the pilot announced to his passengers, "I've got some good news and some bad news. The good news is that we're making record time. The bad news is that we're lost." Many of us live our lives that way. Going ever faster we know not where! And as *The Peter Principle* puts it, "If you don't know where you're going, you'll end up somewhere else!" I love that statement because it implies that even when we don't know where we're actually going, we do know where we would like to go. Specifically, we would like to go to happiness. But if we don't know where our actions are taking us, we'll probably end up somewhere else. That is why it is so important that we (1) keep in mind that our ultimate goal is happiness and (2) examine our actions to see whether that's where they *are* taking us.

Jeremy Bentham's Hedonic Calculus

The preceding statements have been broad and general. Now let's turn to a more detailed application of intelligence to the problem of decision-making. Jeremy Bentham was an important nineteenth-century value theorist and a member of the British Parliament. He was also an interesting character. His body is on display in a glass case at the University of London. He is, at his own request, sitting upright, fully clothed, but with his severed head on the floor between his feet. Perhaps that arrangement makes sense in light of the opening statement of his book *An Introduction to the Principles of Morals and Legislation*. There he says, "Nature has placed mankind under the governance of two sovereign masters, pain and pleasure. It is for them alone to point out what we ought to do, as well as to determine what we shall do."

As can be seen in the preceding quotation and is suggested by the final location of his head, Bentham was a *psychological hedonist*. That means he believed we cannot help but endeavor to maximize the amount of pleasure and minimize the amount of pain in our lives – or, more technically, to maximize the ratio of pleasure to pain in our lives. Bentham was following the path set out by his great predecessor, David Hume, when Hume said, "Reason is and ought only to be the slave of the passions."

Bentham developed a method to help us satisfy our passions or desires more effectively; he called it "the hedonic calculus." "Hedonic" means "having to do with pleasure," and a "calculus" is simply a device for calculating; hence, an "hedonic calculus" is a method for calculating and comparing the pleasures and pains of the alternatives available to us. It is reason as servant to desire.

Why apply the hedonic calculus to the various choices before us? So we can discover which one yields the greatest proportion of pleasure over pain, for that, Bentham believed, is the alternative we want. A special value of Bentham's calculus is that it doesn't just *tell* us that decision-making is complex and difficult – as though experience hadn't already taught us that! Rather, it points out seven specific aspects of every alternative so that we can keep them in mind and "check them off" as we go about the business of decision-making. Attention to these seven aspects first helps us determine how good a thing is in itself, and then it helps us determine how well it compares to available alternatives.

The first aspect Bentham speaks of is *intensity*. If one thing (let's call it A) produces a more intense pleasure for you than another (let's call it B), then all other things being equal between A and B, you should choose A, since it will give you more pleasure, and that is what you want. This point might seem so obvious as to be unworthy of mention. Most of Bentham's seven points will seem obvious when considered individually, but as we go over them all, you should begin to get a new sense of the complexity of good decision-making, so let's continue.

The second aspect is that of *duration*. Some advertisers insist, for example, that the flavor of their chewing gum lasts longer than that of other leading brands. Bentham's point is that if the pleasure of C, whatever it might be, lasts longer than that of D, then all other things being equal between C and D – such as their providing the same intensity of pleasure – we should choose C over D because it provides the same intensity of pleasure *over a longer period of time* than does D. Sometimes the fact that one thing will produce a more enduring pleasure than another is obvious (such as when the choice is between two equally cold sodas, one 16 ounces and one 12 ounces, on a hot day), but the difference is not always obvious – for example, when choosing between cars to buy or neighborhoods to live in. The duration factor is sufficiently important, however, that we should always give it special attention.

Third, we should consider the *purity* of the experience that will be caused by an object. The point here is that sometimes a thing which causes us pleasure now causes us pain later. Hence, we need to determine, so far as we can, how pure (free) of unpleasant consequences those things are that will cause pleasure immediately. The general principle is this: If E causes a purer experience than F (that is, if E has fewer undesirable consequences than F), then all other things being equal – such as E and F causing pleasures that are of equal intensity and duration – E should be preferred to F because it will provide a pleasure of equal intensity and duration, but with fewer undesirable consequences.

Let's say, for example, that you have been looking for a house to buy and have narrowed the alternatives to two. Let's assume that aesthetically they are equally attractive to you and that they are equally sturdy (hence, they will provide an equal duration of enjoyment). But then you discover that the property of one of the houses is more maintenance free than the other – and you hate yard work, such as mowing, raking, spraying, fertilizing, and weeding, and you cannot afford to pay someone else to do it. Clearly, then, you should choose the house with the more maintenance-free yard. It will provide a purer experience for you.

A fourth important aspect of pleasure is *fruitfulness* (Bentham called it "fecundity"). Whereas "purity" is concerned with whether a possible object of choice has *negative* consequences, "fruitfulness" is concerned with whether a possible object of choice has *positive* consequences. Some objects that cause pleasure at first cause pain later; others cause pleasure now and have negligible consequences; still others cause pleasure now and pleasure later. Clearly, the latter are the ones to choose. Let's say, for example, that you equally enjoy playing the piano and playing the oboe. But playing the piano, let's suppose, will enable you to earn some sorely needed money, whereas playing the oboe will not. Hence, playing the piano and the oboe provide pleasures of equal intensity, duration, and purity, but playing the oboe (we're supposing) has negligible additional consequences whereas playing the piano has the important additional consequence of enabling you to earn needed income. Hence, playing piano is more "fruitful" than playing the oboe and is what you should learn to do well if you don't have time to do both. The general principle here is that if G is equal to H in all respects except that G will yield more beneficial consequences than H, then G should be chosen over H, if you must choose between them.

A fifth aspect of pleasure which Bentham pointed out is *number* (he called it "extent"). The point here is that all other things being equal, if J will make a larger number of people happy than K, then choose J and make more people happy at no extra cost! Both prudence and altruism support the wisdom of this choice. Altruistic people will choose J because they are the kind of people who get pleasure from making other people happy. Prudential people *may* not care about the happiness of other people, yet they will still make the altruistic choice. Why? Because if they do pleasant things for other people, those people are more likely to do pleasant things for them. Hence, by increasing the number of people who are benefited by their actions, they are increasing the number of people who are likely to do something nice for them. Also, it is usually more pleasant to live among happy people than among unhappy people, and by choosing J we make the society that surrounds us more pleasant. All other things being equal, then, such as intensity, duration, purity, and fruitfulness, if J will make a greater number of people happy than K, then the smart thing to do is choose J.

A sixth aspect of a pleasure is the *probability* of its occurrence (Bentham called this "certainty or uncertainty"). The point is that if the probability of L occurring is greater than the probability of M, then if all other things such as intensity, duration, purity, fruitfulness, and number are equal between L and M, then we should choose L rather than M because the likelihood that L will come about is greater than that of M (or, conversely, the probability that M will *not* come about is greater than the probability that L will not come about).

The problem being addressed here is that things do not always work out the way we plan for them to. Consequently, all other things being equal, we should choose that alternative which has the highest probability of working out the way we want it to. For example, if you want to buy a lottery ticket in one of two lotteries and the tickets cost the same and the payoff is the same, but you happen to know that 2,000 tickets will be sold in Lottery A, whereas 4,000 tickets will be sold in Lottery B, then the smart thing to do is to buy a ticket in Lottery A, for whereas you *might* get lucky in Lottery B, the chances are *twice* as great that you will get lucky in Lottery A. Hence, buying your ticket from Lottery A doubles the probability that you will achieve your goal of winning the lottery.

Let's take a look at another example of the probability aspect of decision-making. If you were a business person and had to travel by train or plane to complete an important business deal, and if the plane would get you there quicker if it flew on schedule, but snow was threatening and might cause a flight delay which would cause you to lose the deal irretrievably, whereas the train would be slower but would almost certainly get you to your destination on time, then all other things being equal, the smart thing to do would be to take the train and thereby maximize the probability that you will make your appointment.

Stated more generally, if the probability of reaching our goal by L is greater than the probability of reaching it by M, then all other things being equal between L and M, we should choose L and thereby maximize the probability that we will get to enjoy our goal.

The seventh and last characteristic that Bentham pointed out was that of *time-lag* (he called it "propinquity or remoteness"). Here the point is that if N can be achieved sooner than O, then all other things being equal – such as intensity, duration, purity, fruitfulness, number, and probability of occurrence – N should be chosen rather than O. After all, the purpose of making decisions and taking action is to enjoy ourselves, so when there is no good reason to delay enjoyment, we should get right to it! Moreover, it is generally the case that the farther something is in the future, the greater is the probability that some unforeseen circumstance will intervene and prevent us from enjoying it. Some parents, for example, put off enjoying their children until, for one reason or another, it is too late to enjoy them at all. Hence, all other things being equal, we should enjoy things sooner rather than later.

Bentham stopped with seven characteristics. I would like to add an eighth in the name of John Dewey: cost. In this day of easy credit, cost might seem like a part of the purity of a choice since the cost of something may be considered a negative *consequence* of enjoying it: "buy now, pay later" means "pleasure now, pain later." For example, normally one consequence of owning a new car is having to make monthly payments for years. But I am thinking of cost in the pure sense of what we have to give up *before* we can have something. Hence, I believe cost is worth giving separate attention to in distinction from purity.

Dewey pointed out that nothing can be enjoyed without some kind of prior expenditure – even if the cost is no more than the effort it takes to look at something. In that case the cost is usually negligible, but sometimes cost is momentous. Hence, we need to calculate the costs of the alternatives before us, remembering that cost must be calculated not only in terms of money, but also in terms of time, emotional energy, personal relationships, and whatever else is important to you. Hence, if P costs less than Q, then all other things such as intensity, duration, fruitfulness, purity, number, probability, and time-lag being the same, you should choose P since that choice will leave you with more resources for other things.

Perhaps you can think of still other characteristics that should be examined in order to improve decision-making, but these eight should make it clear that decision-making is a complex procedure when it is done well. Each of these eight characteristics seems simple and obvious when considered by itself, but in actual decision-making (such as choosing a profession), we are usually faced by alternatives which exhibit these characteristics all at once in confusing ways. One alternative, let's call it "S," may yield a more *intense* pleasure than another, let's call it "T," but T may yield a more *enduring* pleasure than S; however, S may be more *probable* than T, yet T may be *nearer in time* than S, and so on.

The complexity and difficulty of weighing and comparing eight aspects of two or more alternatives can be mind-boggling. Yet we shouldn't despair, for by means of intelligent analysis of this sort, we are *increasing the probability* that we will make a satisfactory choice. Moreover, whereas this kind of methodical procedure is slow and awkward at first, it is something which – like typing or dancing or hitting a tennis ball – becomes more natural, rapid, and accurate with practice. Hence, it is important for us to work at using the hedonic calculus until it becomes "second nature" to us.

John Stuart Mill's criticism of Bentham

John Stuart Mill knew and was profoundly influenced by Jeremy Bentham. Nonetheless, he thought there was a serious deficiency in Bentham's hedonistic approach to life. Bentham taught that *all* aspects of pleasure can be "quantified," that is, given some kind of numerical value. For example, the *probability* of one alternative occurring might be 70% whereas the probability of another is 90%.

Or with regard to *intensity* we might rank one pleasure at +4 and another at +7 on a scale from –[minus]10 to +10 (with –[minus]10 being agony, 0 being indifference, and +10 being ecstasy). Bentham reasoned that by giving an appropriate number to each of the seven previously discussed aspects of the alternatives before us, we can compute in an entirely objective manner which alternative will give us the most value and, therefore, should be chosen, since the only important consideration among alternatives is the *quantity* of pleasure and pain produced by each.

Mill agreed that the aspects of pleasure that Bentham pointed out can be quantified. He also believed that in addition to the quantifiable aspects of alternatives there is sometimes a *qualitative* aspect that is very important and cannot be quantified. Some pleasures are more dignified and worthy of human enjoyment than are others, and some pleasures are completely unworthy of a human being. Unworthy pleasures, Mill insisted, should be avoided altogether – even if it means more pain and less pleasure! He wrote in a now famous statement: "It is better to be a man dissatisfied than a pig satisfied. It is better to be Socrates dissatisfied than a fool satisfied."[26]

If Mill is correct, then pleasures cannot always be compared to one another on a purely quantitative basis. Consequently, the alternative with the best ratio of pleasure to pain may not always be the best choice. A little pleasure that is worthy of a human being is to be preferred to a great deal of pleasure that is unworthy of a human being. Which pleasures are unworthy of a human being is a topic we will touch on in the chapter on morality. The importance of Mill's point, if it is correct, is that it keeps us from being misled by a purely quantitative approach to decision-making. Yet, when all the alternatives before us *are* worthy of a human being, Bentham's calculus remains a useful and powerful tool.

Intrinsic values and Instrumental values

Finally, in this section on axiology, I would like to familiarize you with the distinction between *intrinsic value* and *instrumental value*. A thing has intrinsic value if it is valued for itself rather than for something it leads to, that is, if it is valued for the immediately felt experience that it elicits within us. The taste of one's favorite ice-cream has intrinsic value. By contrast, a thing has instrumental value when it is valued not for the immediate experience that it gives us, but for its *consequences*. Listerine Mouthwash, for example, has nothing but pure instrumental value. As the Listerine ad says, "It's the taste you hate twice a day!" Its value – unlike that of sweet flavored mouthwashes – is not that of immediate enjoyment, that is, of intrinsic value; its value is instrumental (it kills germs and bad breath).

Probably for most of us another good example of instrumental value is calisthenics. We do them to improve or maintain our health, but in and of themselves they are boring. In general something has instrumental value when it does any of the following four things: (1) when it *gives rise* to intrinsic value later (a course in college that you don't enjoy while taking it may enable you to get a job that you do enjoy), or (2) when it *preserves* the value of something that is important to you (like wax protects the finish of a car), or (3) when it *improves* the intrinsic quality of an experience (as better speakers make listening to your music collection more enjoyable even though the speakers themselves are aesthetically indifferent or unattractive), or (4) when it *prevents* the occurrence of an unpleasant experience or loss (as a capsule of vaccine may be indifferent in itself (it has no taste) yet prevent you from suffering an illness).

Whereas I have separated intrinsic and instrumental value in order to distinguish them clearly, one of the most important lessons of life is that something can have *both* intrinsic and instrumental value. If, for example, you enjoy swimming and have the time and opportunity to do it, then rather than doing boring calisthenics for exercise, you can swim regularly and at one and the same time enjoy yourself *and* get the benefits of exercise. Because intrinsic and instrumental values can exist in one and the same thing, John Dewey urged that we should always "idealize the means" to the ends we are pursuing. That is, we should make the means as enjoyable as possible. Simply because something "has to be done" doesn't mean it has to be unpleasant. We should review the means we are using to do what has to be done and see if some change in the means would enable us to enjoy ourselves more along the way. By means of technology people have been ingenious at making more pleasant (or less unpleasant) those things that "have to be done" – from cleaning the toilet to balancing the checkbook to washing the dishes.

Keeping in mind that something can have either intrinsic value or instrumental value or both, let's expand the notions of intrinsic and instrumental value to include displeasure and indifference as well as pleasure, so either type of value can be positive (pleasant), neutral (indifferent), or negative (unpleasant). Now let's run out the permutations of intrinsic and instrumental values, using a "+" for positive value (whether intrinsic or instrumental), a "–" for negative value (whether intrinsic or instrumental), and an "o" for neutral value (whether intrinsic or instrumental).

Clearly, instrumental values are for the sake of intrinsic values. Hence, from this point of view our objective should be to fill our lives as much as possible with intrinsic values according to our own preferences. Furthermore, when we choose among specific goods we should, as John Dewey taught us, prefer a good that has intrinsic value *and* instrumental value to one which has an equal amount of

INTRINSIC	INSTRUMENTAL	EXAMPLE
+	+	Playing piano will have positive intrinsic value if you enjoy playing it, and it will have positive instrumental value if you make money by doing it.
+	o	The fragrance of scented candles in a department store can provide positive intrinsic value as you pass through but usually has no instrumental value.
+	−	Eating too much delicious food at dinner and waking up with indigestion. Hence, positive intrinsic value but negative instrumental value since what causes lots of pleasure at first causes considerable discomfort later.
o	+	Taking capsule medicine provides no immediate pleasure or pain (provided you're not the type who gags on a capsule!). However, such medicine can have the positive instrumental value of restoring or protecting your health.
o	o	Stepping on a grain of sand as you cross a parking lot. You don't notice it, so it causes neither pain nor pleasure, and it has no significant consequence for you.
o	−	Acquiring unwittingly the habit of being cynical. There was neither pleasure nor pain in acquiring the habit, picked up in childhood from one's parents, but it alienates other people and hence has negative instrumental impact.
−	+	Having a cavity filled. The experience is unpleasant, but it has the positive instrumental value of preventing further pain and preserving the tooth.
−	o	Driving through a papermill town with the windows down. The odor is most unpleasant and has no redeeming instrumental value.
−	−	Being arrested for drunkenness. The experience is unpleasant and so are the consequences. This combination is the "most to be avoided" of the nine possibilities.

Figure 13.1 Permutations and examples of intrinsic and instrumental values

intrinsic value but less or no instrumental value (all other things being equal). Why? Because life is an ongoing process; the quality of our future will be to a significant extent the product of the choices that we make now.

The double good (+,+) will give us enjoyment later as well as now, whereas the single good (+,o) will give us enjoyment now, but not later. In Bentham's terminology, the former good is "fruitful" whereas the latter is not. To give an example, if a person genuinely enjoys memorizing the names and phone numbers of people in the Manhattan telephone directory but enjoys just as much (or possibly a little less!) memorizing words and definitions in a foreign language, then it would probably be better for him or her to memorize the foreign words and definitions, for although the immediate pleasure would be approximately the same, the consequent pleasure of the latter is much more promising. There are few things that can be done with knowledge of the names and phone numbers of the first 2000 entries in the New York City directory, but a knowledge of 2000 words in a foreign language could enable a person to communicate with people to whom that language is native, to read their literature in the original language, to travel more easily in the countries in which that language is native, and to secure a job that requires knowledge of that language.

In brief, if our highest goal is happiness, we will achieve that goal more effectively (1) if we learn to distinguish between what is truly good and what only appears to be so and (2) if when possible we choose things that include instrumental as well as intrinsic value (and "things" include everything: friends, beliefs, careers, hobbies, clothes, cars, etc.).[27]

A summary of concepts in value theory

As a conclusion to this chapter on axiology, here is a list of terms and concepts that should help you think more effectively and communicate more clearly about questions of value.

VALUE (noun): the property of an object because of which it can give one some benefit, that is, can (1) give one pleasure (the taste of good food), (2) enhance one's immediate pleasure (spices), (3) prolong one's pleasure (money – for more of a good thing), (4) preserve a source of one's pleasure (glass over a treasured picture), (5) reduce one's displeasure (a mild anaesthetic), (6) stop one's displeasure (a strong anaesthetic), (7) prevent one from feeling displeasure (a general anaesthetic before an operation), or (8) prevent one from suffering harm (an inoculation).

VALUABLE: what makes something valuable is that it possesses one or more of the properties cited in 1-8 above.

VALUE (verb): to value something means, ordinarily, to desire it because you think it has some value.

EVALUATE: to evaluate, appraise, or assess something is to try to figure out what value it has, that is, what benefit it will have for or harm it will do to us or someone or something.

POSITIVE INTRINSIC VALUE: something has positive intrinsic value if it can directly give one pleasure, for example, the taste of ice cream, the smell of a rose. See (1) above.

POSITIVE INSTRUMENTAL VALUE: something has positive instrumental value if it can indirectly give one pleasure, or reduce, stop, or prevent displeasure or harm, for example, spices, aspirin, and inoculations. See 2-8 above.

NEGATIVE INTRINSIC VALUE: something has negative intrinsic value if it itself is, or would be, unpleasant, for example, the smell of rotting meat.

NEGATIVE INSTRUMENTAL VALUE: something has negative instrumental value if it will, or would, eventually cause displeasure or a loss of something valuable, for example, drinking too much alcohol and getting a hangover, or driving drunk and losing your driver's license.

NEUTRAL INTRINSIC VALUE: something has neutral intrinsic value when it gives, or would give, one neither immediate pleasure nor immediate displeasure.

NEUTRAL INSTRUMENTAL VALUE: something has neutral instrumental value when it will, or would have, no significant consequence for one.

Notes

1. St. Augustine, *The City of God*, tr. Marcus Dods (NY: The Modern Library, 1950), Book 19, Chapter 1, p. 672.
2. According to Aristotle it is not only humankind that is goal-oriented; everything is goal-oriented. Consequently, Aristotle had a *teleological* conception of reality ("*telos*" is Greek for "end" in the sense of a goal or aim). Keep in mind also that human activity can take internal forms, such as thinking and imagining, as well as external forms, such as walking and talking. Furthermore, when Aristotle spoke of humans as "end oriented", he meant "end" in the broadest sense, so as to include not only gaining possession of something external, such as food, but also the achievement of an ability (such as the ability to stand on one's hands or derive square roots in one's head) and the achievement of enjoyment (such as listening to or playing music).
3. However, we are not always clearly aware of the ends that are moving us. Some of our objectives are so natural or so repressed that it takes a lot of reflection for us to become aware of them; sometimes to do so we need the help and perspective of another person.

4 St. Thomas Aquinas, *Treatise on Happiness*, tr. John A. Oesterle (Englewood Cliffs, NJ: Prentice-Hall, Inc., 1964), pp. 60, 65, 66, 127.
5 Blaise Pascal, *Pensées*, tr. A. J. Krailsheimer (NY: Penguin Books, 1966), p. 74, #148.
6 The belief that there is such a thing as human nature and that it does determine or ought to determine our behavior in certain respects is a belief that has been challenged by the French philosopher and playwright Jean-Paul Sartre. We shall look more closely at his challenge in the chapter on freedom and determinism.
7 See George Santayana, *The Life of Reason* (NY: Charles Scribner's Sons, 1953), pp. 462–3.
8 Another valuable definition of happiness is this one by Immanuel Kant: "Happiness is the complete satisfaction of all our desires, extensively, intensively, and protensively." By "extensively" Kant meant that nothing that we desire is left out (which is similar to Aristotle's "completeness"). By "intensively" he meant that our desire is satisfied to just the right intensity of feeling – not too little and not too much. By "protensively" he meant that each desire is satisfied for just the right length of time – not too brief and not too long.
9 For Nietzsche's vision, see, for example, Walter Kaufmann's *Nietzsche* (NY: Meridian Books, 1956), pp. 223ff.
10 John Dewey, *Reconstruction in Philosophy* (Boston: The Beacon Press, 1948), p. 179.
11 To Kant the "*summum bonum*" (Latin for "the supreme good") is the union of virtue and happiness. Virtue (morality) is a greater good than happiness, but the combination of these two is greater than either by itself.
12 See Oliver A. Johnson, ed., *Ethics*, 3rd edn. (NY: Holt, Rinehart and Winston, Inc., 1974), p. 250.
13 Literally speaking, the hedonistic paradox is about pleasure rather than happiness. "Hedonistic" and "hedonism" come from the Greek word "*hedone*," which means "pleasure." Hence, the hedonistic paradox, literally speaking, is that the more *pleasure* we seek, the more *displeasure* we get. There is an important question as to whether happiness and pleasure are the same thing. Recall that "*eudaemonia*" is the Greek word for happiness, and you will understand why some people speak of "the eudaemonistic paradox," according to which the more intensely we seek *happiness*, the more *unhappy* we become. However, "the hedonistic paradox" is a much more common expression than is "the eudaemonistic paradox," and many people equate pleasure with happiness, so I will use the former phrase to cover both paradoxes. In a fuller discussion of happiness we would explore differences between pleasure and happiness and corresponding distinctions between the two paradoxes.
14 Schopenhauer was one of the first western philosophers to be heavily influenced by the Buddhist teaching that life is suffering. This is evident in his theory of the cause of unhappiness and in his way of dealing with it. However, Schopenhauer's philosophy is pessimistic in a way in which Buddhism is not.
15 Schopenhauer argued that we must not kill our craving for happiness by committing suicide. Suicide is a way of saying that we care so much about happiness that we cannot or will not live without it. Hence, to commit suicide is to allow the craving for

happiness to win out over us, and that is what we must not do. First we must come to see our craving for happiness as the destructive thing that it is. Then, according to Schopenhauer, we must destroy *it* – rather than let it destroy *us*.

16 See Whitehead's *The Function of Reason*. "Axiology" is pronounced: ax-ee-Ó-logy.
17 John Dewey, *Experience and Nature* (NY: Dover, 1927), p. 412.
18 In this day of easy credit, the distinction between cost and consequence is blurred. One can enjoy many things without paying a cent prior to enjoyment. Parting with one's money becomes a consequence of enjoyment rather than a cost; the only cost is "signing on the dotted line," that is, making a legally binding promise to pay the money in the future. Since the prospect of paying in full on the spot is sometimes painful enough to prevent one from going through with a purchase, easy credit is a clever way of making a purchase painless – at least temporarily.
19 Boethius, *The Consolation of Philosophy* (Indianapolis, IN: Bobbs-Merrill, 1962), p. 33.
20 The prophet Isaiah warned against delusory goods by asking, "Why spend your money for that which is not bread, and your labor for that which does not satisfy?" Paul of Tarsus urged that we "test everything" and "hold fast to that which is good." See Isaiah 55:2 in the Jewish Bible and I Thessalonians 5:21 in the Christian New Testament.
21 Sometimes bad judgment is caused by the attraction of low cost, rather than the promise of high pleasure. I saw a sign in a shop that warned against deceptively low costs: "Bitterness over poor quality lingers long after the pleasure of low cost has vanished." The point is that bitterness is likely to be part of a package in which low cost is the result of poor quality.
22 *Nicomachean Ethics*, 1178a. There are important differences between Socrates and Aristotle regarding the nature and power of reason, but they both exalt its role in human life.
23 Plato, *The Symposium*, in *The Collected Dialogues of Plato*, ed. Edith Hamilton and Huntington Cairns, p. 557.
24 *The Upanishads*, tr. Juan Mascaro (Baltimore, MD: Penguin Books, Inc., 1965), pp. 57-8.
25 See Philip Brickman and Donald Campbell, "Hedonic Relativism and Planning the Good Society," in *Adaptation-level Theory*, M. H. Appley, ed. (NY: Academic Press, 1971), pp. 287–300. The idea of the hedonic treadmill should remind you of Schopenhauer and the hedonistic paradox. There is, however, an important difference. Schopenhauer thought that all pursuit of pleasure is ultimately self-defeating. Brickman/Campbell and the *Upanishads* recognize that there *are* self-defeating pleasures, but they also believe there are fulfilling pleasures. We just need to learn to recognize both and respond appropriately.
26 J. S. Mill, *Utilitarianism*. Bentham's famous statement to the contrary was, "Pushpin is as good as poetry." That is, if a simple children's game (pushpin) gives as much pleasure to someone (all things considered) as does reading the finest poetry, then there is no reason why that person should choose poetry over pushpin. To think

otherwise is a mistake because, according to Bentham, value is a quantitative function of pleasure and pain, and what causes pleasure and pain is relative to the individual. Hence, for the hypothetical person mentioned above, pushpin and poetry are equal in value.

27 Here is another issue, consideration of which can help you clarify your thinking about happiness. Socrates said that we can think we are happy when we are not. Alan Watts said we can think we are not happy when we are. Are both of these states of affairs possible? They are if we make the following distinctions. Happiness in its fullest sense, what some would call "true happiness," consists of the combination of *subjective happiness* and *objective happiness*. Subjective happiness consists of feeling happy or thinking that oneself is happy, even though one's feeling or thinking that one is happy might be being caused by factors which if one fully appreciated them would make one feel or think oneself unhappy ("He wouldn't feel happy if he knew what's really going on"). Objective happiness consists of having circumstances such that if one fully appreciated them one would feel or think oneself happy, but one may not fully appreciate these circumstances and so may not feel or think oneself happy ("If she understood the situation, she would feel happy"). Subjective happiness without objective happiness means that one feels or thinks oneself happy but is in circumstances such that if one fully appreciated them, one would not feel or think oneself happy and might even feel or think oneself unhappy. Objective happiness without subjective happiness means that one does not feel or think oneself happy but does in fact have circumstances such that if one fully appreciated them one would feel or think oneself happy. True happiness, then, consists of feeling or thinking oneself happy because of circumstances which if they are or were fully understood do not or would not destroy that feeling. Therefore true happiness consists of the combination of subjective and objective happiness. False happiness consists of subjective happiness without objective happiness. What do we call objective happiness without subjective happiness? Unappreciated good fortune?

Reading Further

Risieri Frondizi, *What Is Value?*.
C. I. Lewis, "Book III: Valuation," *Analysis of Knowledge and Valuation*.
Jean-Paul Sartre, "Existentialism," in *Existentialism and Human Emotions*. Sartre argues that values are created, not found.

Chapter 14: Ethics and Morality

- Good and Bad; Right and Wrong; Self-interest and Morality
- Different meanings of "right" and "wrong"
- Ethical Nihilism
    - Casey in *Grapes of Wrath*
    - Metaethics and Normative Ethics
- Ethical Relativism
    - Morality can change; not universal in authority
- Individual Relativism
    - The moral authority of conscience
    - Criticisms
        - Too unstable to fit the facts
        - No basis for objectivity
- Social Relativism
    - The moral authority of society
    - The stability of morality
    - Criticisms
        - A society can be immoral
        - The persistence of guilt
- Ethical Absolutism
    - Explanation and Justification
    - Conscience
- **Theocentric Theories of Ethics**
- Divine Command Ethics
    - The moral authority of God
    - Criticisms
        - Why does God will what God wills?
        - An epistemological problem
- Perfect Being Ethics
    - *Criticisms*
- **Anthropocentric Theories of Ethics**
- Rationalistic (Deontological) Ethics
    - Morality vs. Prudence

- Immanuel Kant on reason and morality
- **Altruistic (Utilitarian) Ethics**
  Jeremy Bentham's principle of utility
  Criticisms
     Greatest happiness vs. Greatest number
     A disturbing implication
- **Universal Eudaemonism and Moral Happiness**
  Happiness, freedom, and rationality
  The principle of universal eudaemonism
  Mill, Russell, and Dewey
  Moral Realism
- **Why be moral?**
  John Rawls and the veil of ignorance
- **The need for moral education**

"I slept and dreamed that life was beauty.
"I woke and found that life was duty ... "

Ellen Sturgis Hooper (1816–1841)[1]

Good and Bad, Right and Wrong, Self-interest and Morality

The good, as we have been discussing it, is ultimately a matter of individual judgment based on the combined values of the felt-quality, the costs, and the consequences of a thing. Because you and I *feel* differently about certain things, we believe there is nothing wrong with you judging a thing "good" – such as a certain kind of music – and me judging the same thing "bad," or vice versa. This means that given the opportunity, we may construct very different lifestyles. With respect to matters of enjoyment we find individual differences understandable and the freedom to differ desirable. As John Stuart Mill put it in his essay, *On Liberty*, "The only freedom worthy of the name is that of pursuing our own happiness in our own way."

Yet we are not willing to let other people do *anything* they want to do in their pursuit of happiness, and we can distinguish two very different reasons for objecting to actions that others might take. On the one hand we object to some actions because they cause us pain or deprive us of pleasure. In other words, we object on the ground of *self-interest*. Let's say, for example, that you own a home in a relatively secluded area and that what you enjoy most about it is the peace and quiet. The city, however, decides to build a ballpark right next to you – with loudspeakers and floodlights for evening games! You might well go to a session of the city planning board and register your opposition to the park, but you would do so on the ground of self-interest, for when you bought the house you knew (let's assume) that the land next to you could be developed legally as a park – you just thought it would never happen. Hence, your reason for objecting would not be that what the city is proposing to do is wrong; rather, from your personal point of view, it is simply undesirable.

On the other hand, let's suppose that instead of the city wanting to build a ballpark next to you, a wealthy, powerful person sees your property, falls in love with it, and makes you an offer you'd better not refuse. Then – though you might do it only to yourself – you would object that what this person was doing was just not right; it was not only undesirable from your point of view, it was *morally wrong*! Now we are at the crux of a knotty problem: What is the difference between *the right* and *the good*, between *the wrong* and *the bad*, and how are they related to one another?

The good, which we have focused on until now, is the basic concept in *axiology* (the study of nonmoral values). The right, to which we turn now, is the basic concept in *ethics* (the study of moral values). Together these two disciplines, axiology and ethics, constitute the more inclusive discipline called "theory of value." Hence, another way to put the questions asked above is: What is the relation between axiology and ethics? To return to our example, it is clear that the aforementioned gangster would consider acquisition of your property a "good" thing (because it would give him pleasure). You, however, would consider it not only a "bad" thing (because it would deprive you of pleasure) but also a "wrong" thing! Now what, if anything, would make extortion of your property not only "bad" (undesirable) but also "wrong" (immoral)?

Before answering that question, let's answer another question which will help focus our discussion more sharply. Earlier we asked, "What kind of a thing is it that is true or false?" The answer was, "An assertion." Now let's ask, "What kind of a thing is it that is morally right or wrong?" There are several possibilities – such as intentions, character, and consequences – that need to be considered on a more advanced level, but the best initial answer, I think, is "an action." It is actions that are, were, or would be morally right or wrong.

An action is a unit of intentional behavior that produces expected consequences. Unintentional behavior – such as pulling the sheet off your bedmate when you, deep asleep, roll over – is not an action and so is not morally right or wrong; also, the unintentional consequences of intentional behavior are not part of an action – if you back your car over a squirrel that you had no reason to think was there, then killing the squirrel was not an action and so is not subject to being morally right or wrong (backing up the car was an action; crushing the squirrel was not – unless you did it intentionally!).

Different meanings of "right" and "wrong"

Now let's return to the question, "What *makes* an action not only *bad* but also *wrong*?" The English language rarely makes it easy to answer such a question because nearly every English word has more than one meaning. We say, for example, "That's the wrong fork to use for your salad," "You gave the wrong answer to that question," and, "You turned the wrong way at that intersection." In the first case a claim about etiquette is being made; in the second case a claim about truth is being made; in the third case a claim about the means to an end is being made. What makes something a matter of "wrong etiquette" is the prevailing custom of a group; what makes something a "wrong answer" is that it does not represent reality correctly; what makes something a "wrong means" is that it won't work at all or that it won't work as well as an alternative that is available. But in using the wrong fork, giving the wrong answer, or taking the wrong turn, you are not necessarily doing anything morally wrong.

So what makes an action *morally* wrong? Many answers have been given to that question, and all are controversial, but there seems to be almost no one who has not had the feeling of an action *being* morally wrong. Consider, for example, the following story from *Time Magazine* (July 25, 1977, p. 17). During July of 1977, the weather was sweltering in New York City; moods were turning mean among people who had no way to escape the heat. Then, as though things were not bad enough already, a power failure struck. Many people, especially younger and poorer people, began pouring out of their dwellings, breaking the windows of neighborhood stores and looting them. Soon looters were stealing from one another.

One teenage girl had such a haul of clothes and radios that she couldn't carry them far without stopping to rest. While she was resting some boys came by; seeing her plight they offered to help her. She was pleased until the boys proceeded to run away, stealing from her the goods that she had just stolen. Soon thereafter she told her story to friends and said of what the boys had done, "That's just not right. They shouldn't have done that." How irrepressible our sense of moral right and wrong must be to manifest itself in such circumstances! That girl wasn't just *angry* at those boys for taking what she had stolen; she felt that what they had done to her was *wrong*.

John Hick has said that that feeling, "The sense of moral obligation, or of 'oughtness,' is the basic datum [or fact] of ethics."[2] Obviously there are many value judgments into which moral feeling does not necessarily enter – for example, judgments about the goodness or badness of food, hairstyles, and odors. So where does that sense of moral right and wrong come from when it does arise? And what is its authority? More generally, what is the authority of the standards by which we judge things to be morally right or wrong? And where do those standards get their authority? There are many answers to these questions, but they can be gathered into three basic types: ethical nihilism, ethical relativism, and ethical absolutism.

Ethical Nihilism

One answer to the question "What gives morality its authority?" is that, in a sense, there is no answer. This is the position of ethical nihilism. "Nihil" in Latin means "nothing." To an*nihil*ate something means to destroy it so completely as to reduce it to nothing. The ethical nihilist argues that there is nothing to morality. There is nothing that makes anything morally right or wrong. Therefore, there is nothing morally right or wrong. The so-called standards of moral right and wrong have no more authority or reality than does Santa Claus or the Easter Bunny.

This point of view is stated exquisitely by a fellow named "Casey" in the first part of John Steinbeck's novel, *Grapes of Wrath*. Casey had been a roving preacher, but he found that being even a roving preacher was incompatible with his freewheeling sex life. After he left off preaching, Casey concluded the following about moral right and wrong: "There ain't no sin and there ain't no virtue. There's just stuff people do."

Casey's statement seems to be a clear expression of ethical nihilism: people may *think* that some things are morally right and other things are morally wrong, but that is only because they are ignorant of the fact that there is no one and nothing that can make an action morally right or wrong. Some people harm other people; some people help others; some people give to others; some people steal from them; yet none of it is morally right or wrong; it's just stuff people do. To be sure, you may not like what some people do, and they may not like what you do, but that doesn't make either of you morally wrong; it just makes you different from one another.

Some people react with horror to ethical nihilism. They envision that if lots of people became ethical nihilists, society would fall into anarchy and chaos. That is not very likely, however. Ethical nihilists, like everyone else, do not want people stealing from them, raping them, enslaving them, and so on. Hence, most ethical nihilists favor the establishment of *laws* to protect them from such abuse, but the laws they favor are simply based on their preferences, not on what they think is morally right or wrong.

To understand the difference between ethical nihilism and the other ethical positions we are about to explore, it is important to understand the distinction between metaethics and normative ethics. *Normative ethics* is the endeavor to discover, clarify, and justify the principles of behavior that we ought to live by. Hence, it proceeds from the belief that there are such principles, rules, or norms. *Metaethics* makes no such assumption. It starts by asking questions about the nature of ethics – just as metaphilosophy starts by asking questions about the nature of philosophy. Metaethics asks what, if anything, makes an issue a moral issue. It asks what it means for a rule to be a *moral* rule rather than a non-moral rule. It asks whether there are such rules. It asks what, if anything, could possibly justify a rule as something that all humans ought to abide by.

Three basic parts of the general field of ethics are *metaethics*, *normative ethics*, and *applied ethics* (which applies the findings of normative ethics to issues concerning medicine, the environment, scientific research, law, government, business, the media, and other areas of life). The ethical nihilist believes in doing ethics in the form of metaethics, but in answer to the questions "Morally what ought we to do, and why ought we to do it?" the ethical nihilist argues that there is nothing that we *ought* to do because there is no good reason to think that we have an obligation to do or not do anything. Hence, ethical nihilism is a *metaethical* position about normative ethics. It says that normative ethics, the search for *what* is morally right and wrong and *why* it is morally right and wrong, is a futile enterprise. The ethical nihilist argues that the questions of normative ethics have not been answered satisfactorily after centuries of effort because its questions have no answers. Therefore, neither do the questions of applied ethics – since they depend on normative ethics for answers. To adapt an analogy from another context: doing normative ethics, searching for what is morally right and wrong, is like feeling around in a pitch dark cave for a black cat that isn't there.

Most people, however, cannot shake their feeling that there is a cat in the cave. At the very least, most of us seem to have *feelings* that some things are morally wrong and others morally right (remember: even the young looter felt morally offended when what she had stolen was stolen from her). Therefore, notions of moral right and wrong certainly seem to most people to have meaning, application, and some kind of authority. The job of *normative* ethics (as distinguished from *meta*-ethics) is to identify and articulate those principles of behavior that we ought to live according to, and to figure out and set forth the reasons for the authority of those standards.

Ethical Relativism

People who believe in the possibility and importance of normative ethics divide into two basic groups: *ethical relativists* and *ethical absolutists*. Ethical relativists and ethical absolutists, unlike ethical nihilists, believe that there *are* moral rights and wrongs. However, ethical relativists believe that moral rights and wrongs are not absolute because they are anchored to things that can change, such as social customs and traditions, and individual feelings and judgments. Ethical relativists also believe that because moral principles are based on such things as individual feelings and judgments, and social traditions and customs, there is no single ethical position that applies to everybody. Individuals differ from one another and societies differ from one another, so it is natural that moral principles are going to differ from individual to individual and society to society. Ethical absolutists, by contrast, are convinced that moral right and wrong are unchanging because they are anchored to something that does not change (what that might be we will examine soon). In contrast to ethical relativism, ethical absolutists also hold that what is morally right and morally wrong applies equally to everyone.

In brief, according to *ethical relativism* moral right and wrong are *subject to change* and *do not apply to everyone*, whereas according to *ethical absolutism* moral right and wrong are *not subject to change* and *do apply to everyone*. It is important to note that ethical absolutists do not say that everyone *agrees* on what is morally right and wrong. They know that people disagree about moral issues. Their point is that whatever really is morally right is universal in the sense that it applies to everyone; it is what everyone *ought* to live by.

Now let's begin looking at specific theories of normative ethics by first examining two relativist positions and then examining several absolutist positions. Keep in mind as we move along that the meaning and authority that moral statements have depends on what it is to which they are anchored.

Individual Relativism

Some kinds of statements are objectively true (as in mathematics), and therefore everyone should agree to them. Moral statements, however, are not objectively true or false, according to individual relativism. Moral statements are true only relative to individual judgment. A moral statement is true only when an individual believes that he or she should abide by it. Hence, it is the conviction of the individual which *makes* a moral statement true or false for him or her. After all, if an individual wants to know whether it would be morally wrong for him to do something, what can he do except search his own mind and heart and then answer for himself? To be sure, he can seek the opinions of others and discuss moral questions with them, but he cannot answer for anyone else, and no one else can answer for him.

Moreover, because the moral judgments of different individuals are based on different personal histories, feelings, situations, and thought processes, different individuals may arrive at different (even very different) moral conclusions, but that does not mean that one person is right and the other wrong. What is wrong is to violate one's personal moral convictions, whatever they are. Moral convictions are among the deepest expressions of who one thinks one is and how one thinks one should live; hence, to violate one's conscience is to violate one's deepest sense of identity and integrity. However, there is no such thing as one's moral convictions being wrong. They are what they are. With further experience and reflection they may change, but they are never right or wrong in some objective sense.

Hence, when an individual relativist says, "Capital punishment is morally wrong," what she is reporting is that she thinks or feels that at that time it would be wrong for her to support capital punishment, but she is not saying or implying that it would be wrong for anyone else to do so. Again, there are no objective truths in morality; there are only truths for the individual. No action is morally right or wrong in and of itself and for everyone. More abstractly, according to individual relativism any moral statement uttered sincerely by an individual, for example, "X is morally wrong" or "Y is morally obligatory," means only that that individual believes it would be wrong for him or her to do X or not to do Y. The individual relativist does not, however, think that his or her moral convictions have any authority over anyone else. The moral convictions of an individual have authority over that individual only. Other individuals have their own moral convictions, and according to individual relativism, that is what each should follow. To be sure, sometimes an individual relativist desires that other people do Y or not do X, but other times she is indifferent as to whether other people do what she thinks it would be immoral for her to do or not do, and at no time (as long as she is being consistent) does she think other people are acting immorally by not doing what she thinks she ought to do and would like for them to do.

Individual relativism is very attractive to some people because of its emphasis on the responsibility of the individual and its tolerance of diverse opinions. Other people, including people who value tolerance and individual responsibility, think individual relativism has serious problems. Let's look at three purported problems. First, some critics say that when an individual says, "X is morally wrong," he typically means more than simply that he thinks it would be wrong for him to do X. When we say, "X is morally wrong" (for example, "Infant torture is morally wrong"), we typically mean that *other* people, as well as ourselves, ought not do X (torture infants). Hence, individual relativism doesn't account for what we usually mean when we make moral statements.

Second, people's standards of morality are not as fickle as they would be if individual relativism were true. The emotions, thoughts, and feelings of individuals about moral issues ebb and flow a great deal, sometimes from day to day, even hour to hour – especially in today's open, pluralistic societies – so if the thoughts and feelings of the individual were all that morality is based on, we should expect moral standards to shift around a lot – but they do not, and that suggests that moral standards are based on something beyond the individual. To be sure, say these critics, moral standards do change over time, but they do not jump around the way we should expect them to if they were based on nothing more than the thoughts and feelings of each individual.

Third, to say that something is morally obligatory or morally wrong is to say that there is something that people ought to do or ought not to do – regardless of their personal feelings or opinions. The oughtness of such a statement means that there is an external authority, an authority outside the individual, because of which the individual should do or not do a certain thing. Individual relativism cannot account for this *authority* of moral principles over the individual, and so, again, *individual* relativism is an inadequate account of morality.

Social Relativism

But what about *social* relativism? Wouldn't that provide moral statements with external authority and more resistance to change than individual relativism can provide? Yes, because according to social relativism, when an individual says, "It is wrong to torture an infant," his or her statement is anchored to the convictions of a *society* of people of which he or she is a member. Hence, according to social relativism a proper reading of "It is wrong to torture an infant" is something like this: "The people in our society are opposed to infant torture." Such convictions become expressed in the teachings, laws, customs, traditions and institutions of a society. They thereby shape the sentiments of future generations and help provide the stability that we usually see in the moral standards of a society.

The moral anchor might be hooked to the convictions of some group other than society in general, of course, such as a religious group, but it has to be attached to the convictions of a group and not merely to the convictions of an individual. Morality is a social phenomenon, not an individual phenomenon. It has to do with how people treat one another. An individual shipwrecked on a deserted island could do nothing immoral. The possibility of morality would not exist on that island until a second person came ashore.

Notice that since morality is a social phenomenon rather than an individual phenomenon, an individual may not *like* all the positions of the group with which he or she identifies most closely. The Mormon Church, for example, teaches that drinking alcohol is wrong. An individual Mormon, however, may dislike this tenet of Mormonism. Yet if he were a social relativist, he would not say, "The Mormon Church is wrong and I am right," since according to social relativism, individual convictions do not constitute a morality. Personal opinions are just that and nothing more. Morality does not begin until shared convictions become established standards for the behavior of the members of a group. Apart from group agreement there is no morality – only individuals with their personal opinions and preferences. This means that if an individual personally likes the moral convictions of a group, it is not the case that his or her morality agrees with the group's morality. According to social relativism there are no individual moral standards; the only moral standards are group standards, which the individual does or does not accept and to which he or she does or does not conform.

If an individual does not like the moral positions of a group of which she is a part, she can console herself that the group's position might change or that there are other groups to which she might go that think or feel differently. If, however, she desires the good will of her current group (because of family ties or business contacts, for example), she may attempt to conceal from the group that she does not like their position, since moral differences often result in social rejection.

Moreover, if she likes strongly enough something which the group disapproves, she may indulge in it on the sly – knowing that it is wrong (in the social relativist's sense of "wrong"), but knowing also that she certainly does enjoy it. Will she feel guilty when she does such a thing on the sly? She may, because of the way she has been reared. Guilt, according to social relativists, is an emotional reaction that is conditioned into children to divert them from doing what the conditioner disapproves of. According to social relativism, then, adult feelings of guilt are merely vestiges of childhood rearing. Because social mores change, those feelings may no longer be appropriate, and in any case should be unnecessary for the adult. Adults should pay attention primarily to the mores of their society and only secondarily to feelings of guilt, as possible signs that one is violating the mores of one's society.

In brief, a social relativist will not care about the moral convictions of a group unless (1) she personally shares those convictions, or (2) she believes they are in her best interests, or (3) she desires the goodwill of its members. If she has children, her attempts at moral education will consist of efforts to inform her children as to what behavior is approved and what is disapproved by the group which she believes will be most important in her children's lives (even though she herself may not share some of the convictions of that group). If her children ask, "But Mother, why is drinking alcohol wrong?," and if the mother doesn't personally disapprove of drinking alcohol, she will say something like, "Because people disapprove of it, and if you do it, they will think poorly of you and you will get into trouble with them." As a social relativist, she will have no higher court of appeal than that.

If rather than a personal question, she is asked an "academic" question (for example, "What do Fiji Islanders think is morally permissible and morally wrong?"), normally nothing can be inferred about her personal position from what she says. She is simply describing the moral norms of a group to which she does not belong. Yet if someone asks her, "What do *you* think is right and wrong?," she will, as a social relativist, translate that to mean, "What are the moral positions of the group with which you identify most closely?"[3] Even from her answer to that question, however, you cannot infer to what extent she likes the morality of her primary group. To find out whether she shares the moral convictions of her primary group, you must simply ask: "Which of the positions of your primary group do you share and which do you not?"

Whatever her answer, it will reveal that she, unlike the individual relativist, does not think that her personal opinions constitute a morality. Until such opinions are shared by others and help give them unity and guidance as a group, they remain mere personal preferences. There is no such thing as individual morality. To be sure, it is the actions of individuals which are moral or immoral, and so it is the individual who acts morally or immorally – but only as judged by the standards of a group. The opinions of the individual as an individual have no moral authority. Indeed, if in obeying the group she acts contrary to her conscience, she is nonetheless doing the morally right thing. Conscience should be obeyed only when it is expressive of the moral standards of society.

Earlier we heard criticisms that individual relativism cannot account for the objective authority of moral principles or the relative stability of moral codes and does not account for what we mean when we make moral statements. Social relativism was presented as an improvement upon individual relativism. Now we must ask whether social relativism is adequate. In some ways it seems more adequate than individual relativism. A phenomenology of moral experience reveals that we experience moral right and wrong as being based on something more authoritative than our individual opinions – something that stands over us and against us as an objective authority, whether we like it or not. Also, moral standards are relatively stable and slow to change, so they must be based on something less subject to fluctuation than the opinions of the individual. Social relativism sees morality as anchored to group consensus, so it can account for the feeling that moral right and wrong are rooted in something outside us and have a kind of external authority over us. That external authority is the authority of the society in which we live. Moreover, because moral beliefs find expression in social practices, traditions, and institutions, it is understandable that moral beliefs tend to be stable and long-lived, rather than fluctuating with the feelings and opinions of the individual.

Moral codes can and do change, however. Indeed, one of the greatest virtues of social relativism, according to its supporters, is that it provides relative stability for the moral code of a society (which is necessary for social order) while permitting changes in the moral code as the experiences, insights, and circumstances of its members change (which is necessary to accommodate progress and prevent stagnation or rebellion).

Still, some critics, including individual relativists and some absolutists, insist that conscience and guilt are such fundamental moral phenomena that any adequate theory of morality must give a more satisfactory account of them than social relativism does (we will return to this issue soon). Furthermore, say absolutists, social relativism does not account for the widespread conviction that the fundamentals of morality are unchanging, and that, for example, it always has been, is, and always will be wrong to torture an infant.

If some group comes along which believes in torturing every first-born infant to death, and which also believes that it is wrong to relieve the suffering of people over age sixty, then in that group (according to social relativism), torturous infanticide will be right, and relieving the sufferings of elderly people will be wrong – no matter how or what you as an individual feel or think about these matters.

There may, of course, be another group that disagrees with the group just described, so you could say to the first group, "See, I told you that torturous infanticide is wrong!" But your pronouncement would mean nothing more than that there is a second group which treats infants differently than the first group does. If there is no higher moral appeal than group agreement, then the sentiment of neither group can or will have authority over the other group, and neither group can be morally wrong.

But, the critics say, don't we sometimes make moral judgments about group standards? Don't we sometimes say that what a group does or approves of is wrong – if, for example, the group is engaging in or permitting such things as slavery or racism or religious genocide or gender discrimination or environmental destruction? By such judgments don't we mean not simply that we think it would be wrong for us, individually, to do those things or that the group to which we belong forbids them, but rather that they are *absolutely wrong*? Don't such judgments presume and imply that moral standards transcend personal and group conviction?

It would be amusing, of course, if the individual relativist slipped into the mentality of an absolutist by saying that infant torture is "really" wrong and then tried to defend her position by appealing to nothing more than her own opinions. It would be amusing because the practitioner of infanticide would say, "Okay, you don't approve of what I do. So what? We are both individual relativists. You have your opinions; I have mine. Your convictions have no authority for me, and mine have none for you. So go away!" A social relativist would receive the same kind of response were he to talk as though the moral positions of *his* group were somehow more authoritative than the moral positions of *other* groups.

The point is that according to relativism, no individual, as an individual, and no group, as a group, has more authority than any other individual or group. Consequently, say the critics, if progress is to be made toward justification of the claim that torturous infanticide or slavery or racism or gender discrimination or religious genocide or environmental destruction is absolutely morally wrong, then we must appeal to something other than the fact that that individual or that group thinks that way.

Let's explore some of these points further by means of an example. Suppose we discover in a foreign country a group of social relativists who engage in torturous infanticide. Wouldn't we say to that group, "What you are doing is morally wrong!"? If we did, then they, as good social relativists, might reply, "What's wrong for you has no bearing on what's wrong for us. Your group has its morality; our group has a different morality – plus an ample supply of its own infants. So mind your own business!"

Now, if we, too, are social relativists, we will have to agree that we have meddled. "Sorry," we might say, "you're right. We have no right to insist that you live by the morality of our group. Because *we* disapprove of what you do to infants, we forgot momentarily that morality is relative to the group. Please accept our apologies for meddling; return to your sacrificial altar. Oh yes, and here is your infant back."

But is that what we would actually say? Or would we suddenly be convinced of the bankruptcy of social relativism and abandon it? If so, then instead of saying, "Sorry 'bout that," we might say something like, "*Etiquette* is group relative, but *morality* is not! What you are doing is morally wrong for anyone to do. Morality doesn't gain its authority from the consensus of a group; it bears on all groups and all individuals at all times, in all places, no matter what the individual thinks about it at the time and no matter what the consensus of the relevant group is. Infant torture, like racist lynching, is wrong no matter when, where, or by whom it is performed; therefore, you should stop it and never do it again – no matter how much you personally approve of it and no matter how large the group which approves of it. The immorality of a racist lynching does not vary inversely with the size of the mob!"

Ethical Absolutism

The preceding statement is characteristic of the widespread conviction that morality ultimately has to do with what is wrong always, everywhere, and for anyone. The contrasting folkways and customs of different societies are often not moral matters – such as how people dress or wear their hair or greet one another. Sometimes, however, a society's folkways or customs are morally significant and might be immoral – consider the old Chinese practice of binding the feet of certain women from infancy so that their feet were rendered diminutive and useless, or the old American practice of slavery.

Still, we are a long way from *proving* or *warranting* the universal authority of a particular moral code. All we have done so far is remove some misrepresentations of what morality is about and come up with the realization that, according to ethical absolutism, when we say that something is *morally* wrong, we do not just mean that it is contrary to our personal convictions or that it is disapproved of by some group; rather, we mean that it ought never to be done by anyone, including ourselves.

Further, we mean that if a thing is really immoral, then it remains immoral even though our feelings toward it change from feelings of disapproval to feelings of approval. To be sure, I may make a mistake in my understanding of what is moral and immoral, and I may grow out of that mistake, but what was right was right all along, and what was wrong was wrong. I just misunderstood it; the mistake was in me, not in the objective moral facts.

Consider a parallel case. The best scientists once thought the sun went around the earth; now the best scientists believe the earth goes around the sun. Hence, beliefs about the sun have changed, but the truth about the sun did not change. The earth was going around the sun while scientists thought it was not, as well as after they changed their minds. Similarly, if the truth is that slavery is morally wrong, then slavery was morally wrong while people believed it was morally okay, as well as after they came to see that it is morally wrong.

The last point should help us understand an important difference between individual relativists and those absolutists who agree with them that we should never violate our conscience. Individual relativists believe there are no objective moral truths, so they do not think our conscience can be mistaken; it is simply an expression of our moral convictions at that time. By contrast, moral absolutists believe there are objective moral truths, so they think we can be mistaken in our moral feelings and convictions. This raises an interesting problem for those absolutists who believe we should never act against our conscience. We should not do so, they say, because our conscience is expressive of our convictions as to what is morally right, and we should always do what we think is morally right; not to do so would be to degrade morality from its proper importance to something that can be brushed aside for the sake of pleasure or convenience. This means, according to these absolutists, that even if what one thinks is morally right is really morally wrong, one should still act according to one's conscience. To do otherwise would be to reject the seriousness of morality. It can be unfortunate and even tragic when what a person thinks is morally right is really morally wrong, but at least that person remains morally serious, so if she later discovers her moral mistake, then, being a morally serious person, she will begin to do what is really morally right and will try to make amends for past mistakes.

Many people, perhaps most people, have had the feeling or conviction that a certain act was morally wrong in the absolute sense explained above. The question we must ask now is whether that feeling or conviction can ever be *justified*. It can be *explained*, of course, but there is an important difference between explaining something and justifying it. To explain a thing is simply to tell how it came to be as it is or why it occurred as it did. Such an explanation is a causal explanation; it does not necessarily imply approval or disapproval of that which is being explained. By contrast, to *justify* a thing is to give reasons for thinking that it is, was, or would be good.

Sometimes, however, an explanation undermines the possibility of a justification. Bertrand Russell, for example, *explained* feelings of moral conviction in terms of the will of society worked upon its individual members. Russell reasoned like this: no one wants their personal property to be stolen; so society nurtures in its young members the strong feeling that they ought never to steal. If the job is done correctly, the young are never given a *reason* why it is wrong to steal (one can always work one's way around a reason!). Rather, a very unpleasant reaction to the very thought of stealing is cultivated in children (a combination of guilt and anxiety). Such children normally won't steal because their emotions paralyze them at the thought of such a thing. If you ask them *why* they don't steal, they simply say, "Because it's wrong," and that's that. If this explanation of conscience is correct, then such children are the victims of emotions bred into them by their society. They are "dupes of the herd," and for most people the dupery continues into their adult years.

Here, then, is one *explanation* for feeling that something is immoral. People who have that feeling have no *reason* for feeling that way; they just feel that way. The people who made them feel that way have *their reasons* for causing them to feel that way, but their reasons may have nothing to do with morality and may even be immoral! A slave owner, for example, might, for the sake of his own convenience and profit, cultivate in his slaves a feeling of guilt toward any thought of disobedience to him. Such a procedure, if it works, seems clearly more effective and economical than giving the slaves reasons for obedience – reasons always invite discussion and criticism.

However, when people come to believe that morality is merely a tool by means of which the herd (or the head of the herd) controls its members, then morality loses its grip on those people. Believing that their feelings of guilt were artificially bred into them, those feelings lose their authority for them and evaporate. "Why," they will ask, "should I continue to act according to the proddings of conscience, given where those proddings came from? Knowing what I know now, I can no longer take the proddings of conscience seriously as signs of what I ought to do and not do. Rather, starting now, I will no longer be a dupe of the herd. I will act for the sake of my *own* best interests. Because of the way I have been reared, it may be unpleasant at first to act against "my" conscience (the 'my' is a cruel joke played upon me by society – it is not *my* conscience; it is *society's* conscience!), but if I go ahead and ignore my conscience, gradually my knee-jerk guilt will fade away and bother me no more. Then I will be free at last from the tyranny of morality!" Obviously this is an explanation of morality which does not justify it; rather, it undermines it.

It should be added that even if no one has consciously, intentionally, and with an ulterior motive, imbued feelings of guilt within us, conscience will still lose its clout for anyone who becomes convinced that their feelings of guilt are simply the outcome of a natural process of socialization. If guilt feelings are the result of an unintentional, rather than an intentional, process, that hardly confers greater authority on them. The appropriate response would be: "It's just something that happened. Now that I realize how guilt feelings come about, I am henceforth liberated from taking them seriously. They are simply tools by which other people – sometimes intentionally, sometimes unintentionally – have tried to control me."

Where do we go from here? Have we now reached the truth about morality or is there something more to be said? Ethical absolutists believe there is much more to be said. Our feelings of right and wrong do not just go away after we are confronted with the possibility that they are the products of intentional or unintentional social control. To be sure, *some* of those feelings may go away. We may have been socialized to feel guilty about expressing anger and then have been liberated from that guilt by coming to see that it was the result of training by authoritarian parents and teachers who wanted no resistance. The point that Bertrand Russell and others have made about morality being a product of socialization is an important point that increases the extent of our self-awareness. Our feelings of guilt *may* sometimes be tools by which others have been controlling us and which we should overcome, but it doesn't follow that *all* guilt feelings are simply a result of socialization.

Any good thing can be misused for personal gain. Consider, for example, the feeling of knowing that something is true. Con artists try to instill that feeling into us with regard to whatever it is they are selling, and they succeed with nearly all of us some of the time. But that doesn't mean we should conclude that our feeling of knowing something is never justified. It means only that we should be very careful, especially in certain circumstances, in trusting our feeling that we know something.

Similarly, the fact that people sometimes manage to get their way with us by instilling and manipulating guilt feelings in us does not mean we should conclude that our guilt feelings are never justified. Sometimes our feeling of knowing something is the result of knowing it, and sometimes our feeling that we ought not to have done something is the result of our having done something that we ought not to have done. Also, what about our feeling that it is wrong to torture an infant? Is that just a matter of socialization? If we think it is not, how can we *justify* our belief that it is morally wrong? Or justify our belief that sometimes feelings of guilt *reveal* that we have done something morally wrong? Let's examine five answers to these questions. Each answer is absolutistic, instead of relativistic, because each claims to ground morality on something other than individual or group sentiment – something that stands with authority over every human being at all times in all places.

Before we examine those positions, it is important to realize that an ethical absolutist is not necessarily a dogmatic person. The ethical absolutist does believe there are definite universal rights and wrongs, but he or she does not necessarily believe that he or she knows for certain which specific things are absolutely morally right or wrong. To be sure, *some* ethical absolutists are dogmatic in their ethical beliefs. They think they know for certain what is morally right and morally wrong. Other absolutists, however, are open to discussion of any of their ethical beliefs. They think there *are* absolute moral rights and wrongs, but they also think that we, as human individuals, are limited, prejudiced, and fallible. Consequently, they believe we need to seek moral right and wrong in discussion with one another and stay open to the possibility that no matter how clear things seem to us now, we may not yet have it right. Still, they are absolutists because they think there *is* something to get right. Hence, unlike individual relativists, they think an individual's moral convictions can be mistaken, and unlike social relativists, they think a society's moral positions might be mistaken.

Now let's look at those five absolutist positions. To begin I will provide you with descriptive titles of these positions to help you get a sense of the orientation of each. The first position is "the divine command theory." The second position is "the perfect being theory." The third position is "rationalistic." The fourth position is "altruistic." The fifth position is "eudaemonistic." The first two theories – divine command ethics and perfect being ethics – are theocentric theories, for reasons stated in the next paragraph. The last three theories are anthropocentric theories, for reasons that will be stated later.

Theocentric Theories of Ethics

"*Theos*" in Greek means "God," so "theocentric ethics" literally means "God centered" ethics. Theocentric ethics holds that in order to find out what is morally right and wrong we must look to God – a unique, powerful, perfectly wise being who is the creator, sustainer, and guide of the universe. (In chapter 17 we will explore the topic of the nature of God, and we will address the question of the existence of God. For our purposes in this chapter I will without discussion assume the concept of God that is usually at the base of theocentric ethics.)

Emil Brunner, a protestant theologian from the first half of the twentieth century, expressed the position of theocentric ethics when he wrote in his book *Our Faith*:

> Moral seriousness is respect to the voice of conscience. If there is no God, conscience is but a complex of residual habits and means nothing. If there is no God, then it is absurd to trouble oneself about right – or wrong. It all comes to the same ultimate chaos. Scoundrel and saint are only phantoms of the imagination.[4]

If Brunner is correct, then ethics cannot be done without taking God seriously, for only God has the authority to make anything morally right or wrong. If there is no God, then there is just you and me, and neither of us has the authority to make anything morally right or wrong, nor does any other human being.

By implication Brunner was saying that we must choose between theocentric ethics and ethical nihilism. If we reject the one, we must accept the other. Why? Because only an absolute being could establish absolute standards. Hence, if there is no God, then there can be no absolute rights and wrongs. Or, as the great Russian novelist Fyodor Dostoyevsky put it: "If there is no God, then all things are permitted." Now let's look at two very different versions of how we might find out from God what is morally right and wrong.

Divine Command Ethics

According to the divine command theory of ethics, it is the will of God that *makes* an action morally right or wrong. If murder is morally wrong, it is because

God commands us not to do it. If charity toward the poor is a moral obligation, it is because God directs us to help them. In brief, nothing is morally right or wrong until God commands us to do it or not do it. Whatever God commands us to do, we ought to do; whatever God commands us not to do, we ought not to do. Whatever God has not given us a command for, we are permitted to do or not do according to our personal preference.

Why should the will of God determine moral right and wrong? Because this world is God's world; it was created by God and God alone, so God and God alone has the right to tell us how to live in it, and how to treat one another and God's other creatures – just as someone who opens her magnificent flower gardens to visits from the public has a right to say that visitors must not litter or pick flowers and must make a donation to a local charity for homeless people.

In order, then, for us to know what is morally right and morally wrong, God must reveal to us what God's will for us is. Religions throughout the world believe that God's will for us as humans and as individuals is revealed through prayer, through living prophets, and through scriptures which record revelations from God to men and women in the past.

Critics of the divine command theory ask, "What is it about God's willing something that *makes* it morally right?" Is God's will what we ought to do simply because God is *all-powerful*, so that we'd better do what God says or suffer terrible consequences? That might, indeed, be a good reason for doing what God says, since the long arm of divine law would surely be inescapable, and the wrath of God could be horrific. But the fact that God has the *power* to enforce what God wills would not seem to make moral what God wills anymore than the power of a gang leader to enforce what he or she wills would make what he or she wills moral! Hence, it is not the power of a person that makes moral what he or she wills – even though that person be divine.

Nor can it be the case, critics say, that because God created the world, therefore whatever God wills for the world is binding on its occupants. Imagine a parallel. Just because a brilliant, powerful woman created a new society on an uninhabited planet would not mean that whatever she willed its citizens to do would be morally right. Just as we can imagine such a person in such a situation commanding people to do something immoral, we can imagine the creator of the world commanding us to do something immoral.

Consider, for example, the gut wrenching Biblical drama of God's command to Abraham to slay Isaac, his beloved and only child by his beloved wife Sarah. After this horrifying command, Abraham, still under command from God, traveled several days with Isaac to the designated place of sacrifice, tied him up, placed him on a sacrificial altar, grasped a dagger, raised it up, and was about to plunge it into Isaac as God had commanded. Only then, as Abraham was about to strike, did God tell Abraham to stop – but it was clear that Abraham would have obeyed the command of God all the way (see Genesis 22).

What was Abraham thinking? Did he believe that what God had asked him to do was regrettable but not wrong? Did he believe that what he had been asked to do was wrong, but that he owed God obedience after all that God had done for him? Did he believe that what he was doing was wrong, but he feared to disobey God? Should he have "stood up" to God and refused to go through with the sacrifice for the reason that it was wrong? Was that what *God* was hoping Abraham would do? Did Abraham *feel* that what he was doing was wrong, but *trust* that God would not ask him to do anything that was really wrong?

Lest the ethical significance of this story be dismissed because, after all, God stopped Abraham at the last moment (which can be interpreted to mean that God would never *really* command such a thing), keep in mind that God did not stay the hands of those Israelites whom, according to the Bible, God commanded to enter into Amalekite villages and slay every man, woman, child, and beast (see I Samuel 15). Indeed, we are told that God became angry at King Saul, who ruled over the Israelites at that time, because the Israelites did not do the job to perfection. One person, Agag, king of the Amalekites, and the best of the animals were spared temporarily when they should have been slain immediately with all the others.

Søren Kierkegaard, the great Danish existentialist philosopher, had no problem with the rightness of such commands from God. In "Problem I" of his book *Fear and Trembling*, he claims that our duty to obey God (no matter what God commands) transcends our duty to abide by ordinary morality. Hence, a command from God (such as the one to Abraham or to the Israelites who attacked the Amalekites) calls for what Kierkegaard described as "a teleological suspension of the ethical." That is, whenever we are confronted by a divine command which contradicts the prevailing moral code or even our own sense of morality, we should suspend our adherence to conscience and ordinary morality in order to fulfill a higher duty, namely, obedience to God.

In order to justify Kierkegaard's teleological suspension of the ethical, let's return to the possibility that Abraham *felt* that God's command to sacrifice Isaac was wrong, but *trusted* that God would not really command him to do something wrong, and so he proceeded. Such trust in God is usually justified by pointing out that God is not merely the creator of the world. God is also a supremely perfect being who is perfect in wisdom and goodness, as well as in power and knowledge. Hence, it would be incompatible with the very nature of God for God to command us to do anything morally wrong, so no matter what God commands, we can do it with confidence – even if it conflicts with our own moral judgment.

That makes sense as far as it goes. But even on the assumption that God is perfectly good and wise, there is a serious epistemological problem with divine command ethics. That problem is focused in the following question: "How can we know for certain that a specific command is really a command from God rather than a delusion from the devil, our culture, our subconscious, or something else?" God is not like a human being to whom we can simply turn and ask, "Did you really give this command?"

"True," a believer might say, "but the Bible is God's Word, so we can turn to it to see what God's commands are." Yet that strategy, too, has an epistemic problem. It presumes we can know that the Bible, rather than the Bhagavad-Gita, the Koran, or some other piece of religious literature, is really the word of God. Perhaps it is, but, again, we cannot determine that it is simply by asking God and getting a decisive answer.

Perfect Being Ethics

Does the preceding critique mean that God is irrelevant to ethics? Not necessarily. It does seem that any command purporting to be from God – whether it is in the Bible or any other piece of religious literature or comes from a living person, a dream, or a voice in our head – should be examined carefully to see whether it is morally acceptable or not. This means that in order to decide whether a purported divine command is *really* the will of God *we* must investigate it and make a decision as to whether it is.

In order to do that, we must have in advance some criterion of morality that we can bring to bear upon purported commands from God.[5] If by our criterion we conclude that a command is immoral, then we must conclude that it is not from God – assuming, now, that God is not merely the creator of the world but is also perfect in goodness. If we conclude that a command *would* be moral to act upon, then we can accept that it *may* be from God, but we no longer need to know that it is from God in order to know that it is morally permissible or obligatory – we have already determined that it is.

But where can we get such a criterion? Because it is a criterion by which we judge even the commands of God, it might seem that we must turn away from God in order to find it, but *perfect being ethicists* argue otherwise. Their thesis is that what is morally obligatory is *what a supremely perfect being would require that we do*; what is morally wrong is what a supremely perfect being would forbid us to do; what is morally permissible is what a perfect being would tell us we may do or not do, as we wish.

An imperfect being might by meanness or ignorance give us immoral commands. A supremely perfect being would not. Hence, the way to figure out the standards of morality is to consider what a supremely perfect being who is creator of the world *would* command us to do. It seems clear, for example, that as the creator of the world a perfectly good being would care about the world and its creatures and would command that we protect and promote their well-being, neither harming nor destroying them without sufficient justification.

Note that perfect being ethics does not ask, "What *does* God command?" It asks, "What *would* God command?" That is where it separates from divine command ethics. According to divine command ethics, in order to know what is morally right and wrong we must find out what God *has* commanded. According to perfect being ethics, things aren't that simple. As we noted earlier, we cannot simply ask God whether a command is from God. Consequently, every command (including commands claiming to be from God – some of which may actually be from God) and every purported moral standard should be evaluated in relation to what a perfect being would require, forbid, and permit.

If a purported command of God is something that a perfect being would forbid to everyone, then it must not be obeyed, even if it is in the Bible or comes from a voice that claims and seems to be the voice of God. If a purported command of God is something that a perfect being would require of everyone, then it ought to be obeyed whether we want to or not, as obedience to a perfectly good being can only be good. If a purported command of God is something that a perfect being would neither require nor forbid, then we should feel free to do it or not do it, as we wish.

In brief, before we can be confident that a purported command from God is really from God, we must determine whether it is plausible that it could be from a perfect being. If it is not plausible, we should not treat it as a command from God. Hence, our basic knowledge of moral right and wrong does not come from knowledge of the revealed commands of God; it comes from an understanding of what God (a perfect being who is the creator of the world) *would* will.

Note that if this God exists and commands us to do something, then if we know that the command is from God, we ought to trust it even if it seems immoral. Why? Because we know that it comes from a perfect being and therefore could not be immoral. (This is analogous to a pilot who, flying in clouds, feels like she has rolled over and is flying upside down, but who doesn't try to right the plane because she trusts her instruments, which say that she is top side up.) The difficulty, however, as pointed out earlier, is in knowing that a command *has* come from God. If we cannot determine that, then we must try to determine whether it *could have* come from God.

Ancient Greek philosophers puzzled over some of these problems, as can be seen in Plato's dialogue, *Euthyphro*. Therein Socrates poses the following dilemma to a priest: Is the Good good *because God wills it*, or does God will the Good *because it is good*? The divine command ethic says that what is good is good because God wills it. That seems unacceptable, as we have seen, because if some powerful force which claimed to be God were to will certain things, such as infant torture and racial slavery, we might *say* those things were good, and we might *comply* with them, but we would probably also believe that they were not good, were immoral, and hence could not have come from a being worthy of the name "God" – a perfectly good and wise being from the concept of whom we get our moral bearings.

Consider the plight of our ancestors who thought that if they did not make human sacrifices, the gods would punish them with natural disasters. Out of fear they made such sacrifices until someone was courageous enough to say, "Enough!" Fortunately, the God of the great religions today emphasizes justice and compassion. But did that God win the hearts of millions *despite* the morality that was commanded? Or was it *because* the morality which God commanded struck those millions as being true to their understanding of what a perfect being would will that they were won over? The latter, according to perfect being ethics, is closer to the truth.

Perfect being ethics is attractive to many people who believe in God because it saves them from the implication of divine command ethics that if it were to seem that God commanded us to do something terrible, then we simply ought to salute and do it. Also, perfect being ethics provides believers with a way to critique purported commands from God without being disrespectful to God. After all, if there is a God, then it is God who has given us both the responsibility to decide whether something is moral or immoral and the mind by which to make that decision. However, there are problems with perfect being ethics. Let's look at several, which will provide us with a transition to anthropocentric, rather than theocentric, ethics.

First, some say the idea of a perfect being is sufficiently vague that it leaves too much room for disagreement among perfect being ethicists. Just as there are deep disagreements between religious people as to the nature of God, there are going to be deep disagreements about what a perfectly good being who is creator of the world would will, and there will be no way to resolve those disagreements, so if ethics is to be of any practical use, we need something more clear and definite to base it on.

Second, perfect being ethics – unlike divine command ethics – does not say that we can do ethics only if we believe that God exists. Perfect being ethics allows that we can find out what we need to know ethically simply by reflecting on the *concept* of a perfect being – whether we think that such a being exists or not. Nonetheless, most agnostics and atheists think there is no good reason to bring the idea of God into the picture when doing ethics. They argue that we can do ethics and be moral without believing in God or even giving a thought to the idea of God.

Third, they add, it is *important* that ethics *not* be tied to the idea of God, lest it alienate agnostics and atheists from engaging in ethics and taking morality seriously. Ethics is a human enterprise. It is the endeavor of us humans to figure out how we ought to live our lives together. It is a human effort to help humans flourish and live their lives with dignity. It is the endeavor of humans to work out principles of behavior that we are all willing to live by and be held accountable for.

Clearly, in order for such principles to have universal or even widespread relevance and authority, they must be established by mutual agreement, not by fiat. Hence, it is crucial that ethics be carried on in a way that invites and encourages *all* people to engage in it with one another by way of rational dialogue. Therefore, what we need to attend to is the nature of humankind, not the nature of God. Toward that end, let's examine three absolutist theories that are anthropocentric, that is, are human-centered rather than God-centered. (*Anthropos* in Greek means "human.")

Anthropocentric Theories of Ethics

Rationalistic (Deontological) Ethics

One effort to give an absolute yet non-theistic foundation to ethics is what I earlier called "rationalistic ethics." The modern pioneer of this position was Immanuel Kant (1724–1804), who is considered by many to be one of the five or six greatest philosophers in western history, along with Socrates, Plato, Aristotle, and Descartes. Kant's rationalistic ethic is known in philosophical circles as "deontology" or "deontological ethics."

Kant's position is that there is an absolute moral right and wrong, but we do not come to know what it is by looking to conscience, society, or God; we do it by rational reflection on the concept of duty and the nature of humankind. Kant's point is that if we *think* about morality, we will realize that its central concern is with (1) what we *ought* to do, (2) what *everyone* ought to do, and (3) what everyone ought to do simply *by virtue of being human*.[6]

The essence of *morality*, as distinguished from *prudence*, is to always act only according to principles that we believe everyone should act according to, and to never violate those principles, in spite of inclinations, pressures, and temptations to the contrary. To be moral is to do what is morally right simply and solely because it is morally right. What is morally wrong, for example, infant torture, is wrong for you, me, and everyone, regardless of our abilities, circumstances, or feelings.

Prudence, by contrast, does not dictate the same behavior for everyone because its ultimate end is personal happiness, not morality. Prudence is simply intelligence applied to helping the individual achieve his or her personal goals in his or her personal circumstances. What is prudent for one person may not be prudent for another because of different desires, abilities, or circumstances. When we live by prudence, however, we are the pawns of our passions, fears, and desires. When David Hume said, "Reason is, and ought only to be the slave of the passions, and can never pretend to any other office than to serve and obey them," he was thinking of reason in its role as servant of desire.[7]

Kant, by contrast, insisted that the role of reason is not merely to figure out how to satisfy our desires. Reason can and does play that role at times, but it is also capable of comprehending what is absolutely morally right and wrong. The moral principles to which reason leads us are what Kant called "categorical imperatives." An imperative is a command to do something. A *hypothetical imperative* says, "You ought to do X *if* you want to achieve Y." A *categorical imperative* says, "You ought to do Z because you are a rational agent." There is no "if" about it. By disclosing moral principles to us, reason provides us with a vantage point from which to resist the push and pull of our fears and urges so as to live with moral integrity. (In the chapter on axiology you may want to review Kant's argument that moral integrity is the value that we ought to seek above all others, including happiness.)

There are certainly reasons to agree with Kant that there is a rational element to ethics, that it involves thought and not merely emotion, that it involves principles of behavior that apply to everyone, and that *ethics* consists of trying in rational dialogue with other people to discover those principles of behavior. *Morality* consists of trying to live according to those principles as best we can.

Two specific ethical principles or categorical imperatives that Kant thought reason leads us to are the following. First, we should act only according to principles that we are willing for all people to act according to. This captures the idea that moral principles are universal principles that apply equally to everyone. If I act according to a principle that I am not willing for you to act according to, I am acting as though I have a right to do something that you do not have a right to do. As humans, however, we are all the same and should have the same rights. For example, if I act according to the principle "Take other people's property when you want it," but I am not willing for other people to act according to that principle and take *my* property when they want it, then I am acting irrationally. I am not treating equals as equals. I am acting as though I am different from others and therefore have a right which they do not have, when in fact I am not different from them in any way which should confer that right on me and deny it to them. Hence, I should not act according to that principle. (Kant's point might also be understood as a version of "The Silver Rule": Do not treat others in a way that you would not want them treating you. The Silver Rule is subtly different from the Golden Rule, which says: Treat other people the way you would like for them to treat you.)

A second categorical imperative that reason reveals to us, according to Kant, is that we should never treat a person as a means only. To use something as a means only is to use it without regard for its own interests or responsibilities. It is okay to use a screwdriver as a means only; it has no interests or responsibilities. It is also okay to use a person as a means as long as you have their free consent – that is what you do when you hire a taxi driver to convey you from one place to another. But in addition to being a taxi driver, the taxi driver is a free agent who is morally responsibility for deciding for him or herself what is good and right for him or her to do, so his or her responsibility for making such decisions should be respected – just as your responsibility for making such decisions should be respected by others. For this reason it is okay to hire a taxi driver but not to kidnap him or her. In the latter case we would be taking away from the taxi driver his or her responsibility as a free, intelligent agent to decide how to live his or her life. We have no right to do that to him or her, and he or she has no right to do that to us. According to Kant, then, reason reveals that we should keenly respect and protect one another's freedom to make our own decisions as to how to live our lives – as long as we are living them in ways that respect the freedom of others.

Altruistic (Utilitarian) Ethics

Immanuel Kant's writings on ethics continue to be profoundly influential on ethical thought. Nevertheless, some thinkers have been very dissatisfied with Kant's approach. They say Kant's highly abstract approach to ethics has produced a bloodless theory which ignores the simple fact that above all humans are pleasure seekers. Therefore, they say, Kant's theory is not going to motivate us in a sustained way to be moral. It may captivate us temporarily with its romantic vision of moral dignity, but in the long run romantic visions which are not grounded in fact get vaporized by the harsh glare of reality. If we are going to have an ethical theory which is not only a good idea but is also capable of motivating and sustaining our commitment to it, it is going to have to be based on the simple fact that above all else we are pleasure seekers.

The most influential statements of this approach to ethics were written by Jeremy

Bentham, whose hedonic calculus we studied in the chapter on axiology, and by John Stuart Mill, who was deeply influenced by Bentham. In the tradition of David Hume's statement that "Reason is and ought only to be the slave of the passions," Bentham reasoned that inasmuch as everyone wants to have as much pleasure and as little pain as possible, and in as much as no human has more of a right to pleasure than others do, what we all ought to do is to act so as to promote the greatest happiness of the greatest number of people.[8] I have called this position "altruistic" because of its generous commitment to the promotion of the happiness of as many people as possible – without regard to race, ethnicity, religion, social class, wealth, and so on. Recall from the chapter on axiology that Bentham was a member of England's House of Parliament. He cared about people, and he was concerned to develop an ethical theory that would constrain Parliament to act for the welfare of the people in general, rather than just the members of Parliament.

In philosophical circles the ethical position of Bentham and Mill is known as "utilitarianism." The most famous statement of this position is Mill's classic essay, *Utilitarianism*. According to utilitarianism, whether a specific action is right or wrong is determined by whether it increases the greatest happiness of the greatest number of people or decreases it. If it increases it, it is *right* and we ought to do it; if it decreases it, it is *wrong* and we ought not to do it. If one action increases the general happiness *more* than another, then it is *better* than the other, and we ought to choose the former over the latter. If one action decreases the general happiness more than another, then it is *worse* than that other, and we ought to choose the latter over the former.

This position is called "utilitarianism" because of its emphasis on the importance of the *utility* (usefulness) of an action or object for producing pleasure or preventing pain. Bentham's utilitarianism possesses the virtue of giving us an uncommonly clear standard by means of which we can determine that something is morally right, wrong, or indifferent, depending on whether it increases, decreases, or leaves unchanged the pleasures and pains of the people affected.

Bentham's hedonic calculus gives us a tool for calculating the utility of the various alternatives available to us. Note that the hedonic calculus can be used to determine what would be the *moral* thing to do, that is, what would make for the greatest happiness of the greatest number of people, and it can be used to calculate what would be the *prudent* thing to do, that is, what would make for the greatest happiness of a particular individual (usually yourself) or a group smaller than the greatest number that could be benefitted. Like Kant, though, Bentham would say that when there is a conflict between the moral thing to do and the prudent thing to do, we ought to do the moral thing.

Bentham's position has been very attractive because of the centrality that it gives to happiness and because of its altruistic regard for the happiness of the greatest number of people. Still, classic utilitarianism, as formulated by Bentham and Mill, contains several problems. Let's look at four. First, if action A will not please me, but will please two or more other people, whereas B will please me but not please other people, the logic of altruistic ethics implies that I ought to choose A rather than B. But why should I choose A instead of B when B is what I want? The utilitarian reply is something like, "It'll all even out, and you'll be better off in the long run." To which the critic replies, "In the long run I'll be dead!"

A likely reply by the utilitarian is, "Well, you may be right. Not everyone benefits in the long run, and you may be one of the unfortunate ones. But *most* people will come out better if we all act according to the principle of utility." To this the reply might be, "Thanks, but if what life is all about is getting pleasure and avoiding pain, as you yourself say, then I don't care to take my chances in a utilitarian lottery. I'll fend for myself and take my own chances, thank you."

We must ask, then, "What reason can utilitarianism give for choosing in favor of other people and against oneself when the odds of getting a satisfactory payoff for oneself are 50/50 or less?" If the utilitarian says, "Well, you ought to do it because it's the moral thing to do," then the critic responds, "But that just begs the question. The question is, 'Why should I be moral if what life is about is maximizing pleasure and minimizing pain?'" Ironically then, after accusing Kantian ethics of having a motivational problem because of being too abstract and not reaching us where we live (in the pleasure zone), it turns out that utilitarian ethics has a motivational problem because of its fundamental conviction that life is all about maximizing pleasure and minimizing pain.

Second, we sometimes get into a situation in which we must choose between making a few people very happy or making a lot of people a little happy. This dilemma is problematic for utilitarianism because Bentham insists that alternatives be quantified. Hence, if option A will give 1,000 people a pleasure of level +2 (on a scale ranging from –[minus]10 to +10), whereas option B will give 200 people a pleasure of +10, we can see that by multiplying the number of people in each case by the pleasure which will be produced, both options will produce 2000 pleasure units. Hence, A and B are equal in utility.

If the 200 people have control of the situation, they'll almost certainly prefer to make themselves ecstatically happy rather than make the majority a little happy, and they will be able to justify their action by Bentham's principle – which favors *the greatest happiness* of the greatest number, and there is no greater happiness than a +10. By contrast, if the majority of 1,000 rules, it will argue that utilitarianism calls for the greatest happiness of *the greatest number*, and 1,000 is a much greater number than 200. So should A or B be chosen? Utilitarianism seems to have no answer – thereby setting up a power struggle between the majority and the minority of a society.

Third, efforts to quantify and compare future pleasures and pains seem destined to inspire hopeless controversies. There is no objective way to weigh and compare the pleasures and pains of different people, and it is just natural for people to think that their own pleasures and pains are more intense and important than the pleasures and pains of other people. Consequently, in trying to make decisions "by the numbers," it is not likely that people will be honest about how painful and pleasant various options will be to them, and it will be all too easy for those in power to rationalize that what *they* want to do is what will produce the greatest happiness of the greatest number.

The fourth and most troubling criticism is that Bentham's utilitarianism does not rule out the possibility of recommendations that seem to be immoral. For example, a racist majority could use the principle of utility to justify its desire to eliminate a minority race by genocide. To be sure, such an action wouldn't make the minority happy, but the majority could argue that it would make them deliriously happy and eliminate forever the recurrent social problems purportedly caused by the minority race.

It is conceivable, then, that a weighing of all the pleasures and pains which would be involved in eliminating the minority would dictate that for the sake of "the greatest happiness *of the greatest number*," the minority should be done away with! Yet, clearly, to do so would be immoral. Right? Consequently, utilitarianism, at least as stated, is shockingly flawed. It is an ethic that is initially attractive to many because of its concern to maximize the number of people who are happy, and because it counts the pleasures and pains of the lowest members of society as equal to those of the highest members of society. Nonetheless, it has loopholes that could be used by cunning people to justify something very different from what Bentham intended.

Universal Eudaemonism and Moral Happiness

Yet can we do better? Is it possible to articulate an ethical position that will capture our deepest intuitions regarding what is morally right and wrong, give us clear, practical guidance, and inspire all or most of us toward mutual commitment to it? Because of the labors of thinkers like Plato, Aristotle, Kant, Bentham, Mill, and – in our time – Elizabeth Anscombe, Philippa Foote, Christine Korsgaard, and John Rawls – I believe we have made much progress toward that goal. Indeed, I believe that the fundamental elements and ideals of ethics have been laid out rather clearly by men and women working over the centuries, but those elements have yet to be joined in just the right way and brought into a clear, sharp focus. Let's look at one final effort to do this.

Universal eudaemonism brings together a Kantian conception of the *universality* of morality and a Benthamian conception of the *motivation* of morality. You may recall that *"eudaemonia"* is the Greek word for happiness, so from the name of this position you can infer that it is concerned with universal happiness. To understand the sense in which universal eudaemonism is concerned with universal happiness, and why, we need to understand two of its basic convictions.

The first conviction is one we are familiar with: The desire that drives and unifies our lives is the desire for happiness. The second conviction is that we are *free, rational* creatures. Because we are *rational*, we are able to *understand* ideals and principles of behavior; because we are *free* we can, at least to some extent, *direct* our lives by ideals and principles. To say that as humans we are rational can be confusing. Eudaemonism does not mean that humans are rational in the sense that they always *act* rationally. It says that humans are rational in the sense that they are *capable* of acting rationally. Consider a similar distinction: a person can be intelligent without acting intelligently. Because we are rational we can, for example, *understand* the ideal of moderation, and then in our freedom we can choose not to do too much or too little; we can understand the principle of fair play and then act accordingly.

This means our behavior is not necessarily determined by *habit, appetite, impulse, temptation, tradition,* or *salesmanship*. In spite of the power of these and other forces, we can resist them successfully – as difficult as that might be at times. We can resist the impulse buying for which a great deal of advertising sets us up, and we can resist impulses to perform rash actions out of anger, lust, fear, fatigue, etc. What enables us to resist these forces? *Our ability to conceive of an ideal and conform our behavior to it.*

But what is the ideal to which we ought to conform our behavior? Here is the proposal of universal eudaemonism: What we ought to do is that which is morally right, and that which is morally right is *that which is compatible with and supportive of the possibility of universal happiness*. Conversely, that which is morally wrong is that which is destructive of or harmful to the possibility of universal happiness.

The ideal, then, is universal happiness, and morality is that which is supportive of it. One action is morally *better* than another when it will contribute *more* than the other toward the creation, maintenance, or improvement of a system of relations, social and natural, that fairly supports all people in their pursuit of happiness. Conversely, one action is morally *worse* than (not as good as) another when it contributes less to such a system than does the other.

We should consider a principle a moral principle only when it is a principle that we believe *everyone ought to follow because we believe it is necessary for the possibility of universal happiness* (or, alternatively, for the fair pursuit of happiness by everyone). Whenever we think that a principle of behavior is a moral principle, we ought to discipline ourselves to follow it. "Discipline" is an important word here because *to be moral* is to live a life that is wholly disciplined with respect to observance of those principles that we believe to be moral – that is, with respect to those principles that we believe everyone, including ourselves, ought to live by because they are necessary to the creation and maintenance of a system of relations, natural and social, within which everyone can fairly pursue happiness.

Such a system of relations is, of course, a guiding ideal, a star by which to navigate, a direction in which to sail, not a port within reach. It is a standard by which to judge where we've been, where we are, and where our actions are taking us. It is not an ideal that we can actualize fully in the foreseeable future, or perhaps ever. Nonetheless, according to eudaemonism it is the ideal by which we ought to guide our actions because it seems clear that *everyone wants to be happy* and *no one has more of a natural right to happiness than does anyone else*. Consequently, we all ought to live in ways that are respectful of those two facts. Note, however, that eudaemonism does not say that we have an obligation to make others happy or that anyone has a right to be made happy by others. Our *right* is to a fair opportunity to pursue our own happiness in our own way; our *obligation* is to live in ways that make it possible for all others to have that same opportunity.

Some of these ideas were expressed eloquently by John Stuart Mill in chapter 1 of his essay *On Liberty*:

> The only freedom which deserves the name, is that of pursuing our own good in our own way, so long as we do not attempt to deprive others of theirs, or impede their efforts to obtain it. Each is the proper guardian of his own health, whether bodily, or mental and spiritual. Mankind are greater gainers by suffering each other to live as seems good to themselves, than by compelling each to live as seems good to the rest.

Regarding the last point, Bertrand Russell quipped, "God save us from people who want to make us be happy!" John Dewey elaborated the point in this way: "To foster conditions that widen the horizon of others and give them command of their own powers, so that they can find their own happiness in their own fashion, is the way of 'social' action. Otherwise the prayer of a freeman would be to be left alone, and to be delivered, above all, from 'reformers' and 'kind' people."[9] All of these statements express the spirit of eudaemonism, and although Mill is associated with utilitarianism, universal eudaemonism seems to fit better with the broad spectrum of his writings on ethics.

It is, of course, very easy to propose an ideal and very difficult to determine which specific actions will actualize that ideal. As a result, history is rife with ethical disagreements and failed utopian efforts. Consequently, regarding specific actions, such as whether to support a certain candidate for office in the upcoming election, as well as regarding general principles, such as whether to oppose capital punishment or support it, we must distinguish between *what is really right* and *what we think is right*. Making this distinction saves us from ethical nihilism for it allows us to understand that there can be a right and wrong in a situation even when we are not sure what it is – just as scientists, after numerous failed experiments, keep doing research on a difficult problem because they believe there is a truth of the matter even when they do not know what it is.

Belief that there are moral facts is called "moral realism." Moral realism gives us reason to take seriously the search for moral truths and to act on what, in our best judgment, those truths seem to be. The last point (acting according to our best judgment) is one that Socrates emphasized by saying that although we might make a mistake in our moral reasoning, we ought always to try to figure out what is morally right, and we ought always to do what we *think* is morally right – even though we do not know for certain that we are correct. In addition, we need to formulate, review, and, when appropriate, revise or replace our moral beliefs in the midst of (or at least as a result of) the ongoing give and take of rational dialogue.

Note that common moral principles, such as those against stealing, slander, murder, slavery, and rape, can be evaluated and explained in terms of the central thesis of universal eudaemonism. Why do those moral principles exist? Because of the unfair ways in which stealing, slander, murder, slavery, and rape interfere with the pursuit of happiness on the part of the people who are victimized by them. Hence, one way to come to appreciate the value and authority of a specific moral principle is to ask how it relates to the possibility of universal happiness. If a rule of behavior is compatible with and supportive of the fair pursuit of happiness by everyone, then it is a rule to which we ought to conform. If a rule of behavior has no positive relation to the fair pursuit of happiness, then that is a reason to think that it is morally indifferent or morally pernicious.

Note also that because universal eudaemonism makes *everyone's* desire for happiness equally important – not just the desire of the greatest number of people who can be given pleasure simultaneously – eudaemonism avoids the openness to abuse to which utilitarianism is subject. A majority could not use universal eudaemonism to justify abuse of a minority. *Everyone* has an equal right to fair conditions for the pursuit of happiness. Also, because universal eudaemonism is concerned with the conditions necessary for the happiness of everyone, it does not lend itself to a tug of war between those, on the one hand, who want to emphasize the importance of the happiness of the greatest number over the importance of the greatest happiness, and those, on the other hand, who want to emphasize the importance of the greatest happiness over the happiness of the greatest number.

Still, like every philosophical position, universal eudaemonism needs to be critiqued. You might want to see if you can think of a counterexample to it; that is, see if you can think of an example of a rule of behavior which seems to you to be morally obligatory or morally forbidden but which does not seem to gain its moral authority from its compatibility or incompatibility with the possibility of universal happiness. If you think of a counterexample or some other problem with universal eudaemonism, ask whether this means that eudaemonism is woefully wrong and needs to be discarded or whether it is a good theory that simply needs to be improved.

The latter distinction is common in science. Sometimes an hypothesis is shown to be simply wrong and is discarded; other times an hypothesis is shown to be largely right or close to true but in need of correction or improvement. For example, Copernicus rejected Ptolemy's belief that the sun goes around the Earth, but he kept Ptolemy's idea that the movement of the heavenly bodies is circular. Kepler kept Copernicus's idea that Earth goes around the sun, rather than vice versa, but he showed that the movement of the Earth around the sun is elliptical rather than circular. Similarly, universal eudaemonism may be the best ethical theory available yet still need to be improved upon. The same consideration needs to be given to each of the other theories we have examined since each of them has intelligent contemporary supporters who believe that with revisions it is still the best theory available.

Why be moral?

Because utilitarianism says that what life is all about is the pursuit of pleasure, some critics say that it has no way to motivate people to be moral when they think they will be able to achieve pleasure more effectively by acting immorally. Does universal eudaemonism have the same problem? How does it motivate people to be moral? One answer, the moral answer, is this: Everyone desires happiness, and no one has more of a right to it than anyone else (or, to put the point positively, everyone has an equal right to the fair pursuit of happiness), so we ought to respect that fact by the way we live. Now let's look at some additional reasons – prudential rather than moral – why people should commit themselves to the eudaemonistic ideal of moral happiness.

These reasons are of two types, personal and social. *Personally* a moral life will be rewarded by a sense of *integrity* and *self-respect*. When you endeavor to live morally you know you are trying to live according to what you yourself believe to be right. Socrates urges in the *Apology* and the *Crito* that we should spare no effort to avoid doing what we believe to be wrong. He argues that the most important thing is not mere life, but a good life, and a good life is not possible without self-respect. The guilt or self-loathing that follows our betrayal of our personal sense of moral right is a self-condemnatory feeling that spoils the pleasures of life. Socrates also pointed out that if there is a life after this one (here we see ethics link up with metaphysics), then how we live this life may significantly determine what happens to us in the next life (which is a point affirmed by nearly all major religions, east and west).

The endeavor to be moral helps us not only avoid the negative feeling of self-betrayal; it also provides us with positive feelings. There is pleasure in doing what we believe to be right. As Mary Baker Eddy, founder of Christian Science, put it, "Consciousness of right-doing brings its own reward."[10] René Descartes put the point even more strongly: "A good done by ourselves gives us an internal satisfaction, which is the sweetest of all the passions, whereas an evil produces repentance, which is the most bitter."[11] This is not to say that we should be moral only to avoid the feeling of self-betrayal and to enjoy the feeling of acting morally. To do what is right simply because it is right is the true moral motive, as Kant pointed out, so it is far more important than either of these other motives. Still, if it is better that we should do what is moral rather than immoral even if our motives are not purely moral, then we should not overlook the possibility that consideration of these other factors may make the difference in a situation of severe moral temptation. Indeed, in such a situation perhaps morality itself calls for us to cast around for non-moral motives that will help ensure that we do what is moral.

The *social* significance of a moral life is that it contributes to the actualization of a system of relations that enables and allows every individual to pursue their own happiness in their own way. Anyone who takes pleasure in the pleasure of others should be able to take great pleasure in living a moral life once he or she realizes the extent to which a moral life contributes to the happiness of other people, including future generations as well as the living.

As to the behavior of *other* people, we should certainly be in favor of them leading moral lives according to universal eudaemonism since their doing so will support *our* personal pursuit of happiness. Moreover, *our* living morally will at times redound to our own benefit because the system that our moral behavior helps to create will be beneficial to *all* of its members, and that will include ourselves as well as others. Conversely, if we support or permit a system of relations that does not provide for the possibility of the happiness of everyone or discriminates against the happiness of some, we may wind up among those whose happiness is precluded or interfered with by that system. Hence, to act immorally is to act inconsistently with our belief that we have an equal right with others to pursue and achieve happiness.

The psychological impacts of immoral behavior should also be considered. If we support an immoral system and are among those who are victimized by it, we may hate ourselves for having been so stupid as to support such a system. If we are not victimized by that system but other people are, we may fear those others (for they will resent what we have done to them), and we may feel guilty (for we will know that those people do not deserve their plight and that we would not want anyone to do to us what we have done to them by supporting that system).

Here, then, is a summary of the relation of the good to the right and of happiness to morality from the eudaemonistic point of view: Happiness (the good) is the goal of human life, and morality (the right) consists of living according to those principles that everyone must live by in order for everyone to be enabled and freed to pursue their own happiness in their own way to the maximal extent that circumstances permit. On the individual, personal level, morality is a necessary condition of achieving happiness in the fullest sense, because if we betray our own sense of right, then even if we are not victimized socially by those who are victimized by our immorality, we must live with fear, guilt, or both – either of which can spoil the broth of happiness.

If we are ever going to approach significantly closer to the kind of world that universal eudaemonism envisions, then it seems that we need to bring two things together: commitment to pursuit of one's own happiness and commitment to the establishment of the conditions of happiness for everyone. This is the ideal of "moral happiness," and the people who bring these two things together are "morally happy."

If the position developed above is correct, then each of us should support a system of relations that frees and enables everyone to seek their own happiness in their own way; otherwise, we will run the risk of being among those who are penalized by an immoral system. But how can we figure out more specifically the rules we should live by? John Rawls, a Harvard University philosopher, in his influential book *A Theory of Justice*, proposes an intriguing vantage point from which to answer that question.[12]

Rawls says that if we want to figure out those principles that would create a fair and just society for everyone, then we should pretend that we are making rules for a society of which we will be a part – but without any knowledge of which part we will be. From behind what Rawls calls *"the veil of ignorance"* about the position we will have in the society we are constructing, we would be very careful to ensure that the laws of the society would treat us fairly and justly no matter what position we might occupy. The drama of this process of rule-making is heightened, Rawls adds, if we stipulate that our position in this society will be determined by some arbitrary device, such as a roulette wheel. Otherwise we might assume that given our ambition and high intelligence, we would surely rise to the upper echelons of any society of which we were a part. On that humble assumption we might tend to construct rules that favor the higher positions of the society. But if it were stipulated that our position would be fixed for life, would be determined by a purely random method, and could be anything possible, then we would be much more careful to construct laws that would not treat anyone unfairly, no matter what their position or circumstances might be, for they might be ours. By a spin of the wheel we could be short or tall, male or female, heterosexual or homosexual, black or white, asian or Native American, mentally challenged or brilliant, wheel-chair bound or ambulatory, born into a Catholic family or a Protestant family, a Muslim family or a Hindu family, etc.

Rawls' suggestion is a valuable device for helping us discern the conditions which must be established if justice, and therefore moral happiness, are to become more widely enjoyed. A problem with Rawls' approach is that the world is already well underway; we are each already in a well-defined set of circumstances, so we do not have to make our decisions from behind a veil of ignorance. Hence, the motivation problem rears its menacing head again. Some people know they are fortunate enough in terms of abilities and circumstances that for the sake of achieving more power, wealth, fame, or pleasure, they can support an immoral system and almost certainly escape its ill consequences. Moreover, there are sources of intense pleasure that are open to the immoral person that are not open to the moral person. We need, then, to ask whether there are ways to shape or influence the moral character of children so that as teens and adults immoral possibilities simply do not appeal to them – or, if they do, so that they will have the resources and will to resist them.

The need for moral education

If we believe in the ideal of moral happiness, how can we educate people to seek happiness within the bounds of morality? How can we nurture people to be unwilling to pursue happiness by immoral means? How can we foster people who – because they care about other people, take pleasure in their happiness, and desire universal well-being – will not knowingly support an immoral system?

It may sound as though I am assuming that human beings can be trained to do what we want and then, given appropriate rewards or punishments, can be relied on to keep doing it. I do not mean to make that assumption. The broader issue involved in such an assumption (as to whether human behavior is free or determined) is an important issue that we will explore in the next chapter. For now I mean only to emphasize a point that is, I think, accepted by determinists and their critics alike: the more weakly an ideal is held, and the stronger is the pressure to betray it, the more frequently it will be betrayed. Exercising our freedom on behalf of what we believe to be right can be more or less difficult, and the more difficult it is, the less likely we are to do it. Consequently, we need to search for means to instill in people the emotional desire and strength of will to act morally in all circumstances, even the worst. To put this matter another way, we need to see if we can discover ways of instilling virtue into people – virtue being a strong and stable disposition to act morally.[13]

What is going to cause or inspire people to act morally, to choose the path of virtue, to always act in ways that are compatible with the possibility of universal happiness? Any answer to this question is going to be very controversial. If the question were not as urgent and important as it is difficult, I would not suggest an answer. It may, however, be the most urgent question before humankind today because it is largely human behavior that makes this world pleasant or miserable, and our behavior is rooted in our character. Consequently, if we would improve the world according to the eudaemonistic ideal, we must influence the dispositions and the character of people to move in moral directions.

Perhaps this should be done by parental example, or self-discipline, or meditation, or philosophical dialogue, or worship, or classroom education, or some other approach. The "how" is something to be investigated empirically, as well as discussed philosophically. *That* it needs to be done seems beyond question – if we are to have a better world. Consequently, I suggest that the most important thing we can do for people is to help them become moral, because only by everyone becoming moral can universal happiness become an approachable ideal. For these reasons I will offer some answers to the question, "What will influence people to act morally?" However, I claim neither novelty nor finality for my answers; their primary purpose is to stimulate deeper thought and broader discussion.

Three important *non-moral* motives for living morally are *prudence*, *gratitude*, and *love*. Before elaborating on these motives, I want to emphasize again that whether these motives can be intentionally and effectively built into people's lives – and if so, how – is a question that philosophy and the social sciences must investigate jointly. Whether we *ought* to attempt to weave these motives into peoples lives is yet another question that each of us needs to think about, and all of us need to discuss.

Prudence is a motive for behaving morally because, as we have seen, the social system which we create and support by living morally protects *us* as well as others. One of the best ways to ensure that other people will be protective of us is for us to be protective of them. If they know we look out for them, they are more likely to look out for us. Hence, intelligent self-interest (which is what prudence is) often advises that we behave in a moral manner. The flaw in this motive is that by itself it works only when people are convinced that they cannot do better for themselves by acting immorally. Whenever they become convinced that they can do better by acting immorally, prudence dictates that they do so! Hence, prudence helps, but it must be supplemented with other reasons for living morally.

A second reason for acting morally is *gratitude*. The Scottish philosopher David Hume stated that "Ingratitude is the worst of sins." Perhaps there is a worse sin, or another sin that is equally bad, but certainly ingratitude reveals something disturbing about the ungrateful person. Those who feel no sense of thanksgiving, no urge to respond in kind when they knowingly receive a gift with no strings attached, are devoid of a feeling that helps turn individuals into a community. They lack a social dimension and cannot participate fully in human society. At the end of his book *A Common Faith*, John Dewey came as close to poetry as prose can in helping us appreciate how much we owe to others:

> We who now live are parts of a humanity that extends into the remote past, a humanity that has interacted with nature. The things in civilization we most prize are not of ourselves. They exist by grace of the doings and sufferings of the continuous human community in which we are a link. Ours is the responsibility of conserving, transmitting, rectifying and expanding the heritage of values we have received that those who come after us may receive it more solid and secure, more widely accessible and more generously shared than we have received it.[14]

There are two aspects to healthy gratitude, both of which are reflected in Dewey's statement. The one aspect is *thankful awareness* of the undeserved benefits we have received from others; the other aspect is *a desire to reciprocate* that generosity. Our reciprocation need not be directly to those who have benefited us; indeed, that is often impossible because our benefactors are dead or unknown. But we can commit ourselves to the proposition that those who have labored in ways that benefit us will not have done so in vain because we, too, will promote the values for which they labored and from which we have benefited.

A "down to earth" example of reciprocity is the following true story. A man I know, "John," was driving through New Orleans, Louisiana, with his family in the 1960s. The roads were very busy as he came upon a fellow whose car had stalled. Many cars had already passed the fellow, who was of a different race from John, and those were racially tense times, but John stopped and offered to push the man's car down the road to a service station. The man accepted, and they got under way. After the man's car rolled to a stop in the station lot, he got out and, to John's surprise, asked how much he owed him. John told him he didn't want money, that he'd just like him to pass the favor on to someone else. The fellow assured John he would, thanked him, and they parted. They had never met before, would probably never meet again, and wouldn't recognize one another if they did! John's explanation for doing what he did was that he had been in equally dire straits more than once and had received friendly help without charge, so he was just passing on one of the favors he had received from others. That kind of grateful reciprocity, I think, helps enhance the quality of life in a society.

Gratitude is a tricky thing, however. A person can "calculate" what is owed to a benefactor, pay it back, and then – except from the point of view of prudence – be unconcerned with the plight of others until he or she "owes" something again. Such myopic gratitude is oblivious to the fact that we are the beneficiaries of a host of unknown benefactors, and it shows no disposition to be an unknown benefactor. Clearly, this narrow kind of gratitude, if it deserves the name at all, has only the most tenuous relation to the gratitude of which John Dewey spoke.

Gratitude can also sour into resentment. Receiving help can be interpreted by the recipient as a sign of weakness and inadequacy on his or her part; consequently, the recipient may resent his or her benefactor for making him or her aware of their deficiencies and for making them feel dependent. Also, one may resent the feeling that one "ought" to repay the favor which one has received. To be sure, gratitude does not have to disintegrate into these forms of resentment, nor should it, but it can, and so, like prudence, it needs to be supplemented by another motive for living morally: love.

The word "love" has many meanings. I am using it here to mean sympathetic support of other people. It is a matter of enjoying seeing people happy and of taking pleasure in contributing to their happiness, directly or indirectly. This kind of love is a deep and enduring sentiment which helps ensure that people will be moral even when they don't "owe" anything, and even when it isn't personally prudent. The 17th century philosopher Benedict Spinoza revealed himself to be a loving person in this sense when he said, "It is part of my happiness to lend a helping hand."[15]

This kind of love is a sentiment that must be nurtured in people; it is not something with which we are born. Studies by the Swiss child-psychologist Jean Piaget suggest that at birth every child is completely egocentric, that is, self-centered. Only gradually does the child come to distinguish itself from others and to understand itself as part of successively larger groups, such as the family, the neighborhood, the city, the nation, and humankind. Piaget called such growth in understanding a "decentering process."

One of Piaget's interpreters, Gordon Allport, an American psychologist, pointed out that the decentering process "may be arrested at any stage along the way, especially in its affective aspects."[16] Allport's point is that there is an affective (emotional) aspect, as well as a cognitive (intellectual) aspect, to a child's gradual identification with family, community, nation, and humankind. A young person can understand intellectually that he or she is part of a community, yet he or she may feel no affection for certain other members of that community and may even dislike them.

One of our national and international problems is that many adults are cognitively and affectively parts of a nation, race, or ethnic group, but they are only cognitively part of humankind. They do not *feel* as strongly the troubles of those outside their affective groups who are starving to death as they feel the discontent of those within their affective groups who are striking for higher wages. I do not mean to be simplistic about the difficulties of one nation helping another or about the importance of higher wages, but if the citizens of one race, ethnic group, or nation do not *care* about the sufferings of the members of another race, ethnic group, or nation, then no effective means of helping them will even be sought, much less implemented. Consider, for example, the problems caused by lack of love between members of different races, religions, and ethnic groups. Alcibiades, a young friend of Socrates, recognized this problem long ago when he noted that a healthy nation is characterized by "the presence of friendship and the absence of hatred and divisions" among its citizens.[17] Hence, we need to cultivate in people a moral disposition which is rooted in love and fortified by prudence and gratitude.

Moral education as the formation of human sentiments is a twice urgent task. First it is urgent because of the explosive state of the world and its societies; second it is urgent because its foundation is most easily, effectively, and fruitfully laid within the first few years of a person's life. This last belief is shared by many psychologists and was stated clearly by Alfred Adler, founder of the school of individual psychology. "By the end of the fifth year of life," he wrote,

> a child has reached a unified and crystallized pattern of behavior, its own style of approach to problems and tasks. It has already fixed its deepest and most lasting conception of what to expect from the world and from itself. From now on, the world is seen through a stable scheme of apperception: experiences are interpreted before they are accepted, and the interpretation always accords with the original meaning given to life. . . . Even if this meaning [which as children we attribute to life] is very gravely mistaken, even if the approach to our problems and tasks brings us continually into misfortunes and agonies, it is never easily relinquished.[18]

What dividends we might reap if we poured our resources into moral education during those first five years of each person's life! What can then be accomplished so easily grows increasingly difficult as the individual ages.

Witness the failure of so many rehabilitation programs that are largely trying to help people acquire as adults the dispositions that they could have acquired easily as children, but did not. To be sure, we still have much, perhaps almost everything, to learn about the role that genetic inheritance does or does not play in causing immoral behavior, but for now it seems highly likely that neglect of childhood moral education increases the number of people who later need rehabilitation. Hence, it seems foolishly short-sighted to neglect this aspect of our children's development. Moral education is character formation, and character is something with decisive significance for the individual and society. As the Masonic Creed puts it: "Character is destiny." As Aristotle put it more than two thousand years ago in his *Politics*: The better the character of a people, the better their government will be – and vice versa.

There are, of course, exceedingly difficult questions to answer if we take moral education seriously. *Who* should do it? *How* should they do it? What part should the *government* play in moral education, if any? It may be that the most effective moral education takes place not in formal institutions, but in affective groups, such as families, neighborhoods, and communities.

So far on the topic of "moral education" I've emphasized (1) cultivating love in people, (2) making them appreciative of what others have done for them, and (3) pointing out the purely prudential aspects of being moral. But effective moral education must also (4) make people disposed to behave intelligently. Bertrand Russell, the most famous British philosopher of the twentieth century, said in his essay *What I Believe*, "The good life is one inspired by love and guided by knowledge." I think he was correct to put love first, for intelligence which is not subordinated to eudaemonistic love has an exceedingly large capacity for evil – as we can see in the lives of people like Hitler and Stalin. Conversely, love which is not guided by intelligence has an exceedingly large capacity for unintentionally causing disasters that are well-meant but disastrous nonetheless. Thus, I concur with Russell: love and intelligence must be yoked together for a good life.

Thinkers like John Dewey have done a valuable job of helping us understand what intelligence is and how to cultivate it. Dewey says in *How We Think* that to be intelligent is to be disposed to respond in a special way to puzzlement, frustration, or blockage of activity. It is to respond to such experiences by *clarifying* the problem, *gathering* relevant data, *figuring out* possible solutions, *trying out* the most promising solution, and then *evaluating* the satisfactoriness of that solution – starting all over if the solution acted on isn't satisfactory.

Being intelligent in Dewey's sense is not something we are born with. We are born with an innate *capacity* to be intelligent, but intelligence in Dewey's sense is something that must be developed. It is a disposition, a way of behaving, a way of responding to problems, so we must *learn* to behave and respond intelligently.

Just as clearly, responding intelligently to problems has an important bearing on our happiness. To be sure, by virtue of good fortune we may be happy without behaving intelligently. But if we do not learn to behave intelligently, chances are slim that our happiness will endure. Either our unintelligent behavior will eventually prove disastrous, or our good luck will vanish and we will not have acquired the ability to cope intelligently with unfavorable circumstances.

It is also possible, of course, that we might be intelligent without being loving. In that case, we probably won't be happy. People will be wary of us because of our high intelligence and calculating attitude, and we will be deprived of shared enjoyments, which according to John Dewey are the most precious of pleasures. In his own words, "Shared experience is the greatest of human goods."[19]

In conclusion, love and intelligence must be yoked together for a good life, so by means of moral education we should endeavor to nurture people who are both loving and intelligent. Such people, moral people, will seek to create, sustain, and improve a system of relations that tends to make people moral and prepares and frees them to pursue their own happiness in their own way – within the bounds of morality. So, seek happiness! But in a way that leaves intact (and hopefully nourishes) those conditions that make happiness possible for everyone.

The preceding ideals are almost laughably lofty given the continuing violence of human history, right up into the twenty-first century, but moral progress has been made. Witness the diminution of slavery and child labor, the increase of rights for women and minority members, and the increase of humane treatment for prisoners of war and the mentally ill. If we are going to continue making moral progress, we need a conceptual star by which to guide ourselves and measure our progress. Perhaps universal eudaemonism is that star. If it is not, we should ask, "Why isn't it, and what should be?" And whatever the ideal is by which we should guide our lives and institutions, how can we move toward it most effectively? There are perhaps no more important questions for us to address than these.

Notes

1 Ellen Sturgis Hooper, "Beauty and Duty," *The Shorter Bartlett's Familiar Quotations*, Kathleen Sproul, ed. (NY: Permabooks, 1953), p. 177.
2 John Hick, *Faith and Knowledge*, 2nd edn. (Ithaca, NY: Cornell University Press, 1966), p. 111.
3 An obvious problem that arises from living in a modern society is that a modern society is usually constituted of numerous overlapping groups, some of which have conflicting moral standards. Consequently, an individual might, for example, be a citizen of the USA, which says that abortion is permissible, and a member of the

Roman Catholic Church, which says that abortion is not permissible. An important question is whether social relativism has a decisive way to resolve such conflicts.

4   Emil Brunner, *Our Faith* (NY: Charles Scribner's Sons, n.d.), p. 5.
5   Kai Nielsen, an atheist, and Richard Swinburne, a theist, agree on the need for such a criterion. See Nielsen's *Ethics Without God* and Swinburne's *The Coherence of Theism*, ch. 11.
6   According to Kant, as well as Socrates, we should always listen to conscience when it urges us to *think* about what is morally right and to *do* what we think is morally right. However, conscience, understood as moral *feeling* toward various actions, is not infallible, so any specific recommendation that it makes should be scrutinized by reason. To think or reason about what is morally right involves, for Kant, much more than simply consulting one's moral feelings.
7   David Hume, *A Treatise of Human Nature*, Book II, Part III, Section III.
8   There is technical terminology that can be helpful here. The belief that humans by their very nature cannot help but seek above everything else to maximize their pleasures and minimize their pains is called "psychological hedonism." That is, our minds ("psyches") are such that we are built to seek pleasure ("*hedone*"). By contrast, "ethical hedonism" holds that that we *ought* to seek pleasure above all other things, but we do not have to. These positions seem incompatible with one another because if we are free to put other things ahead of pleasure, then psychological hedonism is false, whereas if we are not free to put other things ahead of pleasure, then ethical hedonism is false (since "ought" implies "can"). Nonetheless, both of these positions seem to be expressed powerfully in the opening sentence of Bentham's *An Introduction to the Principles of Morals and Legislation*: "Nature has placed mankind under the governance of two sovereign masters, *pain* and *pleasure*. It is for them alone to point out what we ought to do, as well as to determine what we shall do."
9   John Dewey, *Human Nature and Conduct* (NY: The Modern Library, 1957), pp. 293–4.
10 Mary Baker Eddy, *Key to the Scriptures*.
11 René Descartes, *The Philosophical Writings of Descartes* (Cambridge: Cambridge University Press, 1985), vol. I, pp. 351–2.
12 See John Rawls, *A Theory of Justice* (Cambridge, MA: Harvard University Press, 1971), pp. 136–42.
13 In ancient Greek philosophy the idea of a virtuous life was not the idea of grinding, humorless commitment to duty, as a contemporary, popular conception of virtue has it. Rather, it was the idea of a life characterized by those forms of action – courage, justice, moderation, and wisdom – that are conducive to the achievement of true happiness.
14 John Dewey, *A Common Faith* (New Haven: Yale University Press, 1934), p. 87.
15 *The Chief Works of Benedict Spinoza*, Vol. II (NY: Dover Publications, Inc.), 1951, p. 6.
16 Gordon Allport, "Normative Compatibility in the Light of Social Science," *New Knowledge in Human Values*, Abraham H. Maslow, ed. (NY: Harper & Row Publishers, 1959), p. 144.

17 Plato, *Alcibiades I*, sec. 126, in *The Dialogues of Plato*, trans. Benjamin Jowett (NY: Random House, 1937), vol. 2.
18 Alfred Adler, *What Life Should Mean to You* (NY: Capricorn Books, 1958), pp. 12–13.
19 John Dewey, *Experience and Nature* (NY: Dover Publications, 1958), p. 202.

Reading Further

For an example of individual relativism see Bertrand Russell's *Religion and Science*, chapter IX.

For social relativism see *Folkways* by William Graham Sumner. Social relativism is also called "cultural relativism" and "social subjectivism."

For a classic statement of rationalistic ethics or deontology, see Immanuel Kant's *Groundwork of the Metaphysics of Morals* (some translations say "Foundation" or "Grounding" rather than "Groundwork"). For a striking contrast to Kant's and Socrates' reverence for conscience, see chapter XXIX of Thomas Hobbes' *Leviathan*. Hobbes regarded reverence for conscience as a disease that tends to promote anarchy and destroy society.

The classic statement of utilitarianism is John Stuart Mill's *Utilitarianism*.

Plato, *Euthyphro*. Does God will what is good because it is good? Or is what is good good because God wills it?

On theistic ethics see Linda Zagzebski, "The Virtues of God and the Foundations of Ethics," *Faith and Philosophy*, 15/4 (October 1998), 538–553.

Plato, *Meno*. Can virtue be taught? Socrates answered "no." Also see relevant sections of Aristotle's *Politics*. For a radical affirmation that people's behavior can be "shaped" see B. F. Skinner's *Walden Two* and *Beyond Freedom and Dignity*.

Part IV Metaphysics

Chapter 15: Freedom and Determinism

- Can we do what we ought to do?
- Words, Concepts, Positions, Justifications, and Criticisms
  Philosophical Anthropology
- Objective freedom and Subjective freedom
  Thomas Hobbes and Brand Blanshard
  Jean-Jacques Rousseau
- Libertarianism
  The nature of metaphysical libertarianism
  Jean-Paul Sartre and St. Augustine
  Inclination vs. Determination
  Responsibility: physical, moral, and legal
    Charles Starkweather and accomplice
- Universal Determinism
- Theistic Determinism
  Fatalism
- Naturalistic Determinism
  The emergence of science and the spread of naturalistic determinism
  Justifications of naturalistic determinism
    A necessary presupposition of science
    Unreasonable to make an exception of humans
    Human behavior is increasingly predictable
    Feeling free is not being free
    Conducive to peace of mind
- Soft Determinism (Compatibilism)
  Feelings, beliefs, and desires as causes
  Determinism and freedom are compatible
  Free means voluntary
  Determinism and moral responsibility compatible
  Rehabilitation vs. retribution
- Hard Determinism (Incompatibilism)
  Reasons and causes
  Genetic endowment, environmental history, and present circumstances
    Analogies: organ; computer

Incompatibilism: neither free nor morally responsible
- **Criticisms**
  Feeling free and the principle of credulity
  Feeling morally responsible
  Redefining "freedom" is not acceptable
  Free behavior can be predictable
  Extenuating circumstances favor libertarianism
  Science without determinism
  Surely mental events are causes
  Determinism and human dignity

Can we do what we ought to do?

The last chapter concluded with a statement of our moral obligations as human beings. Such a statement assumes that we are free to choose whether or not to fulfill our moral obligations, since surely we cannot have a moral obligation to do what we *cannot* do. As Immanuel Kant put it: "Ought implies can." If we have an obligation to do something, then it must be the case that we can do it; conversely, if we absolutely cannot do something – if we are not free to do it, then it cannot be the case that we ought to do it or have an obligation to do it (unless, perhaps, we are responsible for having lost our objective freedom – as when we are unable to drive safely because we have consumed too much alcohol).

Are we free to be moral? That is the question we will focus on in this chapter. With that question we have begun something new. We have turned from asking *value* questions (What is good? What is right?) to asking *existence* questions (What things exist? What are their properties?). We know that humans exist, but is freedom of will one of their properties? These kinds of questions are metaphysical questions. Metaphysical questions are concerned with the following issues: (1) what is really real, as distinguished from what we think is real but is not, and think is not real but is (for example, is God really real or only a figment of people's minds?), (2) what are the real properties of objects (for example, is the human mind physical or non-physical?), and (3) which properties of a thing are essential and which are non-essential (for example, it is essential that a triangle have three sides, but it is not essential that it be red rather than blue; it is essential that space be capable of containing a physical object, but it is not essential that it contain that particular physical object). Many metaphysical questions have to do with clarification of the concepts of God, the mind, the soul, personhood, individuality, matter, substance, causality, space, time, and numbers, as well as with questions as to whether these things exist, and if they do, in what way they exist.

Before we proceed, recall that every philosophical question eventually leads to every other. Therefore, instead of beginning with questions about value and moving on to questions about metaphysics (as we are doing), we could have begun with questions about metaphysics and followed their lead to questions about value. If, for example, we had begun with the metaphysical question as to whether we have freedom of will, and had concluded that we do *not*, then we may well have turned next to metaethics and concluded that we have no moral obligations, can have none, and so are not morally responsible. By contrast, if we had begun with the question of freedom and concluded that we *are* free, then it would have been appropriate to ask next, "What then, if anything, ought we to do with our freedom?" That question would have taken us into the questions about morality and happiness that we just finished exploring. Instead we are proceeding the other way around. Having investigated positions as to what we ought to do if we are free to do it, we will now explore what it means to be free in the relevant sense, and what reasons there are for thinking that we are or are not free to do what we ought to do.

Words, Concepts, Positions, Justifications, and Criticisms

Before proceeding to that inquiry, we need to explore an extremely useful cluster of five concepts: a word, a concept, a position, a justification, and a criticism. (The concept of a concept is the idea of what a concept is.) These are closely related ideas, but each is different from the others. Once you understand how they differ from one another and how they relate to one another, you will be able to do philosophy in a much more satisfying and sophisticated way, so let's examine each.

A *word* is, nearly always, a sound, a written character, or a movement of the hands (as in sign language) that is used to mean or stand for something. In writing we indicate that we are referring to a word, rather than to what the word means, by putting quotation marks around it. For example, "word" is a short word; "paragraph" is a long word; "tall" is shorter than "short."

Words take different forms. A *proper noun* usually refers to a specific individual; a *general noun* usually stands for an abstract idea. "Koko" refers to a specific individual who is a gorilla. When the meaning of a noun is a general idea rather than a specific individual, then the word stands for a *concept*. The meaning of the word "triangle" is the general, abstract idea of any shape that is created by three straight lines when each intersects the other two. Note that when we use a word without putting quotation marks around it, we are speaking of what the concept is about; we are not speaking about the word. Consider, for example, the following statement: a square has more sides than a triangle. That statement is not about the words "square" and "triangle"; it is about what those words are used to mean.

A *position*, as distinguished from a concept, consists of taking a stand with regard to a concept. For example, if I say that triangles exist, I am saying that I think there are actual things that fit the concept of a triangle. That, of course, is not a very controversial position. If I say that God exists, I am taking a controversial position; I am saying that I think there is something real that fits the concept of God. Similarly, if I say that God does not exist, I am doing the same kind of thing – taking a position with regard to how I think the concept of God relates to reality.

If I simply say, "God," then I am not taking a position; I am simply uttering a word and perhaps suggesting a concept to take a position with regard to. There are, of course, many different kinds of reaction that we can have to a concept. We might find it to be interesting or confusing or unfamiliar or funny. Metaphysically interesting reactions to a concept typically consist of (1) taking the position that there is or is not something real which fits a certain concept (for example, there is/is not life after death; humans do/do not have souls), or (2) taking the position that something that we think is real does or does not have a certain property (for example, the human soul is eternal; matter is not eternal; humans are free agents).

A *justification* of a position consists of a reason given in support of a position. If I say that racism is morally wrong because all people have a natural right to be treated with dignity, then I am providing a reason for thinking that racism is morally wrong (even if racism is legal and commonplace where I live). If, by contrast, I say that racism is not morally wrong because there are no such things as natural rights, then I am providing a reason for thinking that racism is not morally wrong (even though it may still be personally repugnant and against the law). In both cases I am providing a justification, although for opposite positions.

Note that a justification in this sense is part of an argument: the position being justified is the conclusion of the argument. The justification, that is, the reason being given in support of the position, is the premise or evidence in the argument. Just as evidence or a premise may or may not succeed in supporting a conclusion to the extent that an arguer claims it does, a justification, that is, a reason given in support of a position, may or may not justify or warrant belief in that position. Whether it does is something we should decide only after evaluating the total argument of which the justification is a part.

A *criticism of a position* is a reason given for thinking that that position is false; a *criticism of a justification* of a position is a reason given for thinking that that justification is not as strong as someone has claimed. Recall Kant's criticism of the position that happiness is the highest human good. Kant said that if we were forced to choose between happiness and morality, we ought to choose morality above happiness; therefore happiness is not the highest human good. As with justifications, a criticism may or may not be a good one.

A masterful understanding of these five concepts and their relations to one another will help you appreciate more fully what we have done in the preceding chapters, and it will help you understand and organize more clearly what we will do in the remaining chapters. With regard to each new section of material, ask yourself as you read along: What are the key *words* here? What are the *concepts* which are meant by those words? What *positions* are being taken with regard to those concepts? What *justifications* are being given for those positions? What *criticisms* of those positions and justifications are being made? Keeping these concepts and questions in mind, let's proceed with our study of metaphysics and begin with questions about ourselves.

Objective freedom and Subjective freedom

An important first step in examining the issue of freedom and determinism is to realize that there are two very different concepts of freedom. Both are important.

Philosophical Anthropology

Metaphysics is the study of the nature of reality and its parts. Humankind is one part of reality. It is that part which is often of most interest to us, and it is a part which we must understand well if we are going to be able to live our lives well. Philosophical anthropology, described long ago in the chapter "What is Philosophy?," is the philosophical study of the essential nature and universal situation of human beings. Cultural anthropologists study the non-universal traits and situations of human beings in, for example, Borneo, Boston, and Botswana. Obviously it is important for those who would do philosophical anthropology to be familiar with the facts that cultural anthropologists have discovered. Just as obviously, cultural anthropologists can do philosophical anthropology, but when they do, they are doing something which goes beyond what they do when they do cultural anthropology. Among the many metaphysical questions that can be asked about human beings, we will focus on two of the broadest and most captivating – each of which has significant practical implications. The first question (Do we have freedom of will?) will be taken up in this chapter. The second question (What is the mind and how is it related to the body?) will be taken up in the next chapter.

To distinguish them clearly is to avoid much confusion. I will call them "objective freedom" and "subjective freedom" because we each experience ourself as an *object* and as a *subject*. Because of our bodies, we each experience ourself as one object among other objects in the world. Some objects are bigger than we are, some are smaller; some are faster, some are slower; some are more able than we are in certain respects, others are less able in those same respects; some are imprisoned, some are not.

In addition to experiencing ourselves as objects, we experience ourselves as subjects with a mental life. We experience ourselves as having sensations, feelings, emotions, thoughts, beliefs, memories, and desires, and as making choices. However, we do not experience ourselves as merely objects or as merely subjects. We experience ourselves as both, and that two-sided experience of ourselves has given rise to two very different conceptions of freedom. In the next chapter we will see that this dual-aspect experience of ourselves has also given rise to numerous questions about how the mind and the body are related to one another. For now we will focus on two kinds of freedom.

"Objective freedom" refers to a person's freedom as one object among other objects. (This kind of freedom might also be called "physical freedom.") Objective freedom consists partially of what Thomas Hobbes called "the absence of external impediments" and what Brand Blanshard called "the freedom to do what we choose." More precisely, we have *objective* freedom to perform an action whenever we have (1) the *ability* to perform it, and (2) the *opportunity* to perform it. Without either the ability or the opportunity, we are not objectively free to perform an action.

I've heard that a fellow whose hand had been crushed asked the doctor as she took off the cast, "Doc, will I be able to play the piano now?" The doctor replied, "Yes," and the fellow said, "Great! I couldn't play before the accident!" The moral, of course, is that we are not free to do what we are not *able* to do, even if we have the opportunity to do it. By the same token, though we are able to do something, if there is someone or something *preventing* us from doing it, we are not objectively free to do it, since we do not have the opportunity. Even if a person is able to play piano, he is not free to do so if there is none around or if the only one available is locked up tight.[1]

Objective freedom is the kind of freedom we are usually talking about when we discuss political freedom – which can include the external, outward freedom to live where we wish, to work at whatever type of job we qualify for, to dress as we please, to speak out and vote for whomever we prefer. Jean-Jacques Rousseau, an eighteenth-century French philosopher, was speaking of this kind of freedom when he made the following statement in his most famous work on political philosophy, *The Social Contract*: "Man was born free, yet everywhere he is in chains." Clearly Rousseau was referring, at least in part, to our lack of social or political freedom, which is one form of objective freedom.

Parenthetically, when we speak of a lack of political freedom, we are usually speaking of ways in which a government takes *opportunities* away from its citizens. Sometimes, however, a government takes *ability* away from some of its citizens – as when it forces sterilization on women so they are no longer able to bear children, and when it injects drugs into political dissidents so they are no longer able to think and speak clearly.

Subjective freedom bears a certain resemblance to objective freedom, but there is a deep difference. Subjective freedom has to do primarily not with the opportunities that are available to us or with our abilities to perform specific actions (such as playing a piano). It has to do not with the way the world is outside of us, but with the way the world is inside of us – yet not in a physical sense of "inside." (Again we are anticipating a "mind/body" problem: In what sense is the mind 'inside' the body? These problems will be taken up in the next chapter.)

If we are not *subjectively* free, then we are not able to choose whether or not to do something that we are *objectively* free to do. "Subjectively free" in what sense? In the sense that we could choose to do A (a specific action, whatever it might be) or to refrain from A, or in the sense that we could choose to do either A or B. Rousseau – who was speaking about *objective* freedom in the last quotation – is just as clearly speaking about *subjective* freedom in the following statement:

> Nature commands every animal, and the beast obeys. Man feels the same impetus, but he realizes that *he is free to acquiesce or resist*; and it is above all in the consciousness of this freedom that the spirituality of his soul is shown. For physics explains in some way the mechanism of the senses and the formation of ideas; but in the power of willing, or rather of choosing, and in the sentiment of this power are found only purely spiritual acts about which the laws of mechanics explain nothing.[2]

Perhaps an illustration will help. Consider the following. We are likely to believe that we are free to go either way in a trivial choice such as whether to ask for a Pepsi or a Coke at a picnic. In such a matter, we believe the choice is entirely up to us and that nothing makes us choose the one way rather than the other. We are less likely, however, to think that we are free regarding non-trivial choices such as whether to cut the throat of someone we love or to refrain from it.

A friend once defended to me his belief that whatever he was objectively free to do he was also subjectively free to do. So I asked him, "You mean you believe you are free to go home and cut your wife's throat?" Without hesitation he answered, "Yes." He was certainly *objectively free* to do so: he was physically *able* to find and wield a knife, and he would have plenty of *opportunities* to cut his wife's throat. But I knew he was a gentle person who loved his wife and their child, so I wondered whether he truly was *subjectively free* to perform such a heinous act. To be sure, he may have felt like he could do it, but there is a profound difference between *feeling free* to do something and *being free* to do it. My friend may have felt free to slit his wife's throat, but had he attempted to do so just to see if he were able to, perhaps he would have discovered that he couldn't actually do it. "Couldn't" in what sense? Not in the sense that he couldn't have wielded the knife, and not in the sense that someone would have prevented him, but in the sense that he would not have been able to follow through with his intention, no matter how hard he tried. More generally, he may have found that he was mistaken in thinking that he was subjectively free to do whatever he was objectively free to do. (I am happy to report that his wife is still alive and well!)

Libertarianism

My friend and those who hold the same position, such as the French philosopher and playwright Jean-Paul Sartre, are called "libertarians." According to the strongest version of libertarianism, whatever we are objectively free to do, we are subjectively free to do. To be free in the subjective sense means (1) to be able to do something *or to refrain from it*, and (2) to be the sole cause of doing it or refraining from it. To be the sole cause of an act means that nothing outside of you or inside of you makes you do it. You do it simply because you choose to do it; you could have chosen otherwise. St. Augustine expressed this position when, thinking about his past actions, he wrote to himself in his *Confessions*, "I knew just as surely that I had a [free] will as that I was alive. I was absolutely certain when I willed a thing or refused to will it that *it was I alone who willed or refused to will.*"[3] The point of the libertarian position is that nothing makes, forces, causes, or determines us to choose and act as we do. We choose and act freely. That does not mean we never have a reason for what we do. It means our reasons are not coercive; we could choose otherwise.

Now for some clarifications. According to libertarianism we are not subjectively free to do what we are not objectively free to do. If we do not have the ability or opportunity to do something, then we are not subjectively free to do it. However, because of our subjective freedom we are free to decide whether or not to *attempt* things that we are not objectively free to do. In our subjective freedom we can choose to attempt to lift a weight that we know we are not able to lift. More often than not, however, when we attempt to do things we are not objectively free to do, it is because we do not realize we are not objectively free to do them – as when we attempt to pick up a coin that our prankster friends have glued to a table (lack of ability), or as when we attempt to purchase food, only to discover that the store is closed (lack of opportunity).

Although Sartre held that we are subjectively free to do anything we are objectively free to do, he would certainly allow that some things are much more difficult for us to do than others. Indeed, sometimes they are so difficult we say, "I just can't!" – though we could. The interesting point here has to do not with acts of physical exertion; obviously it is more difficult to lift a hundred-pound weight than a ten-pound weight, and we cannot by sheer physical strength lift a 10,000 pound weight at all. Rather, the interesting point has to do with socially influenced actions.[4]

Due to your rearing you may, for example, have great difficulty eating raw oysters or live sea worms (they look like oversized centipedes; South Sea islanders love them). Sartre would say, "You may have *difficulty* doing it, due to your contrary inclinations, but you could do it. You just don't." If you were to try to excuse yourself by saying, "I couldn't! I tried, but I couldn't!," Sartre would say that you were speaking in "bad faith." That is, you would be trying to *excuse* yourself rather than acknowledging that you didn't eat that long wriggly sea worm simply because you *chose* not to. You would be trying to escape your responsibility for not eating that "centipede on steroids" by denying the reality of your subjective freedom.

According to libertarianism, when we are not trying to deceive ourselves, we know that no matter how strongly we are inclined *away* from something, we can, nonetheless, do it, and no matter how strongly we are inclined *toward* something, we can, nonetheless, turn away from it. To summarize these points by saying, "Humans are free," is, according to Sartre, inadequate because it is misleading. It suggests that humans *have* freedom – as though it were an attribute that we could lose, like our hair or a habit. Consequently, instead of saying, "Humans are free," Sartre proclaimed, "Man *is* freedom."[5] Sartre is, of course, talking about subjective freedom. Subjective freedom is the nature of a person; if a creature is not subjectively free, then it is not a person, and as long as we are persons, we are subjectively free.[6]

Finally, it is important to note that in order to present the sharpest contrast between libertarianism and determinism, I have been presenting libertarianism in its most extreme form, or what we might call "hard libertarianism." Hard libertarians, such as Sartre, say that whatever we are objectively free to do, we are subjectively free to do. Soft libertarians disagree. They say that sometimes we are subjectively free to do what we are objectively free to do, but not always. Subjective factors sometimes make it impossible for us to do something that we are objectively free to do. Because of our feelings or emotions we may *not* be subjectively free to swallow that big, wriggling sea worm or kill someone we love, and yet at the same time we may be subjectively free with regard to many other things. Most libertarians are probably soft libertarians, and soft libertarianism does seem to fit more closely with the way we experience life. Nonetheless, soft libertarianism has the formidable challenge of identifying when our freedom of choice is nullified by subjective factors and explaining how that can happen. For the sake of simplicity, I will continue to mean hard libertarianism when I say "libertarianism."

The radical freedom espoused by hard libertarianism means we are totally responsible for all of our actions. Such a frighteningly large responsibility causes us to try to deny our freedom, or at least to convince ourselves that it is limited, so that we do not seem to bear so much responsibility. As long as we are human, though, we can escape neither our freedom nor our responsibility. But exactly what does it mean to be *responsible* for an action?

Again we meet ambiguity, so let's distinguish three kinds of responsibility. First there is *physical responsibility*. We are physically responsible for whatever is caused by our behavior. Here are some examples: If I put a watch around my wrist, then I am physically responsible for having a watch around my wrist; if I accidentally knock over a vase, then I am physically responsible for having knocked it over (even if I didn't intend to knock it over, it was my body that caused it to fall over); if someone holds a gun to my head and tells me, against my will, to open a safe, then, if I comply, I am physically responsible for the safe being opened (it was my fingers that opened the safe). In this causal sense of responsibility, we are each responsible for everything that results from our behavior. That seems non-controversial.

The second sense of responsibility is quite controversial. Let's call it "*moral responsibility*". When we are physically responsible for something, we may or may not be morally responsible for it. We are morally responsible, according to libertarianism, only for those actions that we could have refrained from doing. (We are also, of course, responsible for refraining from an action when we could have chosen to do it.)

If we couldn't have done otherwise than we did, then even though we are physically responsible for whatever occurred as a result of our action or inaction, we are not morally responsible for it. It is clear, for example, that Patricia Hearst participated physically in a California bank robbery by members of the Symbionese Liberation Army. There are films and witnesses of her participation, so she was, in part, physically responsible for that robbery. But was she morally responsible for her participation? Could she have done otherwise? Or had she been brainwashed and was no longer in control of her behavior? After all, she had been a law-abiding citizen until she was kidnapped a few months earlier at age 20 by members of the SLA.

If Hearst had been brainwashed, then she was not morally responsible. If she had not been brainwashed, but was commanded to participate or lose her life, then she might be exonerated of moral responsibility due to extenuating circumstances. Or so it would seem. The problem is not as simple as it first appears. Consider the case of Charles Starkweather, a young man who decided to be, in his own words, "the last of the great American outlaws." As a consequence of his decision Starkweather embarked on a life of crime with his girlfriend, and he ended up killing eleven people, including a two-and-a-half year-old child, before he was captured. Ron Mack, a Cornell University psychologist, had the following reaction to one author's sympathetic treatment of Starkweather:

> My concern here is that the notion of personal responsibility not be lost. Clearly there are a variety of factors which influenced Charlie's short-lived career as a murderer. I do not deny the influence of the past, the environment, or of emotional conflict. Charlie did not have the best of parents, had physical difficulties (poor eyesight and bowed legs) which hampered his progress in school and often exposed him to ridicule. He had an intense need for revenge, to be somebody and to show the world.
>
> But these factors in and of themselves do not add up to eleven murders. Charlie did not go berserk, losing the ability to tell right from wrong, and we would be hard put to specify a mental defect which caused him to commit these crimes, unless wanting to be somebody and choosing a career [as an outlaw] are mental defects. After his capture Charlie explained to the authorities that he had done what he did in order to "be somebody" and that he "had wanted to be an outlaw ever since he was a child, but not this big a one."[7]

Starkweather was electrocuted on June 25, 1959.

His girlfriend/accomplice was imprisoned until June 20, 1976, when she was released on parole, after eighteen years as a model prisoner. She was fourteen-years old when she began her spree with Starkweather; he was nineteen. Was she not morally responsible for her actions because of her tender age? Was he not responsible because of his parents, etc.?

The Starkweather story leads us to a third kind of responsibility, *legal responsibility*. The relations among these three kinds of responsibility (physical, moral, and legal) are complicated but important enough to work out clearly. An adult member of a society is, ordinarily, legally responsible for obeying its laws. The law enforcement agency of the society is given the right and has the responsibility to enforce the laws of the society. But note: there are many things for which you are physically responsible for which you are not legally responsible because there is no law about them. For example, ordinarily you are physically responsible for whether you are wearing socks or not, but you are not legally responsible for whichever is the case because there is no law which says that you must wear socks (thank goodness!).

Many laws are expressions of the moral convictions of the law makers, but certainly not all laws are (such as which side of the street to drive on). Moreover, you can be legally responsible for performing actions that you are morally responsible for *not* performing. That is what civil disobedience often involves. In Nazi Germany, German citizens had a legal responsibility to turn in Jews who were trying to escape. In the nineteenth century, USA citizens had a legal obligation to capture and turn in escaped slaves. Some German and American citizens thought they had a moral responsibility to violate that law of their country, and they did.

Finally, a person who lives outside a society and is not a legal member of it has no legal responsibility for obeying its laws as long as he or she remains outside it. But if there are, as ethical absolutists believe, moral laws that are binding on everyone, then we are morally responsible for living according to those laws no matter where we are and even if there is no agency that will punish us if we do not obey them.

Now let's return to the question as to whether anyone is morally responsible in the sense in which the libertarian believes people are. To deal with this question adequately we need to understand the metaphysical position of determinism. We will begin with a general discussion of *universal determinism* and then examine two versions of it.

Although the concept of determinism has been around at least since ancient Greece, most westerners seem to have been libertarians until the rise of modern science. As the physical sciences, especially from the seventeenth century forward, began to show how completely and precisely events in nature are determined by laws of nature, and as the social and biological sciences, especially from the middle of the nineteenth century forward, began to disclose how human behavior is influenced by genetics, childhood rearing, geography, climate, religion, economics, etc., more and more people have been converted to determinism, or at least have been struck by its plausibility. However, the debate between libertarians and determinists is far from over. As we will see, the choice between their positions involves deep, practical differences as to how to understand ourselves and other people, and how to live our lives and relate to other people.

Universal Determinism

According to universal determinism, every event is the necessary result of an antecedent event or events. An event is anything that happens in the universe –

from the grandest thing, such as the Big Bang, to the simplest thing, such as a sneeze. To say that an event happened necessarily is to say that it *had* to happen; something *made* it happen. To say that an event happened necessarily because of antecedent events is to say that given those antecedent events, that other event could not have happened otherwise; it had to happen, and it had to happen just the way it did. (In other words, those antecedent events were *sufficient conditions* of the occurrence of the event being explained.)

Within the position of universal determinism there are two sub-positions that differ profoundly over the ultimate causes of the events in the universe. These sub-positions are *theistic determinism* and *naturalistic determinism*.

Theistic Determinism

Theistic determinism says that there is a God and that *every event is the necessary result of the will of God*. God determines everything that happens in the universe (even though, of course, God may cause things to happen by first creating laws of nature and then using those laws to make things happen, such as sunshine, rain, and the growth of plants). Theistic determinism is more commonly known as "predestination." It is most common among Muslims and among Christians who stand in the theological tradition of the protestant reformer John Calvin (1509–1564).

We must be careful, however, to distinguish among predestinarians. Some Calvinists think that God has not predestined what we will do; God has predestined only whether we will go to heaven or hell. Perhaps we could call these people "soft or partial predestinarians." Other Calvinists believe that God determines everything that occurs, including everything that happens to us and everything that we do, right down to the blink of an eye. People of this persuasion might be called "hard or complete predestinarians." Only hard or complete predestinarians are universal determinists.

Some theists, then, are universal determinists. Most theists, however, especially in Judaism and Christianity, seem to be libertarians. Most determinists today seem to be atheists. For that reason, and because most current philosophical discussions have to do with *naturalistic* determinism, I will focus on it here and encourage you to study theistic determinism elsewhere.[8] However, because theistic determinism is often confused with *fatalism*, let's take a brief look at fatalism before we proceed to our study of naturalistic determinism.

Sometimes when an unexpected event, especially one that is wonderful or tragic, takes place, people say, "It was fate"; that is, it was the will of God or the gods (I add "the gods" because fatalism is prominent in polytheistic religions). The idea of fatalism is that some events are determined and others are not. Those that are fated, we have no control over. Fatalism is like theistic determinism because it, too, involves belief in a supernatural being or beings. The difference is that theistic determinism believes that *everything* has been determined by God whereas the fatalist believes that *only some things* have been determined by God or the gods. Those events that have been fated, we can do nothing about.

Consider, for example, the following story. In 1985 there was still a good bit of public debate about whether people in the front seat of a car should be required to wear a seatbelt. One of the main arguments in favor of this requirement was evidence that seatbelts save lives (that is, the percentage of people killed in car accidents was significantly lower for those who were wearing seatbelts). In response to that point the following "Letter to the Editor" was published in a major newspaper: "I do not believe seat belts save lives. God does. When your time here on earth is up no seat belt, no helmet or whatever will save your life." The assumption behind that letter is that when we will die has been decided by God and is totally out of our control or anyone else's – but there is no suggestion by the author that *everything* has been decided by God.

That way of thinking can also be found on the battlefield. Soldiers are sometimes told that if there is a bullet out there with their name on it, it is going to get them whether they behave in a cowardly or a courageous manner. In other words, if it has been fated that you will die in a particular battle, then you cannot escape, so you may as well behave courageously, and if it has *not* been fated that you will die in a particular battle, then behaving courageously won't get you killed, so you may as well behave courageously. Again the idea is that we are subjectively free with regard to many things (such as whether or not to act in a courageous manner), but we are not free with regard to all things (such as when, and perhaps how, we will die).

One of the most haunting expressions of fatalism is the following story, attributed to novelist W. Somerset Maugham:

> There was a merchant in Baghdad who sent his servant to market to buy provisions and in a little while the servant came back, white and trembling, and said, Master, just now when I was in the market-place I was jostled by a woman in the crowd and when I turned I saw it was Death that jostled me. She looked at me and made a threatening gesture; now, lend me your horse, and I will ride away from this city and avoid my fate. I will go to Samarra and there Death will not find me. The merchant lent him his horse, and the servant mounted it, and he dug his spurs in its flanks and as fast as the horse could gallop he went. Then the merchant went down to the market-place and he saw me [Death] standing in the crowd, and he came to

me and said, why did you make a threatening gesture to my servant when you saw him this morning? That was not a threatening gesture, I said, it was only a start of surprise. I was astonished to see him in Baghdad, for I had an appointment with him tonight in Samarra.[9]

In this case it was fated not only when someone would die, but also where he would die (and probably how he would die). He could run, he could hide, but he could not escape what the gods had fated for him. His desperate choices merely played into the hands of fate.

There are things other than death that fatalists sometimes believe are fated – such as an accident or who you are going to marry – but, again, what distinguishes the fatalist from the theistic determinist is that he or she does not believe that all things in life are fated, whereas the theistic determinist does believe that.

Naturalistic Determinism

Naturalistic determinists believe that there is no God and that *every event is the necessary result of antecedent events and the laws of nature*. Note: *every* event is the necessary result of antecedent events and the laws of nature. There are no exceptions! An event is anything that happens – whatever, wherever, whenever it might be. One event occurred when you were conceived; another occurred when you were born. A more recent event occurred when these words were printed onto this page. That event was the necessary outcome of certain laws of mechanics, chemistry, and electricity that were in effect during the printing process.[10]

An event happening at this very moment is your reading this sentence! That event, too, is the necessary outcome of a host of natural laws, having to do with physiology and with the social processes whereby you became able to read and became motivated to read this book now. It is important to note that naturalistic determinism holds that human behavior is just as thoroughly determined in every respect as is the behavior of chemicals and plants. To use the earlier formula: every bit of human behavior is the necessary outcome of antecedent events and natural laws. The position that all human behavior is determined is sometimes called "metaphysical behaviorism."

Why accept the position that human behavior is determined? Many reasons have been given. Let's look at five. First, determinism is a necessary presupposition of science. If determinism is not true, then the universe is not determined by laws of nature, and if the universe is not determined by laws of nature, then the scientific enterprise, which is a search for the laws of nature, is doomed to failure.

As B. F. Skinner, a universal determinist, says in his philosophical novel, *Walden Two*, "You can't have a science about a subject matter which hops capriciously about."[11] There can be a science of nature only if nature behaves in predictable ways.

If the things which constitute the universe did not behave in lawful ways, then science would be a futile activity, since there would be no laws for it to find. We can't catch fish where there are no fish, and we don't go fishing where we believe there are no fish. Consequently, either we must give up science or we must give up libertarianism (according to which the universe is not lawful). But surely we would be foolish to give up science and return to pre-scientific levels of medicine, agriculture, transportation, and communication just in order to retain belief in libertarianism. Such a move would be like returning to geocentrism because we prefer to think that the earth is the center of the universe. We can no longer take such a move seriously because now we know that geocentrism is false, and we also know that laws of nature have been found throughout nature. That is a simple fact. Consequently, we must give up the belief that we are subjectively free.

Some libertarians respond to the preceding argument by saying, "Of course there are well-established laws of nature! Of course I don't want to reject the benefits of science! All I am claiming is that *human* actions are not determined by the laws of nature. Everything else is, and that gives science plenty of work to do."

The determinist's response, which is a second reason for accepting naturalistic determinism and therefore metaphysical behaviorism, is that it is unreasonable to make an exception of humankind to the obvious prevalence of determinism in the universe. It verges on absurdity to hold that *everything* in this vast universe is determined except human behavior. Given that the behavior of all the stars, planets, comets, moons, metals, chemicals, vegetation, and animals of this world is determined (and who disagrees with that? certainly scientists don't), it is much more plausible to think that the behavior of humans, too, is determined rather than free. Humanity is a mere speck in an otherwise obviously determined universe. Hence, it is highly unlikely that human behavior is the only behavior in the universe that is undetermined.

Sometimes libertarians respond to that point by saying, "Well, yes, on grounds of sheer probability it would be probable that human behavior, like everything else, is determined. However, every rule has its exception, and there are extenuating circumstances which make it clear that humans are an exception to the deterministic rule of the universe. For example, we believe that the behavior of chemicals and plants is determined because we see how predictable they are, but human behavior is not predictable. Therefore, we should not believe that it is determined."

The determinist's response, a third reason for accepting metaphysical behaviorism, is that the last claim is simply false. There has long been evidence that human behavior is predictable, and therefore determined. Insurance and loan companies have survived and succeeded for centuries precisely because human behavior can be predicted to a significant extent. To be sure, some companies have gone out of business because they were not good at predicting who was a good risk for a loan or an insurance policy, but the success and longevity of many other companies shows that they have been very good at making such predictions.

Furthermore, due to the development of the sciences of human behavior, such as psychology and sociology, our ability to predict human behavior is getting better and better. To be sure, we are not as good at predicting human behavior as we are at predicting the behavior of chemicals and plants, but that is to be expected. The social sciences got underway only toward the middle of the nineteenth century, so they have not had nearly as much time to develop as have the natural sciences (which got underway in the early seventeenth century). Furthermore, human beings are much more complicated subjects of study than are chemicals, plants, or anything else we know of, and probability theory itself tells us that the more complicated something is, the more difficult it will be to predict. Moreover, the difficulty of predicting something does not mean that it is not determined. We cannot predict the weather 100% of the time, but we still think it is 100% determined!

The justification of determinism need not be restricted, then, to theoretical considerations, such as the percentage of the universe that humans constitute; the determinist can appeal to the existence and growth of hard evidence that human behavior is predictable and therefore determined. Put all these factors together – the necessity of determinism for science, the implausibility of making humanity the one exception to determinism in the entire universe, the increasing predictability of human behavior – and it is easy to see that the denial of determinism is simply no longer rational.

Still, some people consider the preceding arguments and reject them on intuitive grounds. "Look," they say, "I recognize that external evidence seems to favor the truth of metaphysical behaviorism, but I *know* I am subjectively free. I am directly aware of it. I can feel it. Why should I deny what seems so obvious?" The determinist's response is that how we *feel* doesn't give us the truth about an issue such as this. We may feel free yet be determined – just as we may feel safe yet be in danger.

Benedict Spinoza illustrated this point by saying that a person who thinks he is free is like a rock which falls off a cliff, looks back up while falling down and says, "You know, I could go back up if I wanted to. I just don't want to." Spinoza's point was that no matter what that personified rock felt like it could do, it was still in the inescapable grip of the laws of nature – and so are we.

Nonetheless, many of us still *feel* that we are subjectively free. How does the determinist explain that feeling if it is false? Spinoza presented a two-part explanation. First, he said, when we act we usually see two or more alternatives before us that we are objectively free to do ("Yes, I did A, but I could have done B or C just as easily"). Second, we do *not* see the forces that *made* us choose A rather than B or C. Hence, we *think* we could have chosen B or C when in truth we could not have. To be sure, *if* we had chosen B or C, we could have done whatever B or C involved, but the issue is whether we could have chosen B or C, given all the factors that were influencing us, and the answer is "no." The laws of nature are constantly in effect and are just as effective whether we are aware of them or not. The grip of gravity, for example, is effective long before we become aware of it, and it isn't diminished when we become aware of it (as though to become aware of gravity were to be able suddenly to jump higher!). Neither do the other natural laws that are at work in our lives ever slack off for any reason.

Brand Blanshard, a mid twentieth-century Yale philosopher and determinist, noted that we are usually not aware of the forces acting upon us because when we act (and we are acting most of our waking hours) we are focusing on the consequences of our actions, not on their antecedents. We are focused on the present and the future rather than the past – but, of course, the causes of our present behavior occurred in the past. Understandably, then, we are usually oblivious of the factors that have made us want what we want and do what we do. Hence, our mere feeling that we are free is worth nothing as evidence in favor of libertarianism, and that is a fourth reason for taking libertarianism less seriously and determinism more seriously.

Finally, some libertarians say to determinists, "I concede that your evidence is strong, and I recognize that my feeling that I am free is worth little or nothing as counter-evidence. However, I find it profoundly depressing and disturbing to think that my life is ultimately determined by forces beyond my control, so I'm not going to accept determinism until you give me a proof that it is true!" The determinist's response is not to attempt a proof. Rather, it is to point out that the libertarian has got things quite backwards. He or she has profoundly misunderstood the emotional significance of determinism and libertarianism.

One of the virtues of determinism, and a fifth reason for accepting it, is that it promotes happiness! As Alasdair MacIntyre said in summarizing the position of Spinoza, "Felicity is the knowledge of necessity, for if the mind can accept the necessity of its own place in the whole ordering of things, there will be room neither for rebellion nor for complaint."[12] In Spinoza's own words: "The mind has greater power over the passions, and is less subject to them, in so far as it understands all things as necessary."[13] That is, once we realize and accept that our lives are entirely determined from beginning to end, we will find peace of mind because we will realize that what we are like and what we do, what other people are like and what they do, are matters over which neither we nor they have any subjective freedom whatsoever.

Human misery of the emotional sort largely results from two things: (1) thinking that our past and present would have been better if only we had made better choices, and (2) thinking that how things will go for us in the future is largely up to us. The truth, however, is that we could not have done otherwise in the past, and we have no subjective freedom with regard to what we will do in the future. So shed your guilt and your anger at yourself about the past, your inferiority feelings about the present, your anxiety about the future, and your envy and resentment of others – they could not help being as they were and are, and neither could you. As Spinoza said, "To understand something is to be delivered of it"; that is, to understand a thing fully is to understand why it had to happen exactly as it did. As a popular saying puts it: "To know all is to forgive all."

The point of these statements is that once we understand why things are as they are and happened as they did, we realize that they could not be or have happened otherwise, so we should be forgiving toward ourselves and others, including Charlie Starkweather. By contrast, according to libertarianism we could have chosen differently, so we *ought* to feel bad about what we have done wrongly or stupidly, and we *ought* to feel anxious about the future, as it really is up to us (within limits, of course) whether our future will be great, mediocre, poor, or disastrous.

Soft Determinism (Compatibilism)

Having established the credibility of naturalistic determinism, now we need to explore a disagreement between determinists. A short space back I mentioned Brand Blanshard's point that because we are usually focused on the the future, we are usually unaware of the factors in the past and present that cause our current behavior. But what are those factors? Here there is deep disagreement among determinists. On the one hand are the "soft determinists"; on the other are the "hard determinists." Soft determinists hold that along with physiological factors such as blood pressure, adrenalin level, blood sugar level, etc., a person's *feelings*, *desires*, and *beliefs* enter into the determination of his or her behavior – and these, of course, are also determined by antecedent events and the laws of nature.

Sometimes, for example, a desire to be educated causes us to read a book. But we did not choose our desire to be educated. It was "built into us" by parents, teachers, friends, etc. Nor do we choose our feelings and beliefs about things. They just happen to us as a result of our experiences. We do not choose to like one person and dislike another. Such feelings just happen to us. We spontaneously dislike one person and like another – though, of course, new experiences may cause our feelings toward those people to change over time. Similarly, we do not choose to believe that socialism or capitalism is a better economic system. Our belief in this regard is the end product of a long process of experience, listening, reading, and thinking.

In general, then, we do not choose to feel a certain way, believe a certain proposition, or desire a certain object. Feelings, believings, and desirings just naturally happen to us as a result of our experiences. Hence, just as our choices and actions are caused in part by our feelings, desires, and beliefs, our feelings, desires, and beliefs are caused by yet other natural and social factors that precede and cause them. Thus, all of our behavior and mental life is determined by antecedent events and the laws of nature.

However, according to soft determinism (hereafter "SD"), our behavior is sometimes free as well as determined! That is why SD is called "compatibilism" – because it states that freedom and determinism are compatible with one another; they can both be true of a person at the same time. Now we must seek to understand how this can be, since it seems like a violation of the Law of Non-Contradiction to say that human behavior is both determined and free. The soft determinist argues that there is no contradiction; freedom and determinism are compatible. To be sure, libertarianism has been discredited, but that does not mean we must give up belief in and talk about human freedom. If the expressions "human freedom" and "free choice" didn't refer to something real and important, they would not have been around for so long. We just need to understand them in a way that is compatible with the modern, scientific point of view.

What shall we mean by "freedom" in this scientific era? Two things. First, by saying that a person is *free with respect to some action*, SD means exactly what we would mean by saying that that person is *objectively free* to perform that action; namely, that person is able to perform that action, and nothing and no one is preventing him or her from performing it. Second – and this is the point which distinguishes SD – by saying that a person *acted freely* (or *is acting freely*) SD means that the person did what he did (or is doing what she is doing) *because he wanted to* (*or she wants to*) – and not because he was (or she is) threatened, coerced, drugged, brainwashed, hypnotized, or affected in any way that disordered (or disorders) his (or her) normal set of values and beliefs.

In addition to being objectively free, then, to be subjectively free (in the SD sense, not the libertarian sense) is to do what one does because one wants to do it. To be subjectively free means nothing more than that; it does not mean to have been able to choose differently than one did. Hence, according to SD, whenever people *do what they do because they want to do it*, they are acting freely – even though they could not have chosen otherwise. In other words, a free action is a *voluntary* action; an *involuntary* action is not a free action.

Would the fact that people's actions are determined mean that they are not morally responsible for their actions? According to SD, "No." Whenever people act freely, then even though they could not have chosen otherwise, it is reasonable to hold them *morally responsible*. How can it be reasonable to hold people morally responsible for things they could not help doing? The answer lies in the distinction between voluntary and involuntary behavior. Note the strategy that follows. Just as soft determinists retain the phrase "subjective freedom" or "freedom of choice" but define it differently than libertarians do, they also retain "moral responsibility" for humans, but they understand and justify it differently than libertarians do.

It would not be reasonable to hold someone morally responsible for being congenitally deaf or for throwing up during a bout of stomach flu. Such things do not happen because people want them to. They happen "against our will," as we say. But some other things happen because people *want* them to happen and *act* on their wants. It is those things – *voluntary* behaviour as distinguished from *involuntary* behaviour – for which it is reasonable to hold people morally responsible. Why? First, they did those things because they wanted to – not because they were forced or tricked into doing them. Second, even though in the case of past actions, voluntary and involuntary, the individual could not have done otherwise than he or she did, if we do not like what they did, then *we* can do something in the *present* to change the individual's desires or feelings or beliefs and thereby change his or her free actions (voluntary behavior) in the *future*.

Here are some examples of how such changes work. If a child plays hooky from school because he *believes* that he will not be caught, but we catch and punish him, he is not as likely to play hooky again because we have changed his belief that he can get away with it. If a friend sunbathes a lot because she *desires* "coppertoned" skin, we may cause her to lose her desire if we show her reports which document that extensive exposure to the sun can cause skin cancer and premature aging of the skin. If the adolescents in two towns in the same county have an unhealthy dislike for one another, we might change their *feelings*, and therefore their future behavior toward one another, by combining them into a single team to compete against the kids in the next county.

To review, it would be absurd to hold a person morally responsible for collapsing whenever their ascending aorta is blocked off. There is nothing they can do about it. A human can't help but lose consciousness when the blood flow to the brain is blocked. By contrast, it is reasonable to hold people morally responsible for their voluntary actions. Voluntary actions are a natural expression of the normal feelings, desires, and beliefs of a person. If we like those actions, we can praise or reward them so as to encourage more of them. If we do not like them, we can act so as to change the person's feelings, desires, and/or beliefs so that he or she does not repeat those actions in the future.

Note, however, that it would be wrong to *blame* or *punish* people *because* of what they did in the past (no matter how heinous); all things considered, they could not have done otherwise. Consequently, the emphasis of soft determinism is on *rehabilitation, not punishment,* of criminals. The past cannot be changed and couldn't be helped. Consequently, it would be entirely unreasonable to punish people, as the libertarian would, on the ground that they could have done otherwise. Libertarianism leads to the *retributive theory of justice*, which holds that criminals should be punished because they knew better and could have done differently. SD says we now know that such a position is unscientific and cruel. However, *future* behavior can still be shaped, so we ought to endeavor to rehabilitate, not punish, people who have engaged in criminal behavior. As Paul Andrews said in response to an edition of *Time* magazine on prisons in the U.S.A.: "Our genes and our environment control our destinies. The idea of conscious choice is ridiculous. Yes, prisons should be designed to protect society, but they should not punish the poor slobs who were headed for jail from birth."[14]

Hard Determinism (Incompatibilism)

Having seen how the soft determinist reconciles determinism, freedom, and moral responsibility with one another by defining key words differently than the libertarian does, now let's examine *hard determinism*. The hard determinist agrees with the libertarian that determinism is not compatible with human freedom and moral responsibility. Hence, the hard determinist, like the libertarian, is an *in*compatibilist. Why does the hard determinist believe that determinism is incompatible with subjective freedom and moral responsibility? Here is a metaphysical explanation which some, but not all, hard determinists give: Contrary to what the libertarian thinks, we are never subjectively free to do something or refrain from it, and contrary to what the soft determinist thinks, feelings, desires, and beliefs are not appropriately called "causes" of our behavior. Only physical things (things which can be investigated by science) can be identified appropriately as causes, and the words "feelings," "desires," and "beliefs" do not identity physical things.[15]

More specifically, according to a hard determinist like B. F. Skinner, the

Harvard behavioral psychologist, the causes of our behavior are Genetic Endowment, Environmental History, and Present Circumstances. Every bit of animal behavior (including human behavior) is the necessary product of these three sets of factors – each of which is physical; none of which is mental. Your genetic endowment – which includes your native intelligence in all its aspects (mathematical, musical, artistic, etc.), the condition of your eyes, ears, and so on, and your athletic possibilities – determines what you are *capable* of doing as a physical organism. Some of us are endowed for great accomplishments in sports; others are endowed for great accomplishments in art; most of us are endowed for modest accomplishments in several things.

Your environmental history (which includes all of the social as well as natural influences upon you – all of which are purely physical) *shapes* your behavioral tendencies in one direction rather than another. For example, if English is the only language that you have acquired from your social environment, then you respond in English when spoken to. If you were raised orthodox Muslim and are of that conviction, then you refuse to eat pork or drink alcohol when it is offered to you. These are obvious examples, but behavioral dispositions get much more subtle and include all of your behavior right down to your posture, accent, gait, and gesticulations (some of which, like posture and gait, may be rooted in genetics rather than environment).

What you actually do in specific circumstances is the result of those circumstances triggering the behavioral dispositions that have been built up in you by your environmental history. It is as though you are a musical organ upon which nature and society play. The sounds that an organ makes are a product of the stops that are pulled and the keys that are pressed. (The key determines the note that is sounded, for example, middle C; the stop determines which instrument the note sounds as though it is coming from, for example, a trombone or a flute.) Hence: (1) the construction of the organ + (2) the stops that are pulled + (3) the keys that are depressed >>*cause*>> (4) the sounds that are emitted.

Similarly, the *behavior* that you emit is the product of the influence of your present circumstances on your present behavioral dispositions. Your body (an organism capable of certain behaviors because of its genetic endowment) corresponds to the organ (a musical instrument capable of certain sounds because of its structure and the materials of which it is made). Your environmental history has predisposed you to emit certain behaviors rather than others – just as pulling certain stops rather than others on an organ prepares it to emit certain of its sounds rather than others. Finally, your present circumstances cause *those* predispositions to come into play rather than others – just as pressing the keys of an organ with these stops pulled rather than those causes certain sounds to be emitted rather than others.

To use another analogy, your body is somewhat like a computer that has been programmed by past inputs to respond in certain ways to present inputs. Your genetic endowment (hardware) determines what your body (central processing unit) *can* be programmed to do. Your environmental history (software) determines *what* your body has been programmed to do. Your present circumstances (present inputs) trigger your body to do what it is presently programmed to do in response to such stimuli.

GENETIC ENDOWMENT + ENVIRONMENTAL HISTORY + PRESENT CIRCUMSTANCES →
CAUSE → PRESENT BEHAVIOR

Skinner was once asked whether he found hard determinism depressing. He replied, "Not at all." He went on to say he was convinced that all of his own behavior was the inevitable result of his own genetic endowment, environmental history, and present circumstances, and that he did not find that disturbing in the least. Skinner's great predecessor in behaviorism, J. B. Watson, expressed his confidence in environmental determinism by writing, "Give me a dozen healthy infants . . . and I'll guarantee to take anyone at random and train him to become any type of specialist I might select – doctor, lawyer, artist . . . [or] thief."[16]

Now let's look at some of the reasons given by hard determinists to justify their belief that determinism is incompatible with subjective freedom and moral responsibility. The soft determinist argues that it is reasonable to hold people responsible for what they do freely (that is, for what they do *because* they want to do it) – even though what they want to do, and therefore what they do freely, is determined. According to hard determinism, it is never the case that we do what we do because we *want* to do it. We do what we do because of our genetic endowment, environmental history, and present circumstances – all of which are physical; none of which can be identified as a feeling, desire, or belief.

According to the hard determinist, the soft determinist is snared in the outdated beliefs of "folk psychology." According to folk psychology there are such things as mental events and processes – feelings, beliefs, desires, and hopes – which cause us to do what we do. But, according to hard determinists, science has shown us that the only causes are physical, and among physical things we do not discover feelings, beliefs, desires, or any other mental phenomena. To continue to believe that such things have causal power is pre-scientific (or in the worst case, anti-scientific). Consequently, it is not reasonable to speak of people's actions as being free or of people being morally responsible for their actions. This does not mean that we should not hold people *legally responsible* for their actions. It still makes sense to make laws that influence people to behave as we want them to, and it still makes sense to use education, fines, jail, and perhaps even the infliction of pain, to keep people from repeating violations of those laws.

Furthermore, since a person's behavior is never determined, even in part, by feelings, desires, and beliefs, it is therefore impossible to change anyone's behavior by changing their feelings, desires, or beliefs (as the soft determinist would have us do). All we can get at to bring about change in a person is their body. Hence, the only way to change a person's behavior is by changing his or her genetic endowment, environmental history, or present circumstances – keeping in mind that talk and touch are among the physical ways to bring about change in a person.

In conclusion, it makes no sense to hold a person morally responsible for the libertarian's reasons (since people are not free to choose to do something or refrain from it) or for the soft determinist's reasons (since people never do what they do because of feelings, beliefs, or desires). By showing us that all causes are physical, science has discredited the idea of moral responsibility just as it has discredited the idea of subjective freedom. It no longer makes any more sense to hold a person morally responsible for what he or she does than it would make sense to hold a sunflower morally responsible for following the path of the sun. Neither the person nor the sunflower can help what it does, and neither does what it does because of feelings, desires, or beliefs. If we want to change the behavior of either, we must physically influence the flower or the person in effective ways – and which ways are effective must be discovered by empirical investigation.

Criticisms

Having examined universal determinism (theistic and naturalistic) and two forms of naturalistic determinism (hard and soft), we now need to look at criticisms of determinism so we can evaluate its plausibility as compared to that of libertarianism. The following eight criticisms are especially important.

First, a critic of determinism might return to the simple fact that we feel subjectively free. The feeling of which the libertarian speaks is not the feeling that we are doing (or could do) what we want to do. As we have seen, that feeling is compatible with determinism. The libertarian is talking about our feeling before we do something that we could do it or refrain from it. That feeling is especially strong in situations of moral temptation. In such situations we feel ourselves hanging in the balance between doing something and refraining from it, feeling that we can go either way, and convinced that which way we do go is solely up to us to choose.

The libertarian insists that such feelings should not be rejected casually. If we can't have confidence in our own deepest feelings, convictions, and intuitions, how can we justify having confidence in anything? In the absence of adequate countervailing evidence, it would be irrational to assume that things are not the way they seem to be. If we doubted our deepest feelings and intuitions, we would no longer be able to justify believing or doing anything; we would be frozen by doubt. It is more rational to believe what seems to be the case and to act on it than to be paralyzed by doubt. Yes, our belief may turn out to be false, but if it does, then at least we will know that, and then we can try another possibility. Paralyzing self-doubt leads to no discoveries. Hence, *the principle of credulity* says that it is rational for a person to believe what seems to him or her to be the case unless he or she has a good reason for doubting it.[17] The libertarian argues that all of us – including determinists – have the feeling of being subjectively free, and we do not have sufficient reasons to think that that feeling is false, so we should continue to trust it.

To illuminate the preceding point, consider another of our deep intuitions, namely, that a world of physical objects exists outside us and is not just a figment of our minds. Given the fact that this feeling and our feeling that we are subjectively free are both clear and strong, it follows that if we are going to distrust the one feeling (that we are subjectively free), then we should distrust the other feeling (that there is a world of physical objects outside our minds) – but there is no good reason why we should reject either conviction. Determinism is a *theory*, not a proven fact. The determinist acknowledges that he or she, too, has a feeling of being subjectively free – only he or she denies that feeling for the sake of a theoretical possibility, namely, universal determinism. But why reject a clear, strong, universal intuition for an unproven theory? The libertarian can see no good reason to do so.

Second, in addition to the fact that we feel subjectively free, we feel morally responsible. Speaking from the first person, the libertarian says: "When I reflect on certain things I have done and am not proud of, I feel as though I could have done otherwise at those times. I did not have to do what I did. Consequently, I feel morally responsible for what I did. To be sure, an easy way out of feeling guilt for those things would be to absolve myself of moral responsibility by telling myself that since my behavior is determined, I couldn't help doing what I did, and so, consequently, I needn't feel morally responsible or guilty. But, as Sartre put it, that would be an act of 'bad faith.' I know intuitively that I could have done otherwise. Hence, to accept determinism would be to hide behind an unestablished theory in order to dull or escape my sense of moral responsibility."

A *third* criticism is aimed at soft determinism. We sometimes hold other people morally responsible because we are convinced that they, as we, could have done otherwise. If they couldn't have done otherwise, then we should agree with the hard determinist that it would be unreasonable to hold those people morally responsible. We might not like what they did, and we might try to ensure that they won't do it again, but we shouldn't speak of holding them morally responsible. The soft determinist's redefinition of "moral responsibility" so as to make it compatible with determinism is a maneuver that is neither admirable nor acceptable, says the libertarian. We should have more admiration for the straightforward manner of the hard determinist, who accepts the traditional meaning of moral responsibility and then argues that we cannot be morally responsible because our actions are determined. The hard determinist isn't playing word games; the soft determinist is.

Fourth, the determinist's point that human behavior is in fact predictable to some extent is not a serious problem for libertarianism. If we are physically normal, then if a physician whacks us under the kneecap with a mallet, our lower leg will jerk – whether we want it to or not. That response is determined by our physiology. The libertarian can accept that and many other facts about bodily reactions without having to give up all belief in subjective freedom. Furthermore, if we are people of principle, then our behavior will be predictable *because* of our freedom! That may sound paradoxical, but it isn't. Consider this: If we know that someone is a committed vegetarian, then we will be able to predict her eating behavior with some accuracy: she will reject what appear to be meat items and choose what appear to be non-meat items, but the fact that we can predict her behavior does not mean that her behavior is determined. In her freedom she chose to be a vegetarian for what appeared to her to be good reasons. She could have chosen otherwise. The fact that in her freedom she decided to live as a vegetarian means that her behavior will be predictable in certain respects. However, it will be predictable *not* because she *has* to do what she does, but because she *freely* chooses to do it. Hence, the fact that a person's behavior is predictable does not mean that it is determined.

"Wait a minute!" the determinist might reply. "You can explain the predictability of human behavior in terms of disciplined commitment to a principle if you want, but given the obvious fact that so much of Nature is determined, isn't it unreasonable to make an exception of humans and humans alone?" Consider an analogy. When we have examined 999 out of 1000 apples in a barrel and have found everyone one of them to be excellent, we should expect that the last apple will be of similar quality. It would be irrational to think otherwise. In like manner, since 99.9999999999% of Nature is determined, shouldn't we expect that the last 0.0000000001% (humankind) is also probably determined?

The libertarian's answer is "yes" – but only if there are no significant extenuating circumstances. The libertarian's *fifth* criticism is based on the belief that there are significant extenuating circumstances. Let's return to the analogy of the apple barrel. If as we got closer and closer to the bottom of the barrel, the odor of a rotten apple became stronger and stronger, we could reasonably assume – in spite of the excellent quality of all the apples examined thus far – that at least one of the last few apples would be rotten. Similarly, the apparent fact that all of nature other than humankind is determined does not mean that we have to conclude that therefore the behavior of humans, too, is determined. The four reasons given above constitute "a cumulative argument" of extenuating circumstances for thinking otherwise: our feeling of subjective freedom, our feeling of personal moral responsibility, our feeling that other people are morally responsible, and the fact that freely chosen commitment to a principle of action will result in behavior that is predictable but not determined.

Sixth, the determinist's argument that determinism is a necessary presupposition of science is based on a misunderstanding of science. The job of science is not to assume that universal determinism is true and then to try to prove it; the job of science is to look for and disclose lawful relationships wherever they can be found.[18] Perhaps they can be found in human behavior, as well as in the behavior of chemicals, plants, birds, and so on, but we cannot know that in advance. It must be shown, not assumed. Furthermore, science does not *need* to assume that determinism holds in a certain area before it can go looking for it. It needs to know only that causal relationships have been found in other areas in order to be justified in looking for them in new areas, such as human behavior. But whether determinism applies in those new areas is something to be shown, not assumed.

Consequently, Frazier, the voice of B. F. Skinner in *Walden Two*, is tangled in confusion in the following argument. "I deny that freedom exists at all," Frazier states. "I must deny it – or my program [of shaping human behavior] would be absurd. You can't have a science about a subject matter which hops capriciously about."[19] Frazier is right that we cannot have a science of a subject matter that does not behave lawfully, but that only means that if human beings do not behave lawfully, then we cannot have a science of human behavior. To be sure, there is nothing absurd or inappropriate about scientists *trying* to see whether they can establish a science of human behavior. Until they do, however, there are several good reasons (discussed above) for thinking that they won't succeed. Hence, scientists may have to be content with sciences of the other 99.9999999999% of the universe – and surely that is enough to keep them busy. Science will not collapse without the assumption of universal determinism.

A *seventh* criticism is aimed at hard determinism. If hard determinism is true, then I am not an agent, nor are you. To be an agent is to consciously think through what to do, to decide what to do, when and where to do it, and then to act on those thoughts and decisions. But if hard determinism is true, then I do not decide what I will do, and I do not arrive at my decisions as a result of what I value, believe, hope, and think. If hard determinism is true, then I do not make decisions; rather, decisions happen to me. Furthermore, the real causes of my behavior could be described entirely in the terminology of physics, chemistry, and biology, without any reference to what I value, hope, feel, or believe. If hard determinism is true, then I am (my consciousness is) merely an inconsequential shadow cast by the physical processes that really make things happen, including my mental life and my physical behavior.

Here again we encounter a theory which flies in the face of a deep and nearly universal conviction that our mental life – what we think, value, hope, feel, decide, and so on – *does* give rise to our actions, and that thinking, hoping, evaluating, and so on, cannot be explained in the terminology of the natural sciences. Therefore, following the principle of credulity again, we should continue to honor our deep intuition that we are agents and that our actions are decided by our mental life. To be sure, we should still be fallibilists and admit that perhaps hard determinism is – like heliocentrism was – a position that we shall have to accept eventually. But for now we have no sufficient reason either to give up our conviction that our mental life has an impact on what we do or to think that the language of the natural sciences can take the place of talk about values, feelings, beliefs, hopes, reasons, and other mental phenomena.

Eighth, according to some libertarians, determinism is a demeaning doctrine. It takes away from humans the special dignity that goes with being subjectively free to do what is morally right and to decide the path of one's life. In the words of Rabbi Harold Kushner, author of *When Bad Things Happen to Good People*, "To say 'it is not his fault, he was not free to choose' is to rob a person of his humanity, and reduce him to the level of an animal who is similarly not free to choose between right and wrong."[20]

This way of understanding humans also tends to promote the attitude that it is okay to try to manipulate people's behavior by clever rewards and punishments because, after all, people have no real mental or moral autonomy anyway; their behavior is going to be determined by factors beyond their control no matter what anyone does, so we may as well try to make them behave the way *we* want them to. But, clearly, a position that promotes such a cynical attitude toward humans is based on an inadequate conception of the nature and potentialities, the dignity and responsibilities of humans.

In conclusion, for all the above reasons some critics think that naturalistic determinism is highly questionable and should be rejected in both of its forms, hard and soft.

Notes

1. To get my point across quickly, I have spoken as though objective freedom pertains only to external actions, but it does not. It pertains also to internal actions, such as solving an algebra problem in your head. If you have the ability to solve a certain algebra problem in your head, and you have the opportunity to do it, then you are objectively free to do it – though, of course, you may choose not to do it. But if you do not have the ability to solve that problem in your head (because you do not know how to solve it at all, or you can solve it but only by writing it out), or if you do have the ability to solve the problem in your head but do not have the opportunity (because you are being chased by a maniac with a chainsaw), then you are not objectively free to solve that problem in your head at that time.
2. Jean-Jacques Rousseau, *The First and Second Discourses* (NY: St. Martin's Press, 1964), p. 114. Emphasis mine.
3. *The Confessions of St. Augustine*, tr. John K. Ryan (Garden City, NY: Image Books, 1960), Book 7, Chapter 3. Emphasis mine.
4. Here is a subtle distinction that you might find useful. Sartre's position that whatever we are objectively free to do we are subjectively free to do is what I call "hard libertarianism." St. Augustine defended "soft libertarianism." The soft libertarian says that we are not always subjectively free to do what we are objectively free to do. We can lose our subjective freedom to decide to do or not do something that we are objectively free to do. Augustine believed that free, sinful actions undermine our ability to make virtuous choices in the future. Consider the following analogy: If we begin digging a hole under our feet and keep digging, we can get out of the hole easily at first, only with difficulty as it gets deeper, and not at all after a while. Similarly, Augustine was saying, by sinful choices we can gradually destroy our ability to *choose* options that we are objectively free to perform. In terms of more recent concepts, many people are persuaded that drunkenness, addictions, and childhood abuse, among other things, render some adults subjectively unable to resist certain temptations that they are objectively free to walk away from, even while they remain subjectively free with regard to many other things.

    Hard libertarianism says that such influences make it difficult but never impossible for people to walk away – unless, of course, they are rendered physically unable to walk, but then they are not objectively free to walk away. As long as we are objectively free to walk away, we are also subjectively free to do so, according to hard libertarianism. Perhaps the paradigm fictional example of a hard libertarian is Captain Kirk of the Star Trek television series. In numerous programs he struggles with every shred of his will, to the last moment of consciousness, to resist the drugs and alien forces that try to break his will, whether by terror, force, or seduction.
5. Jean-Paul Sartre, *Existentialism and Human Emotions* (NY: Philosophical Library, 1957), p. 23. Emphasis mine.
6. Sartre's great predecessor in the libertarian tradition was Immanuel Kant, the eighteenth-century German philosopher and scientist. Kant claimed that a human's ac-

tions are not determined by the laws of nature; each human is a locus of freedom in the midst of the deterministic realm of nature.
7   *Ithaca Journal*, Ithaca, New York, June 1976.
8   Some people resist theistic determinism because they find it disturbing to be told that God has determined everything that we do and that happens to us. Theistic determinists reply that we should not be disturbed by this possibility; we should be comforted by it. God is perfectly good and wise, so if God has determined everything, then everything happens for the best, even when we don't understand why it is for the best. By contrast, humans are neither perfectly wise nor perfectly good, so if humans are free and have the power, even to a limited extent, to determine what happens in the world, then the world will not be nearly so good as it would have been if God had determined all. Hence, the argument goes, we should hope that God *has* determined all that happens in the world and history.
9   See Daniel C. Dennett, *The Intentional Stance* (Cambridge, MA: MIT Press, 1987), p. 13.
10 In view of current physics, two qualifications to classical naturalistic determinism should be made. First, according to most cosmologists the laws of nature as we know them did not exist when the Big Bang occurred; rather, they formed very shortly after the Big Bang. Hence, they could not have caused the event of the Big Bang (if, indeed, the Big Bang should be called an event). Second, according to quantum mechanics some events at the sub-atomic level, such as the jump of electrons from one ring to another, are non-deterministic. Nothing causes them; they just happen. Nonetheless, naturalistic determinists argue, since shortly after the Big Bang it seems clear that all events above the sub-atomic level can be explained in terms of the laws of nature and antecedent events.
11 B. F. Skinner, *Walden Two* (NY: The Macmillan Company, 1948), p. 257.
12 Alasdair MacIntyre, "Pantheism," *Encyclopedia of Philosophy*, Paul Edwards, ed. (NY: Macmillan Publishing Co., Inc., and The Free Press, 1967), vol. 6, p. 33.
13 Alasdair MacIntyre, "Spinoza, Benedict (Baruch)," ibid., vol.7, p. 539.
14 *Time Magazine*, October 4, 1982, p. 6.
15 Sometimes the difference between soft determinism and hard determinism is presented as a semantic disagreement over whether it is appropriate to apply the words "free" and "morally responsible" to humans if their behavior is determined – the soft determinist saying "yes" and the hard determinist saying "no." Because we are doing metaphysics in this chapter, I am trying to show that this disagreement can be, and sometimes is, based on important metaphysical differences as to what a cause is and whether what we usually mean by "feelings," "beliefs," and "desires" can count as causes.
16 J. B. Watson, *Behaviorism*, 2nd edn. (London: Routledge, 1931), p. 104.
17 Richard Swinburne formulates and defends the principle of credulity in his book *The Existence of God* (NY: Oxford University Press, 1979), pp. 254–71.
18 There is an important distinction that helps prevent confusion when thinking about science. It is the distinction between science as *process* (or method) and science as

product. If we limit use of the word "science" to the *process* of looking for laws of nature, then of course we can have a science of human behavior (and of anything empirical) because all we mean by a science of something is a scientific investigation of it. But ordinarily that is not all we mean when we speak about, for example, the science of chemistry, the science of biology, etc. We call these disciplines sciences because investigations of their subject matters have disclosed and established the existence of laws that govern their behavior. (Discovery of the laws is the *product* of the process.) Hence, just because scientists are looking into a subject matter does not mean that there is a science of that subject matter in the "product" sense of science. To be sure, there is always a product, that is, outcome, of scientific investigation, but the outcome may not include discovery of a natural law. The outcome may simply be data that do not reveal any lawful connections. Jazz improvisation is a fascinating phenomenon. It can be and has been studied assiduously, but study of it has not produced a science. The academic study of jazz is just that, a study, not a science. A *study* of a subject-matter becomes a *science* when it discovers laws according to which its subject-matter behaves. It is generally agreed that the *study* of human behavior has not yet produced a *science* of human behavior. The libertarian believes it never will.

19   Skinner, *Walden Two*, p. 257.
20   Harold S. Kushner, *When Bad Things Happen to Good People* (NY: Avon Books, 1981), pp. 83–4. G. W. F. Hegel expressed a similar attitude toward determinism in his book *The Philosophy of Right*.

Reading Further

B. F. Skinner, *Walden Two*; *Beyond Freedom and Dignity*.
Richard Taylor, Chapter 5, *Metaphysics*, 4th edition or later.
Robert Kane, *The Significance of Free Will*.

Chapter 16: The Mind/Body Problem

- Dualistic Interactionism
    Dualism
        A human is a combination of two things
        Descartes' definitions of mind and body
    Reasons for being a dualist
        The public/private distinction
        Same activity/Different means
        Personal identity and continuity
        Mind versus body
    Interactionism: 4-way causality
    Criticisms
        The problem of the essence of the mind
        The problem of the existence of the mind
            Extrospection and Introspection
        Implausibility of mind affecting body
        Counterintuitive effect of body on mind
        A physical account of personal identity and continuity
        The spatial relation of mind to body
- Occasionalism
- Parallelism
    Criticisms
- Epiphenomenalism
    Criticisms
- Physical Monism
    What a human is
    The fallacy of reification
    The identity thesis
    Linguistic behaviorism
        A category mistake
    Criticisms
        The essence of mind is intelligible
        The existence of mind is known
            Descartes' "Cogito"

> *Mind is subject, not object*
> Behavior is not what we mean by "good mind"
> The identity thesis does not hold
> Personal identity and continuity again
> Mind/Body causation
> Subjective freedom and physical monism
> What is matter?

- **Psychic Monism**
  Matter is not known directly
  What is a human? George Berkeley
  "Real objects" vs. dreams, illusions, and hallucinations
  The concept of matter is not needed
  No way to prove that matter exists
  > Samuel Johnson
  "Material" and "physical" objects
  The fallacy of reification again
  Criticisms
  > Causation
  > The commonality of perceptions
  > Knowledge of other minds

- **Neutral Monism**
  The dual aspect theory
  What a human is not
  The mystery of the source of mental and physical phenomena
  Immanuel Kant: *phenomenon* and *noumenon*
  Criticisms
  > How could one thing be mental and physical?
  > How could causation operate?
  > Too vague to be helpful
  > Why think there is a *noumenon*?

- **Phenomenalism**

We have thought about whether we are subjectively free and, if we are, what we ought to do with our freedom. Surely, though, whether it is reasonable to think that we are subjectively free, and whether it is reasonable to think that we have absolute moral obligations, or none at all, cannot be settled satisfactorily apart from an understanding of the kind of thing we are. If we are subjectively free, there must be something about what we are that makes it plausible that we are free. Similarly, if we are not subjectively free, it must be because there is something about us that makes it impossible for us to be free. Hence, we turn our attention now to the philosophical study of what a human being is and what it is to be human.[1]

The first question we will focus on is, "What am I?" Notice that the question isn't "*Who* am I?" It is "*What* am I?" The question "*Who* am I?" is a question about one's personal identity and therefore must be answered differently by each of us according to our personal history, physical features, aspirations, and circumstances. But the answer to the question "What am I?" should be the same for all of us, since we are all human. So we need to ask: What kind of a thing is a human? What kind of a thing are you and am I? The answers that have emerged from centuries of thought are diverse and intriguing.

Dualistic Interactionism

Dualism

The first position we will look at is *dualistic interactionism* (hereafter "DI"). Recall that the libertarian believes that human beings are free in one and the same moment to choose to do a certain action or to refrain from doing it. What gives a person subjective freedom, according to a libertarian like René Descartes, is that a person's body is under the control of his or her mind, and the mind is subjectively free to decide whatever it wants, and then to use the body to attempt to achieve what it decides. The body, according to DI, is under the control of the mind in somewhat the same manner that the claw of a claw machine in an amusement arcade is in the control of our hands. Our hands make the claw move right or left, backward or forward until we believe it is centered over what we want to drop it on and pick up.

Think also of an infant learning to walk or an adult patient trying to regain use of a damaged limb. The patient knows what she wants to do, and she can, let's suppose, do it gracefully with her uninjured limb, but has a great deal of difficulty doing it with her damaged limb. Yet she may gradually regain complete facility with her injured limb. For example, she may still play piano magnificently with her left hand but, due to her injury, have to start all over with her right hand. She knows exactly what she wants to do, but she has to relearn how to make her right hand do it. The infant is in a more awkward predicament than the adult because it has no idea *how* to do what it wants to do. Still, infants move very much as though they know *what* they want to do, such as grasp a toy or walk to a parent; they just have not yet learned how to make their bodies do so.

According to DI, such discrepancies between intention and performance are best explained by realizing that a human being is a combination of two distinct parts: a body and a soul, each of which can exist apart from the other.[2] It is the soul, however, which is what the person really is; the body is merely temporary housing – a transient tool by means of which the soul receives information from the world and acts on the world. Socrates expressed this position in a humorous way when, moments before his execution, his friend Crito sorrowfully asked him, "Socrates, how shall we bury you?" Socrates replied, "Anyway you want – if you can catch me." His point was that in this life he was a combination of a body and a soul; at death they would separate, and the only thing his friends could get hold of to bury would be his body, not his soul – that is, not him.

Much later in western history, René Descartes developed his influential position that *the body is an extended unthinking substance, the mind is a thinking unextended substance*, and a human being is a combination of the two, which are joined at conception and separated at death.[3] To say that the soul is unextended is to say that it does not exist in space or have spatial dimensions or properties. Therefore the soul is not physical, and therefore it is not subject to the laws of nature, which apply only to physical things. That explains why we are subjectively free: decision-making is a power of the soul, and the soul is not under the control of the laws of nature.

When Descartes said the soul is "a thinking thing" he meant much more than we usually mean by that phrase. When we use the word "thinking" we usually have in mind only intellectual activities, such as questioning, analyzing, and reasoning. When Descartes said the mind or soul is "a thinking thing" he meant for "thinking" to encompass all mental activities, including awareness, feeling, remembering, imagining, desiring, and deciding, as well as questioning, analyzing, and reasoning.[4] By contrast, matter (the stuff of which bodies are made) has size, shape, and weight, and can be in motion or at rest, but it does not think. It is just mindless stuff.

Reasons for being a dualist

Dualism regarding human nature is a position that has found wide acceptance throughout history and around the world. But what reasons are there for believing it? Primarily certain strikingly different types of experience that humans have. First consider the fact that the body is a public thing. At any given moment it has spatial coordinates and dimensions, and it can be examined by any number of people. The mind, by contrast, has no spatial location and is inaccessible to the public. It is radically private. To be sure, the brain, too, is private in a sense, but unlike the mind, it is not radically private. We can remove the top of your skull and see your brain, but when we do, we do not see your mind, so the mind must not be identical with the brain, and it must not be physical – otherwise it would have a location and we could see it.

Second, there is a radical difference in the way in which the mind and the body engage in four common activities. First, both the body and the mind *retrieve* things, but the body retrieves them by means of the activity of physical grasping, whereas the mind retrieves things by means of memory – which involves no reaching or grasping at all, at least not in the literal sense involved in physical retrieval of objects. Second, both the mind and the body *create* things. The body creates a painting by grasping a brush, dipping it in paint, and touching the brush to canvas. By contrast, the mind creates by means of imagination, which employs no physical implements at all. Third, the mind and the body both *manipulate* things, but consider the difference between the way you go about manipulating a plastic pyramid in order to examine its geometrical features and the way you go about manipulating a pyramid in your mind for the same purpose. In the one case you have to use your hands or some other physical means to manipulate the pyramid. In the other case you do not; you simply turn it around and upside down in your mind – which suggests that these two means of manipulating things are quite different in kind. In brief, the fact that we engage in such similar activities in such radically different ways suggests that we are constituted of two radically different things – the one being physical and the other non-physical.

A third argument in favor of dualism is that without the mind we could not account for our personal identity and continuity. I may lose an arm or a leg or even all of my limbs yet remain the same person. I would remain the same person despite such tragic losses because I am my mind, whereas I only *have* a body – parts of which I can lose without losing my mind. If I were only a body, then when I lost a limb, I would no longer be the same person. But I can lose a limb and remain the same person. Therefore, I must be something different from my body. Christopher Reeve, who played Superman in the movies, did not lose his limbs, but he lost the use of them when a horse-riding accident rendered him quadraplegic. After his accident, in an effort to reassure people about himself and to encourage others in similar plights, Reeve dictated a book and titled it *Still Me* – emphasizing that what he really was was a mind or soul and that that was still alive and well.

Another reason to be a dualist is the following. According to physiologists the cells of my skin, muscles, and bones are continually dying and being regenerated, and the particles that constitute my nerve and brain cells are continually being replaced (though these cells do not die and get regenerated), so after approximately seven years, the substance of my body has been *completely* replaced. Consequently, if I am only a body, then I am never the same person from moment to moment (since microscopic changes in my body are continually going on), and after about seven years, I am a completely different person! But that is not credible. It is obvious that I remain the same person while my body changes. Hence, I must not be identical with my body as a whole or with any part of it.

Another implausible implication of the idea that a human is just a body and nothing more is this: Because every cell of our body is regenerated or replaced within approximately seven years, we could not remember anything that happened to us more than seven years ago. Why? Because we cannot correctly remember having experiences that we did not have. When I say that I remember something that happened to *me* more than seven years ago, I am saying that I remember something that happened to me. However, if I am nothing but a body, then I (this present body) was not around to experience anything that happened more than seven years ago. Hence, any belief I have that I remember something that happened to me more than seven years ago must be mistaken because I (this body) did not exist then!

Such a position seems obviously false. I do remember experiences that happened to me more than seven years ago. Some people remember experiences that happened to them more than seventy years ago. Hence, a human being must be more than a body. There must be something about us that is not completely replaced bit by bit over time, something non-physical that provides the basis of our personal identity-and continuity over years and even decades. According to the dualist, the most plausible explanation of our personal identity and continuity over time is that we are spiritual, non-physical entities, "unextended, thinking things," as Descartes put it, not bodies.

A fourth argument for mind/body dualism comes from Plato's dialogue *Phaedo*. In that dialogue Socrates (who is the main character in most of Plato's dialogues) argues that because the mind resists and drives the body – for example, forcing it away from food despite a strong sense of hunger, forcing it to keep working despite a heavy sense of sleepiness – the mind and the body must be two different things. If they were not, then the mind couldn't resist the passions and inertia of the body; it couldn't resist the impulses of lust and rage; it couldn't drive the body on despite fear, fatigue, depression, and pain. Further, it wouldn't make sense to say that the mind is part of the body, since if it were part of the body, we could locate it, just as we can locate the other parts of the body, and if the mind were part of the body, it should be in harmony with the rest of the body, not opposed to it. But as we just saw, sometimes it is opposed to it. Therefore it must not be part of it.

Interactionism: four-way causality

Having examined several reasons for accepting the dualistic part of dualistic interactionism, now let's see what it means to say that mind and body interact. This can be explained by running out the permutations of mind–body relations: there can be mental–mental, mental–physical, physical–mental, and physical–physical interactions. Let's work through each of these possibilities, beginning with the last. First, there can be physical–physical interactions in a human because one part of the body can influence another part. For example, the physical act of sticking one's finger down one's throat will cause the physical reaction of gagging; the physical act of striking oneself sharply under the kneecap with the edge of one's hand will cause the lower leg to snap forward, if it is functioning normally and hanging freely. Hence, bodily events can cause bodily events.

Second, there can be physical–mental interactions. When you bite your tongue (a physical event), it causes pain (a mental event); the damage is in your tongue, but the pain is in your mind, not in your tongue (remember: there are no mental properties in physical things). If someone massages your shoulders after a period of tension, the physical massaging causes your muscles to relax (a physical state); the relaxing of your muscles causes an experience of relaxation (which is a mental state).

Third, there can be mental–physical interactions. Events in the mind can cause physical events in the body. Actors develop the ability to cause themselves to cry real tears (a physical process) by thinking about sad things (a mental process). Athletes push their bodies to new levels of accomplishment by exerting "mind over matter."

Fourth, mental events can cause mental events. For example, thinking of one thing can cause you to think of another thing. An old priest, I have been told, once went to a party. Finding it rather dull, he livened it up by telling everyone that as a young priest, the first confession he heard was by a man who confessed he had murdered someone! (He had, of course, a religious obligation never to tell anyone *who* the murderer was.) The people were appropriately astonished. They were even more astonished when a latecomer made his way into the gathering, spotted the priest, and said aloud, "Ah, Father, how good to see you! Ladies and gentlemen, would you believe that I was the father's first confessor?"

Notice, I didn't tell you the significance of the latecomer's statement. Your *understanding* of the statements by the priest and the latecomer caused you to infer that the latecomer had murdered someone. Many other examples of mental events causing mental events can be given: imagining something can cause a desire for it; a desire for it can cause a thought of how to attain it; a thought of one thing can cause a memory of another thing; and so on. All of these things (thoughts, imaginings, desires, and memories) are mental rather than physical.

From these examples we see that a human being is a combination of two things, a body and a mind, which interact with themselves and with one another. By means of this conception of what a human is, we can account for our subjective freedom, the radical privacy of the mind, the radical differences between mental and physical activities, the fact of personal identity and continuity in spite of the body's gradual, complete replacement, and the fact that we can resist the inertia and passions of our bodies.

Criticisms

As popular as DI has been, it has never been without its critics, and especially since the nineteenth century, when metaphysical materialism (the position that nothing exists but matter and space) began to eclipse worldviews that include God and the soul. We will explore metaphysical materialism in chapter 18. Meanwhile let's look at five basic criticisms of DI.

The first criticism arises from reflection on the question of the *essence* of the mind or soul. Just what is a mind? If we cannot make sense of a mind that is not physical and exists separately from the body, then we cannot make sense of DI. Some critics of DI argue that we cannot make sense of the concept of a nonphysical thing. They say it is an inherently confused and confusing concept. We know what a body is. It is something that occupies space, has shape, size, color, and location. But what is a mind? According to DI it doesn't occupy space or have a color, size, or shape, and it doesn't seem to have any other characteristic by means of which we can understand what kind of a thing it is. So how do we know what we're talking about when we talk about the mind? For all practical purposes, it is no different from nothing. To use a phrase from Spinoza, "the mind" is just "a word thing." When examined closely it proves to be a word without a meaning, and therefore it should be rejected.

A second criticism of DI focuses on the question of the *existence* of the mind or soul. The point of this criticism is that even if we can make enough sense of the concept of a mind to have some dim understanding of what a mind would be like, there is no way to show that minds exist – neither our own nor others. There are only two ways to show that something exists: by *introspection* (looking inside ourselves by reflection) and by *extrospection* (looking outside ourselves by means of our senses). After we've looked inside ourselves and outside ourselves, there is nowhere else to look, yet neither of these methods reveals the existence of even one mind.

When we look outside ourselves at other people or ourselves, all we ever see are their bodies or our own. We never detect another person's mind by means of our eyes or fingers or noses or tongues or ears. And as the great Scottish philosopher David Hume pointed out, when we look inside ourselves by introspection, we never encounter a mind or soul; all we encounter are specific feelings, such as pain or pleasure, warmth or cold, depression or elation. We may encounter a specific thought of some thing or a memory of someone, but we never encounter a soul or a mind. Indeed, how would we know we had encountered one even if we did? Hence, even if we can make sense of the concept of a mind or a soul, we could not verify the existence of either. Therefore, since we shouldn't believe in what we can't substantiate, we shouldn't believe in the existence of non-physical minds and souls.

A third criticism of DI has to do with *causality*. If mental entities and events have no physical substance or properties and do not exist in space, how can they causally affect physical objects? How can they even make contact with them? It is understandable how a pinprick to the finger can pierce the flesh and send stimulation through the nerves to the brain, and how the brain can discharge stimulation back to the finger, causing it to jerk away from the pin, but how can we fit the mind into this sequence? We can follow the movement from pin to finger to nerves to brain, but we cannot follow it from brain to mind. How could a brain possibly be hooked up to a mind, or vice versa?

We know how the surface of the finger is connected to nerves and the nerves to the spinal column and the spinal column to the brain, but all of a sudden things become mysterious when we say, "and the brain is connected to the mind." Rather than resorting to such a mystery, it is more plausible to conclude that there is nothing beyond the brain; the signals transmitted from our extremities to our brains cause our brains to send signals back to our extremities. Hence, we don't need a non-physical mind to explain human thought and behavior, and whenever we do try to use it to explain them, it turns out that it explains nothing; it merely turns a physiological problem into a metaphysical mystery.

Consider the debacle that Descartes got into when he tried to explain the interaction of the mind and the body. He took a straight forward physiological approach while explaining how stimulations to the surface of the body travel into us along physical paths, but then he claimed that when these stimuli reach the pineal gland in the brain, they cause the pineal gland to quiver in an exquisitely delicate manner which transmits signals to the mind – but he didn't explain how the physical vibrations of the pineal gland could cause non-physical experiences in the mind. Conversely, he held that when the mind wills the body to stand up, for example, the mind causes the pineal gland to vibrate delicately, and those delicate vibrations are then transmitted to coarser and coarser parts of the body until finally the leg muscles contract and cause us to stand. Obviously, this does nothing to *explain* how a non-physical mind can cause the pineal gland to start vibrating in the first place! Nice try, René, but no cigar.

A fourth criticism is based on the counter-intuitive extent to which thought and consciousness are influenced by the body. If dualistic interactionism were true, we should expect that the relation between the mind and the body would be like the relation between a short wave radio operator and her radio. According to DI a person *receives* information from the world through his body and *sends* information to the world through his body. Similarly, the short wave radio operator receives information from the world through her radio and sends information to the world through her radio (either by microphone or morse code).

Let's suppose that a hurricane hits the ham operator's house, causing debris to damage her radio so that she can still receive information but can no longer send information. She is not injured, so she can still think clearly and can understand what her radio picks up, including inquiries from friends as to whether she is okay, but she cannot communicate back to them. If DI is true, that is what it should be like when the brain is damaged. That is, since a person is supposed to be a non-physical mind rather than a body, when the brain is traumatized by a blow to the head, the person should still be okay since a blow to the head won't hurt a non-physical mind. However, until her brain recovers from the trauma she will be unable to communicate back to people who are asking her if she is okay – just as the ham operator can hear and think clearly but cannot communicate back until she repairs her radio. But that is not the way things are. A blow to the head does not just traumatize the brain. It also causes a person to lose consciousness or to be rendered unable to think clearly for awhile – sometimes causing amnesia. But that is not what should happen if DI is true, so DI is probably not true.

A fifth criticism of DI consists of an attack on its claim that the mind or soul is necessary to account for personal identity and continuity. Personal identity and continuity need no more be due to the continuity of some one specific thing within us than the continuity of an institution is dependent on some one enduring thing within it. Ithaca College in New York State once existed several miles from where it is now. Its students, personnel, buildings, equipment, and location are all different from what they were in the college's first half century. But it's still Ithaca College! Why? Because the Ithaca College of *now* grew out of the Ithaca College of yesteryear, just as someone who is an adult today developed from an adolescent who developed from a child who developed from an infant, and so they all are one continuous individual.

One individual is continuous with another individual not because some one thing, like a soul, remains unchanged over time in that individual, but because the later individual developed from the earlier individual – even though no single particle of matter from the earlier individual be left in the later individual. Hence, having a non-physical mind or soul that can remain identical over time is not necessary to being the same person over a lifetime. What I mean, for example, by saying "He is the same person as the childhood friend whom I called 'Gene,'" is *not* "He has the same soul as did my childhood friend whom I called 'Gene.'" Rather, I mean "He is the person, and the only person, who is historically continuous with, who developed from, my childhood friend whom I called 'Gene.'"

The sixth and final criticism of DI that we will consider has to do with the claim of DI that minds do not exist in space and therefore do not have spatial location. The problem is this: If a mind has no spatial coordinates, then it doesn't make sense to speak of a mind being *in* a body. If it were *in* a body, it would have spatial coordinates and we could locate it just like we can locate the heart and liver. But according to dualism, the mind isn't in space at all, so it can't be located in the body. Nor would it make sense to think of the mind being externally attached to the body, since then, too, we could locate it, as we can locate hair on our head or a mole on our neck or a tick on our leg. Indeed, if the mind is non-spatial, there seems to be no way to make sense of the idea of a body *having* a mind at all! What sense could it make to speak of "having" a mind if the mind cannot be in or attached to the body? Yet what would 'have' the mind if not the body?

If dualism is true, at least it offers this consolation: you can never lose your mind because you can't have one in the first place – at least not in the sense of a thing that is in or attached to your body and separable from it. But that is small consolation, for it raises again the enigma of interactionism: If the mind is not in or attached to the body, there does not seem to be any way that it could interact with the body.

In brief, DI suffers from numerous problems having to do with the essence, existence, causality, and spatiality of the mind; moreover, the mind is not necessary to account for personal identity and continuity. Consequently, its critics say, DI should be rejected. It is not a satisfactory account of what a human is.

Occasionalism

Soon we will look at some contemporary alternatives to DI, but first let's look at two intriguing attempts by Descartes' contemporaries to save dualism by giving up interactionism. Nicholas Malebranche (1635–1715) and Gottfried Leibniz (1646–1716) were convinced, by some of the reasons we examined, that there is no way that a mind and a body could be causally connected to one another. However,

Malebranche and Leibniz remained convinced that mind and body both exist and are profoundly different kinds of thing. Hence, they were faced with a problem: how to explain the fact that physical events *seem* to cause mental events (as when physically seeing a clock seems to cause us mentally to remember an appointment) and mental events *seem* to cause physical events (as when our mental concern about someone seems to cause us to physically phone ahead and explain that we will be late).

"Occasionalism" is the name of Malebranche's explanation as to why physical and mental events seem to cause one another in spite of the impossibility that they could do so. According to occasionalism, because mental events and physical events follow one another with such regularity and dependability, we easily become deluded into thinking that they directly cause one another. As we have seen, though, it is implausible to think they could do so.

If there is a God, however, as occasionalists believe there is, then on the occasion of a physical event, such as physically plunging into a cool swimming pool on a hot day, God could cause us to mentally feel a pleasant sensation of wetness and coolness. But note: *the water doesn't cause those sensations;* God does. The water *can't* cause those sensations because they exist in the mind, and there is no way for the body to impact the mind. Hence, when you enjoy the cool wetness of a swim on a hot day, thank God, not the water.

Similarly, if you decide to thread a needle, first you mentally will to do it, and then you physically do it, but because the mind cannot affect the body, the mental act of willing to thread the needle does not cause your fingers to grasp the thread in one hand and move it toward the eye of the needle held in the other hand. Rather, on the occasion of your willing to thread the needle, God causes your eyes and hands to do what God knows you are intending to do, namely, thread the needle.

The general position of occasionalism, then, is that *on the occasion* of the occurrence of certain physical events (especially physical events that stimulate the senses of a living body), God causes characteristic mental events (step on tack – feel pain; light stroke to bottom of foot – feel tickle), and *on the occasion* of certain mental events (especially mental events that involve willing bodily behavior), God causes characteristic physical events to follow (will to stand up – your leg muscles contract; will to look at something – your eyes focus on it).

Parallelism

G. W. Leibniz agreed with Malebranche's rejection of interactionism, but thought that occasionalism painted an unbecoming picture of God: God would have to be pictured as scrambling around, frenetically responding to this physical event, that physical event, this mental event, that mental event, and so on – and billions of such events occur in every moment. It would be more dignified, and therefore more appropriate to the nature of a supremely perfect being, Leibniz thought, if God – who knows all of the future anyway – prearranged things so that ze would not have to intervene on this and that and every occasion to cause the appropriate things to happen. Hence, Leibniz concluded that God does not intervene on the occasion of an event happening; rather, by zer power ze has pre-established from eternity that the appropriate thing will happen when the time comes. (For an explanation of "ze" and "zer" see chapter 17, note 1.)

There is, then, a "pre-established harmony" between the mental realm and the physical realm; the two realms run parallel to one another in the way that God deems most appropriate, but the two realms are not directly causally connected. They are like two separate mechanisms that are synchronized so that even though they operate completely independently of one another, they give the appearance of being causally connected. Consider, for example, a clock and a chiming device that are not connected but are synchronized such that whenever the minute hand of the clock reaches 12, the chiming device chimes the same number of times as the number to which the hour hand of the clock is pointing. If the chiming device were hidden in the clock, we would naturally assume that movements in the clock caused the chiming to begin. For reasons that should now be obvious, Leibniz's theory is sometimes called "parallelism."

Criticisms

Occasionalism and parallelism are intriguing, even brilliant, but seem too problematic or counterintuitive for most people. First, they are both based on belief in the existence of God. Consequently, because atheists deny the existence of God, they must reject occasionalism and parallelism, and because agnostics are unsure as to whether there is a God, they must suspend judgment about them. Second, these positions contradict our commonsense convictions that physical events *do* cause mental events (the doctor's needle causes us to feel pain!) and that mental events *do* cause physical events (our willing to do one more sit-up causes our stomach muscles to contract).

To be sure, the preceding objections still leave us with the question as to how mental and physical events *are* connected, if not by God. Soon we will look at some additional theories about that, but first let's consider the value of having examined occasionalism and parallelism. Perhaps the most instructive thing about these positions is the dramatic way in which they illustrate how criticism of a theory with which people initially feel comfortable (for example, dualistic interactionism) (1) can cause some people (for example, Malebranche and Leibniz) to *modify* it because they believe part of it is wrong but another part of it is correct and should be saved (occasionalism and parallelism save the dualist part of dualistic interactionism but reject the interactionist part), and (2) how criticism of a theory can cause people to *reject* it altogether, and especially when attempts to modify it produce what seem to be even worse problems (for example, some critics think that dualism with interactionism has serious problems, and that dualism without interactionism, as in occasionalism and parallelism, has even worse problems, so they reject both interactionism and dualism).

In brief, if we would be rational, then criticism of a theory requires one of the following responses: (1) that for good reasons we reject the criticism and retain the theory, or (2) that we reject the theory because we think the criticism is sound and fatal, or (3) that we keep the theory but modify it in response to the criticism. Malebranche chose to modify DI by replacing the interactionist part with occasionalism; Leibniz was unsatisfied with occasionalism and replaced it with parallelism. Both were ingenious moves, but few people have been persuaded by either modification. Perhaps the greatest value to be gained from a study of occasionalism and parallelism is a realization of the extent to which trying to save a theory can cause intelligent people to take questionable positions. We sometimes go to great lengths to try to save a belief that we like very much or that our egos are deeply involved with.

In conclusion, in most people's judgment, dualistic interactionism has fewer problems than occasionalism or parallelism, so most dualists today are dualistic interactionists, yet DI still has all of the difficulties that were mentioned earlier, so let's press on to see if we can find a more-satisfactory theory of what a human is.

Epiphenomenalism

According to dualists like Malebranche and Leibniz, mental events and physical events are both real but distinct from one another and unable to affect one another. What seem to be causal connections between physical and mental events are to be explained by positing God as mediator between the two realms of mind and matter. What is a person to do who believes in dualism but rejects both interactionism and the causal mediation of God? How can he or she explain causal connections between physical and mental events? The theory we will examine now is a dualistic theory, but it begins a movement away from the classic mind–body dualism of interactionism, occasionalism, and parallelism.

According to *epiphenomenalism*, mental events are real, and they are not physical, but as critics of interactionism have pointed out, it is implausible that nonphysical things, such as beliefs, desires, and feelings, could affect our brain, nervous system, or muscles. Hence, we should retain dualism, but reject mind-to-body interaction. However, we do not need the God-hypothesis to explain why a prick to the finger causes a feeling of pain. All we need to do is observe the many ways in which physical events cause mental events. There are, for example, simple events such as being jabbed with a pin (a physical event) causing a feeling of pain (a mental event), and there are degrees of correlation which indicate clearly that mental events are directly dependent on physical events. For example, the more alcohol there is in your bloodstream, the less clearly you can think, and the less alcohol there is in your bloodstream, the more clearly you can think. Hence, it can be established both by commonsense and scientific evidence that our feelings are causally dependent on our physical states.

This causal dependence can be extended to the whole of our mental life, including not only our feelings but also our desires, beliefs, dreams, rememberings, and so on. All aspects of our mental life are generated from our bodies somewhat like light in a light bulb is generated by a motor. The intensity of the light increases and decreases with fluctuations in the rpm's of the armature of the motor, but the light does not affect the operation of the motor itself; the light is a by-product of the spinning of the armature. Similarly, our mental events, processes, and states are generated by our bodily processes. The feelings, emotions, desires, intentions, and so on, that are generated by our bodily processes are "epiphenomena," that is, side-effects that do not affect us physically, just as the light from the bulb does not affect the generator, the froth produced by waves does not affect the surging of the ocean, the sparks from the wheels of a speeding locomotive do not affect the speed or direction of the locomotive, and the shadow of an object does not affect the object.

In brief, our bodies are physical, the behavior of our bodies is physical, and the most plausible hypothesis, in view of commonsense and scientific research, is that all the factors that determine our behavior are physical. To be sure, we cannot deny the reality and non-physical nature of mental events, but we should put them in their proper place. A scientific analysis of human behavior relegates mental phenomena to the status of epiphenomena (things that occur but do not cause what happens next).

There are three reasons for taking this position. The first, mentioned previously, is that it is not plausible that something without physical mass or spatial location could affect a physical thing in any way. Such a thing would have no way of exerting influence on a physical thing. Hence, it is implausible to think that mental events cause physical events. The second reason is that the human body is a self-sufficient physical system. We can explain entirely in terms of the physiology and conditioning of the body, including the brain, how physical input from the world (which causes seeing, hearing, smelling, touching, and tasting) results in physical output from the body (such as talking, writing, walking, etc.). To be sure, mental phenomena *accompany* what goes on in the body and what the body does, but they do not explain why those things happen. Intelligible, useful, scientifically respectable explanations of behavior must be entirely in terms of physical factors. (Recall that the version of hard determinism presented in the last chapter says the same thing when objecting to the soft determinist's explanation of human behavior – though a hard determinist is not necessarily an epiphenomenalist.)

Third, there are obvious cases of epiphenomenal mental events in everyday life. Once we realize this, epiphenomenalism begins to seem more plausible. For example, when you get flu, you feel nauseous (mental state) and then you throw up (physical event), but *feeling* nauseous does not cause you to throw up. Rather, something *physical* going on in your body (an invasion of viruses) first makes you feel nauseous and then makes you throw up. The nausea comes before the wretching, but it doesn't cause it. The nausea is an epiphenomenon, a side effect, of what causes the regurgitation. Lovely example, huh? Let's try another. You feel sleepy, and then you go to sleep. Does feeling sleepy (mental state) cause you to go to sleep (physical event)? No. Something physical going on in your body first causes you to feel sleepy and then causes you to go to sleep. For a final example, consider what happens when you touch a very hot object. It hurts and you jerk your hand back. We ordinarily assume that we jerk our hand back because it hurts, but physiologists have discovered that when we touch a very hot object, the nerve impulses from our fingers do not go to the brain before sending signals back to the hand. The intense heat causes the nerve impulse to turn around in the spine and trigger a rapid reflex response – causing us to jerk away our hand *before* we feel pain! Hence, the pain (mental event) couldn't have caused us to jerk our hand away, since we didn't feel it until after the process of jerking our hand away had already begun. Hence, that feeling of pain is epiphenomenal relative to our behavior.

There is, then, no need to appeal to the mental realm in order to explain human behavior. Moreover, were we to attempt to explain behavior by means of mental phenomena, our explanation would cease to be scientific because we would be appealing to phenomena which are not scientifically accessible. Scientific explanations must proceed in terms of things that are publicly accessible and subject to public confirmation. Mental phenomena are not public; they cannot be seen or manipulated, so they cannot be employed in genuine scientific explanations (though, of course, mental and spiritual phenomena often turn up in pseudo-scientific explanations of behavior). Because scientific explanation is the only kind of explanation worthy of serious attention, we should relegate mentalistic explanations of behavior to the museum of yesterday, not the laboratory of today.

Feelings, desires, beliefs, intentions, memories and other mental events, states, and processes should, then, be left out of our understanding and explanation of human behavior. To be sure, this is not the ordinary point of view regarding human behavior, but that is irrelevant. We once misunderstood the relation between the sun and the earth; we got it backwards. We thought the sun went around the earth. Science helped us straighten out our understanding of the relation of the earth and the sun; now it is helping us straighten out our understanding of human behavior.

Here are three additional implications of epiphenomenalism. First, it rules out not only mind-to-body causation but also mind-to-mind causation. Every mental event, state, and process is the side effect of a bodily event, state, or process. This means that there are no mental phenomena left over that could be caused by other mental phenomena. Second, it also means that mental phenomena cannot occur or exist separately from the body, though the body can exist separately from mental phenomena. That is what happens at death, when the physical processes that generate mental phenomena quit functioning, or, metaphorically, when the motor that generates the light breaks down.

Third, epiphenomenalism also implies there is no such *thing* as a mind. There are only specific mental events, states, and processes that come and go – like the individual explosions that make up an extended fireworks display, perhaps creating the illusion that there is a waving flag in the night sky. The fact that it seems like you have an enduring mind is an illusion created by the rapidity with which mental events are generated by the body. This illusion is analogous to the illusion that is caused by a movie projector. It appears to viewers that there are enduring objects sitting still or moving around the movie screen, but that appearance is an illusion created by still shots flashing by so rapidly that they deceive our eyes. Similarly, according to epiphenomenalism our bodies generate mental events so rapidly that we are deceived into thinking that we have an enduring mind, or that our self is a single, enduring thing. But those beliefs are based on illusion just as would be the belief that an object on a movie screen is a single enduring thing.

In brief, though epiphenomenalism is dualistic in affirming that mental phenomena are not physical, it differs from the classic dualism of Socrates and Descartes (1) by denying that mental events can affect the body, (2) by denying that mental phenomena can exist separately from a functioning body, and (3) by denying that there is such a thing as the mind. According to epiphenomenalism, then, the body doesn't *affect* the mind (there is no mind to affect); nor does it *generate* a mind; rather, it generates a flow of mental phenomena. The only legitimate use for the noun "mind" is as shorthand for the flow of mental phenomena from our brains – somewhat as we use "movie" to stand for thousands of still frame shots flowing rapidly past a light that is projecting through a magnifying lens onto a screen. For efficiency in communication it is okay, of course, to use words like "mind" and "movie"; we just need to be careful not to be misled by those simple nouns into misunderstanding what a mind and a movie really are.

Criticisms

Epiphenomenalism has had more supporters than have occasionalism and parallelism, including 20th century supporters like Keith Campbell, but like them it has not enjoyed widespread support. Let's look at some reasons why. First, some critics say that epiphenomenalism's explanation of human behavior is so counterintuitive that it seems absurd, that is, highly implausible. The claim of epiphenomenalism that mental events, states, and processes do not influence our behavior implies that what we think and feel (mental state) when we write (physical process) a love poem has nothing to do with what we say in the poem or how we say it! It implies that the solution we write at the end of a difficult math problem is never the product of what we thought as we worked through the problem! Such conclusions seem obviously wrong – or at least seriously problematic.

The epiphenomenalist might retort, "But sometimes what initially seems absurd becomes accepted as true, such as heliocentrism." The critic cannot deny that, but he or she can give a response similar to one we encountered in our evaluation of determinism. Specifically, we have a very deep conviction that our feelings, thoughts, and beliefs make a difference in our behavior. To be sure, it is conceivable that we are wrong, but why should we believe we are wrong and throw out that conviction for the sake of a theory, epiphenomenalism, that has no more support than it does? Further, wouldn't acceptance of epiphenomenalism lead to a worldview that we simply couldn't live by – a worldview according to which we could never believe that anyone did anything for us because of the way they felt about us, or because of what they believed about us, or because of what they desired for us?

The critic might concede that epiphenomenalism is a clever and somewhat plausible way of dealing with the problems of interactionism; its examples of nausea, sleepiness, and reflex responses should make us more cautious in our claims about the causes of human behavior. But those examples do not even come close to showing that *no* human action is caused by what we feel or believe or desire. Moreover, there are other theories that are at least as plausible as epiphenomenalism yet do not require that we deny what we cannot seriously deny, namely, that our feelings, desires, and beliefs make a difference in our behavior.

As a second criticism, consider that epiphenomenalism is incompatible with libertarianism. (However, it is the perfect philosophical anthropology to underpin *hard* determinism.) If mental events do not influence physical events, then we cannot by free choice make our bodies do anything. If epiphenomenalism is true, then our bodies are entirely beyond the control of our wills. Moreover, if mental events are entirely the product of bodily processes, then, because bodily processes are determined according to the laws of nature, it follows that every human choice is impersonally determined by the laws of nature. Hence, epiphenomenalism requires that we reject as illusory our feelings of freedom of choice and of control over our behavior, but our sense of subjective freedom is far too powerful and important to reject for the sake of a weakly supported theory, so we should stick with our intuition, say libertarians.

A third criticism is less a matter of intuition and more a matter of logic. Epiphenomenalism is correct that it is a mystery as to how mental events could cause physical events, but it is equally mysterious how physical events could cause mental events! How can something that has physical mass and spatial location generate something that has no physical mass or spatial location? Causality in the one direction (physical to mental) is no more intelligible than it is in the other direction (mental to physical). Hence, epiphenomenalism has a mystery in its own explanation of human behavior. Therefore it has no right to reject DI for having a mystery in its explanation of human behavior. Epiphenomenalism is in the same boat with DI. If DI goes down, so does epiphenomenalism. What we obviously need to do is abandon the dualist ship and leave those enigmas behind. They are all caused by mind-body dualism, so let's swim over to some positions that reject dualism altogether.

Physical Monism

The last criticism of epiphenomenalism was that if epiphenomenalists are going to hold that mental events are not physical, yet physical events can cause mental events, then they cannot rule out the possibility that mental events can cause physical events. That is, if they allow that a physical thing can cause a non-physical thing, then it is arbitrary for them to deny that a non-physical thing can cause a physical thing. To affirm causality in the one direction (body to mind) leaves open the door for causality in the other direction (mind to body). Hence, the epiphenomenalist has not closed the door to dualistic interactionism and all of its enigmas. The only way to escape those enigmas is to reject dualism altogether. That is exactly what physical monism does.

A *monist* is someone who believes that something consists of only one kind of stuff or thing, rather than two, as the dualist believes. The word "physical" in the name of the physical monist tells us that he or she believes that the one kind of stuff out of which a human is made is matter. The physical monist argues that a human being is his or her body, nothing more, nothing less. Critics ask, "How, then, did so many people come by the mistaken impression that a human being consists of *two* kinds of thing, a soul and a body?" The physical monist answers, "By means of the fallacy of reification."

To "reify" something is to think that it is an independently existing thing when it is not a thing but is only a property of a thing. (*Res* in Latin means "thing," so to reify something is to "thingify" it.) Beauty, for example, is a property of some things, but beauty is not a thing that exists on its own apart from physical objects. Similarly, wisdom is a characteristic of some people, but it is not a thing that exists separately from people. Unfortunately, many people from Socrates forward have made the mistake of thinking that beauty and wisdom exist *apart* from things, as well as in them.

This is a common and understandable mistake among people who speak European languages. These languages, such as English, are dominated by nouns; most of these nouns refer to things. For the word "dog" there are dogs; for the word "tree" there are trees; for the word "stone" there are stones; and so on. As a result of this feature of our language, whenever we come across a noun, we expect there to be a *thing* to which the noun refers. Hence, when we see or hear the word "mind," we expect it to refer to a thing – just as the word "body" refers to a thing. But whereas the word "body" does refer to a thing, the word "mind" does not; by presuming it does, we commit *the fallacy of reification*; that is, we treat a noun as though it refers to a thing when it does not.

Here is a charming, real-life example of the fallacy of reification. A missionary named Patricia St. John was serving as a relief worker in a poor Muslim town. She started a Sunday evening fellowship for young children, and she always began the meetings by serving cookies and Coca-Cola, which were rare treats for those children. One evening David, a boy about five years old, asked her, "Could we have some fellowship, just you and me alone?" St. John said, "Of course," and made an appointment to meet with him in his home. On the day of the appointment she arrived in David's room with a children's Bible and nothing more. The boy stared at her in a puzzled way. "Where is the fellowship?" he asked. After a period of bafflement, St. John realized the misunderstanding that had occurred. She excused herself, ran to the nearest store for a bottle of Coke, and then spent a happy but totally secular half hour with David. For David, she reported, "Fellowship was Coke – neither more nor less."[5]

In our terminology, David had reified the noun "fellowship." He mistook it for a thing rather than for a kind of human interaction. Many people have made a similar mistake with the noun "mind." But if "mind" does not refer to a mind, then to what does it refer? It is obviously not a meaningless word or it would not have been used by so many people for so many centuries. The response of the physical monist is that all of our mental terms, including "mind," refer to either (1) physical events in the brain or (2) behavior. Let me explain more fully and provide a name for each of these 'planks' in the platform of physical monism.

The first plank is "the identity thesis." The identity thesis holds that mental events, such as feelings of pain or pleasure, are not non-physical. Every mental event is a physical event in the brain. Note: even though the physical monist believes that a human being is a purely physical entity, the physical monist does not deny the existence of mental events. To do that would be absurd. But the most plausible hypothesis, according to the physical monist, is that mental events, like everything else we know of, are physical. (The identity thesis covers, of course, not only *mental events*, but also *mental states*, such as euphoria and depression, and *mental processes*, such as reasoning and imagining.)

By various means, such as phantom limb experiences, we can show that mental events take place in the brain, not in some other part of the body, such as our legs or fingers. In a phantom limb experience, someone who has lost her left leg may feel as though she has an itch in her left foot; someone who has lost his right arm may seem to feel an ache in his right hand. But in neither case, obviously, is the pain located where it seems to be. We *learn* to associate pain and other feelings with the various parts and areas of our bodies. It is even more obvious that intellectual states and processes exist in the brain because in spite of where the neighborhood bully said your brains are located, you can lose your big toe, or some other part of your body, yet still remember people's names and do arithmetic in your head. Hence, mental events, states, and processes are physical events, states, and processes that take place in the brain.

An implication of the identity thesis is that a certain electro-chemical event in the brain is *identical* with remembering the face of a friend, and remembering the face of a friend is identical with a certain electro-chemical event in the brain. Hence, there is no need to explain mental events in terms of a non-physical mind, as does the dualist, and there is no need to think that mental events are non-physical, as does the epiphenomenalist. Mental events, states, and processes can be understood as physical occurrences in the brain.

In this way we avoid the enigmas of dualism and honor our commonsense conviction that interactionism is true. How is this possible? Because now we see, for example, that a desire to eat fresh cherries is a brain state that can be caused by a preceding brain state (perception of a bin of fresh cherries in a grocery store), and that this desire (which is a brain state) can then cause a subsequent brain state (a decision to purchase some cherries) that then causes appropriate actions (picking out cherries and taking them to the checkout counter). But note:

all so-called 'mind-body' and 'body-mind' interactions are physical interactions between the brain and other parts of the body (such as our senses, nervous system, glands, and muscles). Hence, mind-body and body-mind interactions are physical-physical interactions (in addition to which there are, of course, environment-body interactions and body-environment interactions). Physical monism can even make sense of mind-mind interactions: they involve different parts of the brain (such as the part having to do with memory and the part having to do with desire) interacting with one another. Hence, physical monism, unlike epiphenomenalism, honors our intuition that our feelings, beliefs, and desires play an important role in determining our actions, and yet it avoids the enigmas of dualistic interactionism.

The second plank in the platform of physical monism is "linguistic behaviorism." This plank is a response to the dualist's argument that so many people could not have talked about the mind for so many centuries if there were not such a thing. The point of linguistic behaviorism is that statements that ascribe a mind or a soul to a person, whether directly or by implication, are to be understood as referring to *behavior*.

For example, we sometimes say of a person, "She has a good mind." What do we mean by that? We do not mean that she has a good non-physical thing, called "a mind," stuck in her brain. We mean that she learns things quickly or solves problems faster than most people or scores higher on tests or some such thing – all of which have to do with how she *behaves* in response to intellectual challenges. Similarly, if we say of a blues singer, "He really has soul," we don't mean that he has a certain kind of non-physical thing inside or attached to his body. We mean that he performs very effectively that behavior which we call "singing the blues"; his singing causes us to feel a certain way and perhaps to want to dance. If we say that someone "has a good soul," probably we mean that she behaves in an altruistic way; if someone needs help, she gives it readily and generously. If we say that one person is "broad-minded" and another is "narrow minded" – well, you can see the problem with taking such language literally! Such talk is really talk about how those people behave in response to new ideas or ideas they disagree with.

As I mentioned earlier, our noun dominated language tricks us into thinking that talk about souls and minds and spirits is talk about things. It's like the time that a professor told his foreign visitor – who knew little about American life or the English language – that he wanted to show her the college's excellent chemistry laboratory. They walked to the science building and entered a large room where the professor pointed out piece after piece of impressive scientific equipment. "Well, how do you like it?" the professor asked. "Very much," replied the guest. Then she added, "But when are you going to show me the excellent laboratory?" This response by the guest indicated that she had made what Gilbert

Ryle, in his book *The Concept of Mind*, calls "*a category mistake.*" The foreign visitor thought the word "laboratory" referred to a thing, like a microscope, rather than to a collection of things jointly used to carry out scientific research. The professor was first puzzled, then amused, by his guest's question. When he explained the mistake to his guest, she was sufficiently unselfconscious that she chuckled at her error.

The professor then said he was eager to show his visitor the college football team. "Those young people," the professor said, "have a great team spirit." So he took his guest to the football field, pointed out the quarterback, the center, the tackles, and so on, explaining each position as he went. Meanwhile, the team was playing a rousingly successful game against its opponent. When the professor finally finished his explanation, he asked his guest, "Well, what do you think of them?" The guest replied, "I am quite impressed, but where's the great team spirit?" Again the professor was puzzled – until he realized that his guest was expecting that the great team spirit was a thing – like another player or a mascot! Again, Gilbert Ryle would say that the visitor had made a category mistake in her thinking about the noun "team spirit." She thought it belonged in the category of words that refer to an individual, like a player or a mascot, whereas instead it referred to a property of the team.

How about a final illustration? This time the scene is a hospital room where an autopsy surgeon stands next to the bed of an old friend who is on the verge of death. An intern is present to help the surgeon. The surgeon says aloud as she gazes at her old friend, "I'm really going to miss this guy. He has a big heart and a brilliant mind." Within minutes the patient dies. The surgeon and intern wheel him to the autopsy room where the surgeon removes her old friend's heart, holds it up for the intern to see, and says, "I told you he had a big heart. It was enlarged by a childhood disease." The intern replies, "You were certainly correct. Now, please, show me his brilliant mind!" A physical monist would love the reply by the intern. It involves a delicious equivocation on "big heart" and an illuminating contrast between "heart" in its literal sense and "mind" in its literal sense – thereby bringing out nicely the fact that words such as "mind" and "soul" are not ordinary nouns. If they were, what would be meant by, "If one more thing goes wrong, I'll lose my mind" or "He'd sell his soul for a promotion"?

Criticisms

Physical monism has now had the floor long enough. Let's see what its critics have to say. We will begin with criticisms of the criticisms that were made of the dualistic conception of the mind.

First, recall the claim of some critics that the very idea of a non-physical mind doesn't make sense. Dualists respond that there is no serious problem concerning the essence of the mind. We know what a mind is. Perhaps we cannot define it with the precision we would like, but that is true of many of our most important concepts, such as love, justice, and beauty. We do not therefore conclude that there is no such thing as love or justice or beauty! Further, it is clear that a human mind is a thing which experiences, remembers, desires, recognizes, imagines, understands, believes, questions, reasons, hopes, decides, and wills. We have no trouble understanding what each of those things is, so we should have no trouble understanding the concept of a thing (the mind) that has the capacity of engaging in all of those activities.

Second, there is no problem with regard to the existence of the mind. René Descartes pointed out that each of us is essentially a thinking thing. I am a thinking thing, and so are you. It may be an illusion that you have a body, but it cannot be an illusion that you are a thinking thing – and that is what a mind is. Hence, each of us knows that at least one mind exists – our own! Further, whenever you think you exist, you cannot be wrong. Why? Because in order to think that you exist, you must exist!

Descartes' famous formulation of this point is, in Latin, "*Cogito ergo sum*," which literally means, "I think, therefore I am"; that is, whenever I think I exist, I must exist, because only something that exists can think. Therefore, whenever you think you exist, you are correct, and in thinking that you exist, you are warranted in thinking that a thinking thing, that is, a mind, exists. Hence, you can know the existence of at least one mind, namely, your own, whenever you want to.

To be sure, David Hume was correct that you cannot discover the existence of your mind by extrospection or introspection. But that should be no surprise. The mind is always the looker, never that which is looked at; it is always the *subject* that is looking. It is never the *object* of direct awareness; it always transcends the object it is considering. We become aware of the mind's existence by realizing that it must be there in order to carry on and unify the various activities that go on in our non-public, subjective life: thinking, desiring, remembering, recognizing, deciding, imagining, dreaming, willing, etc. It is, then, by means of our awareness of the subjective dimension of our lives that we realize the necessity of something which provides the unity and continuity of that aspect of our lives. This is sometimes called "*the transcendental argument for the existence of the mind.*"

If someone were to refuse to take the transcendental argument seriously because we cannot see the mind directly, that would be as unwarranted as denying that gravity exists because we cannot see it, or as denying that other people have a mental life because we cannot see their feelings, thoughts, and dreams. To be sure, we cannot see gravity or the dreams of other people, but it is reasonable to infer their existence from what we do see and experience. Similarly, from what we experience it is also reasonable to infer the existence of minds.

Third, let's look at linguistic behaviorism. It is true that we judge the intelligence of people by either their behavior or the products of their behavior, such as inventions. We have no alternative. The public side of people, their appearance and behavior, is the only side to which we have direct access. But when we conclude that someone is intelligent, we do not merely mean she behaves in certain ways. We think she behaves in those ways because her body is under the control of her mind; her mind is what is intelligent. Intelligence, then, is not intelligent behavior; intelligent behavior is the expression of an intelligent mind. Hence, the behaviorist account of intelligence is inadequate. We need the concept of mind in addition to behavior in order to say what we mean when we say that a person is intelligent.

The physical monist might respond, "Not so! Computers behave intelligently, but we don't believe that they have minds." That is true, the critic agrees, if by "intelligent behavior" we merely mean "behavior that is effective at achieving some goal." But we have no reason to think that computers behave intelligently because they have minds; whereas in our own case we are aware that it is our thinking that brings about the solution of problems, and which creates and programs problem-solving devices like computers. Hence, the ultimate reality behind the 'intelligent' behavior of computers and human bodies is the mind. Bodies and computers are tools by means of which the mind communicates and implements its solutions to problems.

Fourth, having critiqued linguistic behaviorism, now let's look at the identity thesis. The physical monist might try to counter the importance of the preceding criticisms by saying, "Yes. Of course we think, remember, desire, etc., in a unified way, but it is the brain that carries out and unifies these activities. So the existence and functioning of the mind is just the existence and functioning of the brain, which is physical. Hence, mental events, states, and processes are identical with physical events, states, and processes." The critic disagrees for the following reason.

In order for one thing to be identical with another thing in the philosophical sense of identity, they must have all the same properties – and therefore not be two things, but one thing. For example, the inventor of bifocals and the first postmaster general of the USA were identical because they were both Ben Franklin. Every property that the inventor of bifocals had, the first postmaster general also had because they were both Ben Franklin. By contrast, identical twins are not identical in the philosophical sense because although they are identical in some properties, such as genetic structure and physical appearance, they are not identical in all properties, for example, spatial location and time of birth.

Because the preceding concept is what "identity" means in metaphysics, it follows that if mental events were identical with physical events, then because physical events have spatial properties, such as size, shape, and location, therefore mental events would also have those properties, but they do not. We can see all of the sizes, shapes, and locations of the various parts of the brain, but no one has ever seen the location of the people in someone else's dream, or the size of someone's thought about space, or the shape of the feeling of happiness. It doesn't even make sense to think that dreams and thoughts and states of mind have such properties. Therefore, the identity thesis is mistaken. Mental events are not identical with physical events. Hence, mental events cannot be physical events in the brain.

Fifth, recall the physical monist's argument that in spite of the fact that a human body is completely replaced every seven years or so, a person's identity and continuity over a lifetime can be explained satisfactorily in terms of the continuous development of that person's body over time. Critics of physical monism point out that the public, physical continuity of the body is not the same as personal continuity; personal continuity is based on such things as feelings, memories, beliefs, and values, all of which are private and non-physical. Consequently, our personal identity and continuity must depend not on our bodies but on something non-physical, such as a mind or soul.

That seems true, it is argued, because if you woke up one morning in someone else's body, you would know you were still you rather than the person in whose body you were – though you would, of course, have difficulty convincing other people of that (as would the other person if he/she woke up in your body!). Why would each of you be the same person in spite of having a different body? Because of your sense of self, your memories, your abilities, your values, your plans and aspirations.

To be sure, perhaps this kind of body exchange has never happened; but that is irrelevant. The mere conceivability of this possibility shows that your self is not identical with your body. By contrast, note that it would not even make sense to speak of you keeping the same body but waking up with someone else's mind. A different mind makes a different person. A different body does not. Hence, the mind, not the body, is the basis of our personal identity and continuity.

A physical monist might respond by pointing out that it is not only our bodies that change; our minds change, too. Our feelings about things change, our beliefs change, our values change, our intentions change, and so on. Hence, the mind, like the body, is *not* identical over time; therefore, after seven years we are *not* identical to our earlier self, physically or mentally.

The response of the dualist to this point is that it involves a misunderstanding of the mind. The body is *constituted* of its parts (bones, muscles, hair, organs, etc.). There is nothing more to the body than those parts in organic relations to one another. But the soul is *not* constituted of its feelings, beliefs, desires, etc. Rather, it *has* them, somewhat like a basket has contents: the contents can change while the basket remains the same; or, to use another analogy, the mind is like a lump of clay that has different shapes at different times but remains the same lump of clay no matter what shape it is sculpted into; or, for a final analogy, some people have spoken of the mind as like a chalkboard on which a succession of things can be written, the later ones replacing the earlier ones – though the chalkboard remains the same through it all.

The point of these analogies is that the body of a living human *is* its parts in organic relations to one another and nothing more; our body is not something in addition to our arms, legs, torso, organs, etc., in functional relations with one another. The mind, however, is something over and above the succession of feelings, thoughts, beliefs, desires, intentions, etc., that it has. Each mind or soul is numerically distinct from every other, thereby making each of us different from every other person at any given time, giving each of us our unique identity over time, in spite of bodily and mental changes. In analogy to the changing shape that a single lump of clay takes on over time, the *self* is the changing 'shape' that a soul takes on over time because of its experiences and decisions. The soul at the heart of the self possesses strict identity over time rather than the pseudo identity of natural objects. At every later moment of its existence, the soul is identical to itself at every earlier moment of its existence. To use the basket analogy, the contents change, but the container does not – that is, the self changes, but the unique identity of the soul does not.

Another criticism of DI, you will recall, is that it is incomprehensible how a mental event could cause a physical event, since mental events have no spatial location from which to operate upon physical things, and no physical mass whereby to impact them. According to interactionists, that is an irrelevant criticism because it is based on an outdated conception of causation – which leads to a *sixth* criticism of physical monism.

Analyses of the concept of causation from David Hume onward have shown that we do not even know why *physical* events cause physical events! More generally, we do not know why anything causes anything else. All we know is *that* when some things occur, other things follow regularly. For example, place a piece of blue litmus paper in acid and it turns pink; place a piece of pink litmus paper in a basic solution, and it turns blue. We do not know why this comes about; all we know is that when we do the one thing, the other follows. Why? We'll never know. There is no absolute or conceptual necessity in the laws of nature. That's just the way they are – at least for now, in so far as we have experienced them. Maybe they were different in the past; maybe they will change in the future; maybe even now they are different in some other part of the universe.

If in our experience acid had always turned pink litmus paper blue whereas a base turned blue litmus paper pink, that would be normal to us. We would just say, "That's the way it is. Period." But we shouldn't add, "That's the way it *has to be*." Science doesn't discover *why* things are as they are, or how they have to be. Science discovers and reports how things are so far as we have experienced them and insofar as we can make reasonable projections about them from our experiences. Those are exceedingly important things to do so that, as long as the laws of nature don't change, we know what to expect from the world and from our actions upon it. However, science has no right to say or ability to determine that it is *impossible* for a certain thing to cause another thing, or that it is *necessary* that a certain thing cause another.

Hence, when we say that A causes B, we should not presume that we know *why* A causes B; all we have a right to report is that we have discovered a dependable relationship of succession between A and B – that whenever A occurs, B follows. Scientific reports of causal laws are, then, statistical statements telling the ratio of times when B has followed A. However, such a ratio does not tell us *why* B follows A, *nor* does it tell us that B *will always* follow A or that it *must* follow A.

This means that the contemporary understanding of causation does not require that a cause and its effect be of the same metaphysical type (for example, both physical or both mental). It requires only that between two events there be that type of dependable relation that we call a causal relation. Consequently, the contemporary understanding of causality does not rule out the possibility that mental events cause physical events. To use David Hume's terminology, if there is a "constant conjunction" or a "regularity of succession" between mental events and physical events, then it is just as correct to call mental events causes of physical events as it is to call physical events causes of physical events.

There obviously is such a "constant conjunction" and "regularity of succession" between mental events and physical events. Nothing is more common in our experience. Every waking day of our lives, those of us who are fortunate enough to be whole and well observe that when we will to get out of bed, our bodies get out of bed, and when we will to say to someone, "That's a good point," our vocal chords, tongue, and lips cooperate to make the appropriate sounds. Hence, we have plenty of good evidence for saying not only that physical events cause physical events, but also for saying that mental events cause physical events, physical events cause mental events, and mental events cause mental events – even though physical events are non-mental and mental events are non-physical.

Seventh, note that physical monism entails that your sense of subjective freedom is an illusion. Why? Because if, as physical monists believe, the behavior of matter is completely determined by the laws of nature, and a human is nothing but a material entity, then your behavior must be completely determined by the laws of nature. But as we saw earlier, determinism is an unproved theory. Now we have seen that physical monism is, too. Consequently, we are not required by reason to give up our deep intuition of subjective freedom for either of these theories.

Were we to give up this intuition simply because it is unproved, then we should also (in order to be logically consistent) give up all of our other unproved intuitions. But one of our unproved intuitions is that matter exists! Hence, for a physical monist to argue that we should give up our belief in subjective freedom and accept determinism and physical monism because our belief in subjective freedom is merely intuitive and has not been proved would have the ironic result that the physical monist should therefore reject physical monism because it, too, assumes the existence of something that has not been proved: matter. But it would be better, the critic argues, to stick with our unprovable intuitions than to reject them and wind up with a bizarre world view in which there is no matter, no freedom, and no other minds.

Eighth, and last, a new voice picks up on the preceding point about matter and says that whereas there are, as we have just seen, no serious problems with the essence and existence of a non-physical mind, there are serious problems with both the essence and the existence of matter. We have been assuming all along that we know what matter is. But what is it? The physical monist says, "It is that which exists in space." But what is space? Is it something that exists outside of our minds and independently of them? Or is space something that exists within the mind and depends on the mind for its existence? If the latter is true, then, obviously, matter cannot exist independently of the mind, since matter exists only in space, and space exists only in the mind. According to the next position, psychic monism, that is the truth about matter, space, and mind. Hence, if psychic monism is true, then matter is not what most of us have been socialized to think it is.

Psychic Monism

According to psychic monism, developed brilliantly by the Irish philosopher George Berkeley, there is no such thing as matter – at least not in the sense in which most of us think of it. Most of us think matter is something that exists independently of us and, by stimulating our senses, causes us to have perceptions of physical objects of different sizes, colors, shapes, and so on. But let's think about that. We ordinarily identify as a physical object any object that combines the sensations of visibility (can be seen) and tangibility (can be touched). But seeing something is an experience, and so is touching it. It turns out, then, that a physical object is simply part of an experience. But an experience is something that exists in the mind. Therefore, so do physical objects.

Moreover, all we ever know directly are our experiences. But since, according to the physical monist, matter is not an experience or part of an experience but, rather, is the supposed cause of our experience of physical objects, it follows that we are never directly aware of matter. From this, it follows that we have *constructed* (made up) the idea of matter on the *assumption* that there is something lying behind and causing our experiences of shape, color, texture, weight, and so on. But there is no necessity in thinking that it is matter that causes our experiences of physical objects; there are other and better ways to explain our experience of physical objects, as we will see when we study idealism in chapter 19.

Consequently, we should think of a human being not as a combination of a mind and a body, nor as just a body, but rather as just a mind. Hence, to the question, "What is a human being?," the psychic monist answers, "A mind plus its experiences and volitions – nothing more and nothing less." Its experiences are what happens to it – such as seeing the moon or hearing a rooster crow; its volitions are what it wills or does – such as taking a walk or refusing to lie. Now let's look at two reasons for holding this position. The thrust of the first point is that we do not need the concept of matter; the thrust of the second point is that we could never know that matter existed, even if it did.

As to the first point, consider the fact that we do not believe that the objects in our dreams are constituted of matter. Rather, we believe that the people, animals, and other things in our dreams are pure experiences in the sense that when we wake up from a dream, we believe that the figures in the dream no longer exist in any form. To come at the point a bit differently, we believe that the things in a dream depend for their existence upon our dreaming them, so that when we wake up, they cease to exist. (If in a dream you see a horse grazing in a pasture, when you wake up you don't think that horse is still in the pasture – right?) Hence, physical objects in dreams do not exist independently of our minds, and they are not made of matter, yet they can be as real as any object in waking experience.

You may now be thinking, "But objects in our waking experiences are *real* whereas objects in our dreams are not real." Consider, however, how we distinguish between objects in our dreams and objects in our waking experiences. Think, for example, about a book that you read in a dream as distinguished from the book you are reading right now. By what criteria do we conclude that one object is or was a real object, whereas another object is or was only a dream object? If we do a phenomenological examination of objects in waking experience and objects in dreaming experience, we note a number of differences. First, dream objects often do not behave in the predictable ways in which waking objects behave. Pet a real cat and it remains a real cat; pet a dream cat and it may turn into a bird or a puddle of butter.

Second, dream objects do not stay put like real objects do. In waking experience, your body remains among the same objects, for example, trees, flowers, buildings, rocks, until you walk away, drive away, etc., in some steady fashion, be it swift or slow. In a dream, however, you can be among certain objects at one moment, then suddenly, without any idea of how it happened, find yourself among completely different objects.

Third, in waking experience we find that if we want to return to a location, we can usually do so, whereas in a dream that is almost always impossible. Fourth, dream objects typically last for a shorter period of time than do waking objects – if you meet a person in waking experience, you might strike up a friendship that lasts for years, whereas any relationship you strike up with a person in a dream ordinarily lasts no longer than that dream.

Fifth, dreaming and waking also differ in the amount of control that we have over our bodies and other things – though sometimes we have more control in a dream, sometimes more when we are awake. In waking, for example, we cannot fly by flapping our arms, but sometimes in a dream we can. In dreaming, by contrast, sometimes we cannot make our legs move very fast when we are trying to escape from danger, whereas when we are awake we ordinarily have no trouble running at whatever our usual top speed is.

Notice that in the preceding attempt to identify the differences between objects in waking experience and objects in dreaming experience, the word "matter" was not used and the concept of matter was not needed. Why? Because we do not, in fact, use the concept of matter to distinguish between objects in our waking experience and objects in our dream experience. It is only *after* we identify an object as "real" that we say that it is composed of matter (if we say or think that at all – we usually don't). Rather, we identify an object as real or merely a dream object on the basis of the different *experiential* characteristics that we associate with dreaming and waking – and experiential characteristics all exist in the mind. In brief, we do not use and do not need the concept of matter in order to distinguish between dream objects and waking objects. The same analysis can be applied to the distinctions we make between "real experience," on the one hand, and "hallucinations" and "illusions," on the other.

We are not born knowing the differences between real objects and dream objects. Consider how a child *learns* to distinguish between real objects and dream objects. It does not discover by itself and is not shown by adults that waking objects are made of matter whereas dream objects are not. We don't begin to believe that waking objects are constituted of matter because we find matter in them. We *impose* the concept of matter on waking objects because we believe that concept explains why objects in waking experience behave differently than objects in dreams. Hence, matter is not something that we discover in experience. It is an explanatory concept that we have made up, and as we will see in the section on worldviews, metaphysical idealists believe there are better ways to explain the differences between real objects and dream objects than by employing the concept of matter.

As further proof of the uselessness of the concept of matter, consider the following problem. Suppose that your life went as follows: you dream twelve consecutive hours a day and are awake twelve consecutive hours a day; you begin each dream by dreaming that you are waking up, and you end each dream by dreaming that you are going to sleep, and what happens in each of your dreams is just as normal in every respect as what happens in your waking experience.

Could you tell which half of the day you were dreaming and which half you were awake? The psychic monist says "No." There would be no difference between the experiences. But if it were by means of *matter* that we discover the distinction between dreaming and waking, we should be able to make that distinction even in this extreme case, because if matter causes waking experiences, it would be there when we are awake. But when we are trying to decide whether something really happened or was a dream, we cannot just look to see whether matter caused the one experience but not the other. We are limited to examining the different characteristics of the experiences that we call dreaming and waking, and then we must decide in terms of those characteristics whether what happened occurred when we were awake or asleep. Matter, then, is irrelevant to the distinction we make between objects that we experience when awake and objects that we experience when dreaming or hallucinating. In brief, matter doesn't matter. Experience matters.

The second major point of the psychic monist is that in addition to the fact that the concept of matter is useless, there is no way to prove the existence of matter. Obviously, we never see or perceive matter directly. What we see are people and dogs, trees and stones, lakes and clouds, but not matter. If we look through a telescope or microscope, we see stars, planets, meteors, and moons, viruses, bacteria, crystal structures, etc., but not matter. What we experience by perception, then, are always clusters of physical properties, such as the cluster of color, shape, size, weight, and texture that constitute a medium size red apple.

Matter is said to be that which impinges on our senses and causes us to see apples and stones, bacteria and viruses, but that means that matter (*if* it exists) is always behind our experiences causing them, but is never directly evident in them. Since matter is always behind or outside of our experiences (if it exists at all), it follows that because we can never get behind or outside of our experiences, we can never find out whether it is matter that is causing our experiences.

Why can't we get behind our experiences to see what is causing them? Because every time we try to do so, all we wind up with is another experience. If, for example, you were to try to prove to someone that an apple is made of matter, you could cut it in half, but when you did and looked inside, you wouldn't find matter. You would just begin to have different experiences from those you had before; instead of having the experience of seeing a rounded, red, dry surface, you would now be seeing a flat, white, moist surface. Similarly, if you put a piece of the apple under a microscope, you would just have a new experience. And so it goes with every attempt to get behind experience to find out what is there. All we wind up with are further experiences. The truth to be had from all this is that all we ever know to exist are minds and their experiences and activities. Hence, we should unclutter our conception of what a human is by getting rid of the notion that a human is or has a body made of matter.

Because in the west we are reared to believe in the existence of matter, most of us find the recommendation to reject belief in the existence of matter to be weird at best and absurd at worst. Samuel Johnson, a famous eighteenth century English author, was incensed by what he considered to be the absurdity of Berkeley's position. To Johnson it was *obvious* that matter exists. It is reported that as he walked with his friend James Boswell, talking about this ridiculous argument by his contemporary, George Berkeley, Johnson approached a large stone and kicked it with his foot, saying, "Berkeley, I refute you thus!"

Johnson believed that kicking his foot against the stone, feeling its hardness and its resistance to his foot, perhaps feeling pain as a result, proved that the stone was made of matter. But did it? How could it? Johnson didn't kick the stone open and reveal matter inside it. He didn't roll the stone over and reveal matter underneath it. All he did was engage in a series of experiences that he also could have had in a dream – and surely he would have agreed that a stone in his dream was not constituted of matter. Indeed, what a delicious irony it would have been if Johnson had had his experience of refuting Berkeley by kicking the stone, and then had woken up to realize that he had dreamt the whole thing and not kicked a "real" stone at all!

In conclusion, it should be noted that although psychic monists recommend that we remove the noun "matter" and the adjective "material" from our *metaphysical* vocabulary, they do not recommend that we throw out the adjective "physical."[6] A physical object, as distinguished from a non-physical object (such as the number five or the concept of infinity) is any object that has physical properties, such as shape, color, texture, or spatial location. Obviously such objects exist in dreams and hallucinations, as well as in waking experience, and we all agree that the physical objects in dreams and hallucinations are not made of matter. This means that physical objects are not necessarily made of matter, and since we can never know that the physical objects in *waking* experience are made of matter (for the reasons given earlier), we should not believe that they are. To be sure, we know that *physical objects* exist; we have immediate awareness of them. But we have no good reason for thinking that *material objects* exist, that is, objects made of matter, so we should not believe in them.

What has gone awry is that people have taken what was once a perfectly harmless word, "matter" – which once stood innocently for the stuff of which something is made, such as wood or clay – and then reified it. Their mistake was in thinking that the stuff of which things are made must exist outside of and independently of the mind, but that is no more true in waking life than it is in dreams. As we have just seen, there is no need for such a stuff and no way to prove it exists. Hence, anyone who believes in the existence of matter must do so on the basis of sheer blind faith. It would be more rational to formulate our conception of humans without employing the concept of matter, and then there is nothing for a human to be but a mind plus its experiences and activities. Hence, contrary to what the physical monist claims, it is not the mind that has been reified; we know there are minds; it is matter that has been reified.

Criticisms

We have already noted that many people, because of the ways in which they were reared, have *emotional* difficulty taking psychic monism seriously. But are there any good *reasons* for rejecting it? Let's consider three that have been proposed. First, there is the issue of causality. If matter does not exist, then all physical events must be explained in terms of mental causation, but that hardly seems convincing. The position of a psychic monist must be that when I hit a coconut with a hammer, it is not the case in waking life that the hammer breaks open the coconut any more than it is the case in a dream that a hammer breaks open a coconut. According to psychic monism a hammer in waking life has no more material content than a hammer in a dream – and the same goes for the coconut. Hence, if psychic monism is true, then all physical causation in waking experience is of the same illusory nature as is the causation in dream experiences. In both cases, it just seems like the hammer breaks the coconut, or that your hand causes the hammer to slam against the coconut. But that certainly doesn't seem to be the way things are.

Once again we encounter a philosophical theory that is in gross violation of a commonsense conviction – and contrary to what the psychic monist says, this conviction that in waking life hammers really do cause objects to split open is widespread in cultures around the world, and not merely in the west. To be sure, the psychic monist is correct that we cannot prove that matter exists by just looking to see if it is there, but assuming that it exists makes more sense out of the way our waking experience feels to us than does denying its existence – and what good is a theory that does not fit with the way our experiences feel to us?

Second, there is the fact of *the commonality of perceptions*. When you and I and dozens of other people go to the same movie together, we see the same film. That we see the same film is obvious from the fact that we laugh at most of the same places, get choked up at most of the same places, and afterwards can talk together about different parts of the story. But if psychic monism is true, then there is no matter, so the movie theater, the screen, and the image on the screen did not exist outside our minds. But if the movie was not external to us, if it was not out there staring into our faces, so to say, simultaneously causing each of us to have the same perceptions, then how do we account for the fact that we had so many perceptions in common?

A third criticism of psychic monism has to do with our lack of direct access to other people's minds. If there were no matter, so that people were just minds, then in our social interactions we should be able to know directly what is going on in another person's mind because, according to psychic monism, there is no material body standing between us and the contents of that person's mind. However, the only mind we know the contents of directly is our own. That we do not know directly what goes on in another person's mind is evidence that there is something between us and the other person's mind which prevents us from knowing the contents of their mind directly. The most plausible explanation of such inability on our part is that the other person has a body that stands between us and their mind. That explains why we can gain knowledge of the contents of another person's mind only by listening to what they tell us (speaking is a form of bodily behavior) or by observing such things as the posture, complexion, and non-verbal behavior of that which stands between us and their mind – namely, their body.[7]

Neutral Monism

If we reject dualistic interactionism, occasionalism, parallelism, epiphenomenalism, physical monism, and psychic monism, are there any other theories of mind/body relations worth considering? There are at least two more: phenomenalism and neutral monism.

The *neutral monist*, such as Benedict Spinoza, agrees with the physical monist that there is no such *thing* as the mind, and agrees with the psychic monist that there is no such *thing* as the body. The neutral monist says we should not reify mind or body. Neutral monists agree that there are mental phenomena, such as thoughts, memories, and desires, and that we experience physical properties, such as shape, size, and color. However, they do not believe that mental phenomena are dependent on physical phenomena or that physical phenomena are dependent on mental phenomena. They are convinced by both kinds of monism that it would be impossible for a non-material mind and a material body to causally affect one another.

They believe that the physical and the mental are different aspects or expressions of something else on which they both depend. This third thing or "unknown substance" – which is called "neutral" because it is neither mind nor matter nor anything else that we understand – lies beneath all of our experiences, physical and mental, and gives rise to them. Hence, instead of it being the case that a human is one thing with two parts (a mind *and* a body) or just one thing (a mind *or* a body), a human is one thing with two aspects; that is, a human is constituted of an unknown substance that has a physical aspect and a mental aspect (and so this theory is called "the dual aspect theory" as well as "neutral monism").

The basic justification for neutral monism is that, as we have noted before, we cannot get behind our experiences to discover what is causing them. Hence, the neutral monist argues, we should humbly admit our ignorance. We have only two means to direct knowledge of real objects: extrospection and introspection. But we no more find a mind by introspecting than we find matter by extrospecting. The stunning truth is that we simply do not know where our physical and mental experiences come from.

Indeed, who knows what he or she will think next, much less five minutes from now? Often we start out to think about one thing, wind up thinking about another, and have no idea when or why we switched from the one topic to the other! It is as though there is something beyond us that thinks through us; it is as though our thinking is the expression of something beyond ourselves. To be sure, sometimes before we speak we think up what we want to say and then say it later, but it is impossible to think up ahead of time what we want to think for the first time. Where do our thoughts come from? We don't know. They just bubble up – from whence we know not. Ultimately the same thing is true of our perceptual experiences. Experiences happen. Why? Ultimately we do not know.

In brief, our experiences, both physical and mental, must come from somewhere, but we know not where or why, so the most honest position for us to take is that they come from "something we know not what." Because *something* must be causing our experiences, we know *that* it exists, but we cannot know what it is because we cannot get behind our experiences to apprehend it directly. Hence, a human being cannot know the nature of the substance of which he/she is constituted; a person can know him or herself only in terms of the experiences and actions to which the stuff of reality, whatever it is, gives rise.

Consider a person who can hear and smell, but who has been totally blind and paralyzed from birth and is incapable of tactile sensation. Now place that person in a canoe for her first trip into the Florida Everglades. No doubt the sounds of the birds, frogs, cougars, alligators, etc., and smells of the swamp would be fascinating to her, but consider how little she would ever understand of the swamp that was causing her experiences. We know even less than that about the ultimate reality that has given rise to us and all of our experiences.

Immanuel Kant made a useful and famous distinction here. He distinguished between a *phenomenon* (that which appears to us) and a *noumenon* (that which causes the appearance). We can know a thing only as a *phenomenon*, that is, only by how it appears to us; we cannot know what a thing is like apart from perception, as it is in itself, as a *noumenon*. The neutral monist is saying that you can know your self only at the phenomenal level, not at the noumenal level. You know even less of other people because whereas you know your self through both physical and mental phenomena, you know other people only through the physical phenomena associated with their bodies. (The plural of phenomenon is phenomena, and of noumenon is noumena.) Here, then, we have a theory that avoids the problems of the other theories and that humbly recognizes both the limitations of human knowledge and the impenetrable mystery at the core of human life.

Criticisms

Neutral monism certainly seems correct that the human mind strains at its limits when it deals with metaphysical problems, and it is admirably humble in refusing to claim that it has knowledge about the relations of the mind and the body. As a theory, though, it is not without its own problems. First let's look at a criticism by dualists. Neutral monists say that the physical and the mental are two aspects of whatever a human is. Hence, the physical and the mental are like two views of the same object through two different senses, for example, sound and smell. But this seems impossible because how could one and the same thing be both a mental thing and a physical thing? It is obvious that one and the same thing could have two different physical characteristics simultaneously – for example, a toy could be red *and* spherical. It is also obvious that one and the same thing could have two different mental characteristics simultaneously; for example, a mind could be both remembering something and feeling happy. But to say that one and the same thing could have both physical and mental characteristics simultaneously would be to say that it would be both physical and non-physical, both mental and non-mental, and that would be self-contradictory. To say that something is physical in nature is to say that it is not mental; to say that something is mental in nature is to say that it is not physical. Hence, in saying that a human is one mysterious thing which is both mental and physical, neutral monism falls into self-contradiction and therefore cannot be true.[8]

Second, neutral monists have a problem with causality. Remember that according to neutral monism, there is no such *thing* as the mind or the body. There is only an unknown substance that gives rise to mental and physical experiences. From this it follows that physical events do not cause physical events (since there is no body to cause anything), mental events do not cause mental events (since there is no mind to cause anything), and, obviously, mental events cannot cause physical events or mental events cause physical events since there are no minds and bodies to interact. As we have seen, such claims are startlingly counterintuitive. They imply that physical events and mental events merely accompany one another in various combinations, but never cause one another. Rather, all, events of any type are to be traced back to a mysterious, unknowable substance.

Perhaps this theory is true. But it certainly doesn't seem like an unknown substance causes me to decide to try to score on my handball opponent by angling the ball out of his reach on his left side. *I* make that decision because I want to score a point, I note that my opponent is a bit off center to the right of the court, and I understand that if I strike the ball well, it will be difficult for him to move to the left fast enough to intercept it. Nor does it seem as though an unknown substance, rather than my hand, slings the ball against the front wall to the left of my opponent. Nor when I dig my fingers into the sidewall, trying to scrape off my opponent's return shot (he was faster than I thought!), does it seem as though an unknown substance, rather than the wall, causes my pain. In brief, concerning causation, neutral monism is too counterintuitive to accept without much stronger evidence on its behalf. According to the next criticism, the very nature of neutral monism seems to rule out the possibility that such evidence could be acquired.

Third, neutral monism is unsatisfactory as a theory. How much value can there be in a theory that throws up its hands in the face of a difficult problem and says, "Well, we'll just avoid controversy here by saying that what we are looking at is 'something we know not what' "? The purpose of constructing a theory is to propose something clear and specific so that we can see how well it holds up under scrutiny. We learn more from a clear, specific theory that can be shown to be mistaken than from a theory that is so vague and general that it cannot be evaluated. Hence, neutral monism as a theory is not of much value because there is no way for us to evaluate its central claim that there is an unknown substance which is neither mental nor physical but which is responsible for all that happens in the world.

As we saw earlier, a good theory is not always true, but if it is not true, it is at least fruitful; that is, it at least helps us get closer to the truth by being clear enough that we can discover where it is wrong, and by exciting within us new insights to pursue. But neutral monism, because of its recourse to an unknown substance, is not fruitful and cannot be. It is a dead-end street that we'd better bypass if we want to develop a satisfactory philosophical anthropology.

Fourth, a new and radical voice, that of the phenomenalist, breaks into the conversation and asks, "How do we know that there is any substance at all, known or unknown, at the core of each human?" Perhaps there is no substance behind our experiences, no noumena behind the phenomena. Perhaps experiences just happen and are the only things that exist, so that a human is not a body, or a mind, or a combination of them, or an unknown substance. Perhaps a human is only a series of experiences, a stream of feelings, sensations, memories, desires, decisions, intentions, thoughts, beliefs, and nothing more. Perhaps, as David Hume put it, a human is "nothing but a bundle or collection of different perceptions, which succeed each other with an inconceivable rapidity, and are in a perpetual flux and movement."[9] Perhaps the psychic monist is correct that we have no body made of matter, but is wrong that the mind is more than its experiences.

Phenomenalism

The position at the heart of the preceding criticism is that of *phenomenalism*. It is associated especially with David Hume, A. J. Ayer, and Buddhism. It holds that a person is a mind and nothing more, but the mind is not an enduringly identical thing. Rather, the mind is somewhat like a river or a fireworks display; it is a flow of experiences and nothing more.

We will not examine phenomenalism more fully until chapter 19, but with mention of it we have covered most of the mind/body positions that have had strong advocates. Each of those positions except phenomenalism has been found to have significant weaknesses as well as strengths; were we to pause to critique phenomenalism, we would find that it, too, has weaknesses as well as strengths. Most western philosophers today believe that physical monism or dualistic interactionism is the most adequate (or least inadequate) theory of what a human is; many eastern and some western philosophers think psychic monism is the more adequate theory; furthermore, epiphenomenalism and neutral monism are not without intelligent supporters. Clearly, our rational dialogue about what we are needs to continue.

Notes

1. "Anthropos" is the Greek word for "humankind." "Anthropology" literally means "the study of humankind." Cultural anthropology (which is what we usually think of when we hear the word "anthropology") consists primarily of the empirical study of humans in their native settings, so there are studies by cultural anthropologists of people in Java, Borneo, the rain forests of South America, and the highrises of Manhattan. Whereas cultural anthropology focuses on distinct groups of people and tries to capture what their lives are like and what makes them distinct from other groups, the philosophical study of humans is a search for what is true of all humans. Philosophical anthropologists want to know: What makes a human to be a person? What are the situations that all humans face? What are the experiences that all humans have? Is there a meaning to human life? Obviously, when we do philosophical anthropology it is important to be familiar with and reflect upon the findings of cultural anthropologists.
2. You may be puzzled by the fact that I have switched from talking about minds to talking about souls, and there is a third term, "spirit," that I may as well toss into this can of worms. How are mind, soul, and spirit related to one another? There are no standard answers to that question, so here is what I propose while we are within the perspective of dualistic interactionism. Let's use "soul" to refer to the non-physical part of a human ("body" refers to the physical part). Let's use "mind" to refer to the soul in so far as we are concerned with the soul's cognitive powers, such as memory, imagination, reason, and decision-making. Let's use "spirit" to refer to the soul in so far as we are concerned with its affective capacities, which have to do with emotions, music, art, religion, and values.
3. See Descartes' Sixth Meditation in his *Meditations on First Philosophy*.
4. Descartes' belief in dualism came out poignantly in a statement he made when he knew he was fatally ill. A friend at his deathbed overheard him say, speaking to himself, "My soul, you have been a captive for a long time; now the hour has come when you must leave your prison, this body; you must bear this separation with joy and courage." See Jack Rochford Vrooman, *René Descartes* (NY: G.P. Putnam's Sons, 1970), p. 246.
5. I got this story second-hand. It was said to come from *Ordinary Woman, Extraordinary Faith* by Patricia St. John (Wheaton, IL: Harold Shaw Pubs.: 1996).
6. The psychic monist agrees that there are *figurative* uses of "matter" and "material" that we may use without metaphysical mistake or confusion. For example, sometimes we use "matter" as a synonym for "subject," "topic," or "issue," as in the sentence, "The matter at hand will require more time than we can give it today." Sometimes we use "matter" as synonymous with "make an important difference," as in, "Does it matter?" "Material" is sometimes used to mean "relevant," as in, "That comment is not material to the issue being discussed."
7. I would like to thank a student, Dan Smolen, for this criticism of psychic monism.
8. A self-contradictory statement is a statement that has two parts. One part says or implies that something is true, whereas the other part says or implies that that very

same thing is false. Hence, one part of it contradicts the other part, so it contradicts itself. Consider the statement, "My father and I are the same height, but I am taller than he is." It is impossible that at one and the same time I can be both taller than my father and the same height as he is. By the law of excluded middle, I must be either taller than my father or not taller than my father. By the law of non contradiction, I cannot be both. By the law of identity, which ever it is, that is it – period! Hence, a self contradictory statement cannot be true because it is impossible that contradictory statements can be true about one and the same thing at the same time. To put the point a bit differently, a true statement is completely true whereas a self-contradictory statement is part false and part true. But no true statement is partly false. Therefore, no self contradictory statement is true.

9 David Hume, *A Treatise of Human Nature*, Book I, Part IV, Section VI.

Reading Further

Plato, *Phaedo*; *Alcibiades I* (sections 128–131).
René Descartes, *Meditations*.
Richard Taylor, *Metaphysics* (4th edition or later), chs. 2, 3, and 4.

Chapter 17: Philosophical Theism

- Terminology: polytheism, deism, theism, pantheism, atheism, agnosticism
  Philosophical Worldviews
- Hard and soft beliefs
- Religious Theism and Philosophical Theism
- How can we tell what God is like?
  The observational approach
  The definitional approach
  The inductive approach
- Four arguments for belief in the existence of God
  The principle of sufficient reason
  The cosmological argument: Thomas Aquinas
  The teleological argument: Plato; William Paley
  The argument from religious experience
    Two kinds: Intensive and Suffusive
  Arguments from authority
    The masses
    The intelligentsia

Terminology: polytheism, deism, theism, pantheism, atheism, agnosticism

Theism is a member of a family of positions that express diverse beliefs about divine beings, that is, about supernatural beings or forces of great power and intelligence. These positions include "polytheism," "deism," "theism," "pantheism," "atheism," and "agnosticism." As with most such clusters of positions or concepts, a good understanding of any one of them requires an understanding of how it relates to and differs from the others, so let's consider them together.

A *polytheist* believes that there are two or more divine beings. The ancient Greeks, for example, believed that there were many gods and goddesses, including Hera, Hermes, Zeus, Aphrodite, Athena, Poseidon, Hades, Dionysus, and others, each distinguished by some special power, characteristic, or responsibility.

> ## Philosophical Worldviews
>
> Since beginning our exploration of metaphysics we have focused on relatively narrow issues, for example, "Is human behavior free or determined? How are the human mind and body related?" Now we shall do what we earlier called "speculative philosophy." We shall widen our vision as far as possible and ask what the whole of reality is like.
>
> A person's belief as to the kinds of things that make up reality, and as to how the universe got to be the way it is and where it is going, is in philosophy called "a worldview." This expression can be misleading because in ordinary language "world" usually refers to the planet Earth or to the physical universe. In metaphysics, however, "world" usually refers to the whole of reality, inclusive of whatever exists, including God, angels, abstract numbers, and non-physical dimensions of reality – if any of these is real. Hence, a "worldview" is not a theory that is merely about the Earth or even the physical universe. It is a theory about the whole of reality.
>
> For this reason the central questions of speculative metaphysics are, "What is the nature and content of reality? Where has it come from? Where is it going?" More specifically, "Is reality made up of God, human souls, and material objects? Or only human souls and material objects? Or only matter? Or only God and human souls? Or only human souls? Or only God and matter? Or only God? Or what?" Worldviews are answers to these and related questions.
>
> In this part of our journey we will examine three of the most fundamental and influential worldviews: theism, materialism, and idealism. These worldviews are fundamental in the sense that virtually every worldview that goes by another name will prove upon examination to be a version of one of these three views. Hence, knowledge of these three theories will enable you to better understand and evaluate the specific worldviews that you encounter through reading, conversation, travel, and the media. Now we turn our attention to theism.

The ancient Romans continued the polytheistic tradition of the Greeks.

Some polytheists are dualists; they believe there are two divine beings, no more, no less. These divine beings are thought of as opposites to one another: light vs. darkness, good vs. evil, creativity vs. destructiveness, force vs. gentleness, order vs. chaos, etc. Religions that are dualistic in nature include Taoism (whose yin-yang symbol expresses exquisitely this dualistic view of reality and history), and Zoroastrianism (according to which history is best understood as an ongoing conflict between Ahura Mazda and Ahriman – two divine beings each of which is trying to win each human over to their side and to guide history in their preferred direction; many Zoroastrians, but not all, believe that Ahura Mazda, the force of goodness, will ultimately prevail over Ahriman).

The term "dualism" can be confusing in philosophy because it is used to refer to more than one kind of position. There is mind/body dualism and there is religious dualism, but just because a person is a dualist of the one kind, it does not follow that he or she is a dualist of the other kind. Many mind/body dualists are not religious dualists. For example, Muslims believe that every human has a soul as well as a body, so they are mind/body dualists. However, in saying "There is no God but Allah" they vigorously reject any suggestion that there is a multiplicity or even a duality of divine beings.

A *deist* believes there is one and only one God, who is infinite in knowledge, power, and goodness. This God exists, has always existed, and always will exist. Ze[1] created and sustains the universe, equipped us with intelligence by which to understand how Nature works, and equipped us with an ability to enjoy the good and beautiful things of Nature,[2] society, the body, the mind, and the spirit. God has also given us a moral sense of what we ought to do and ought not to do. When God brings history as we know it to a close, it will not matter whether a person belonged to "the right religion" or any religion at all. Rather, each person will be judged and rewarded or punished according to how morally or immorally he or she has lived.

According to deists, although God created and sustains the world and gave us all of the gifts and capacities mentioned above, ze does not intervene in the world. Between creation and the end of the world, God simply observes the world. Ze does not answer prayers, cause miracles or religious experiences, or make special revelations to humans, *and never has*. Understandably, critics of this position sometimes call it "the absentee landlord" conception of God – though God is not really absent anymore than a parent is absent who watches over a child from a distance but refuses to intervene when the child gets into some difficulty which the parent wants the child to get out of by itself. This view of God flourished among intellectuals in eighteenth-century Europe and America, and is common today among people who are not affiliated with a religious institution but who believe, nonetheless, that there is a God.

A classical *theist* is someone who believes in the kind of God that the deist believes in, but who also believes, contrary to deism, that God has been and is involved with the world – answering prayers, causing miracles, making zer presence felt in religious experiences, revealing zer will to people like Moses or

Muhammad. Christian and Hindu theists believe also that God has personally entered into human history by taking bodily form, as in Jesus or Krishna.

A *pantheist* rejects the position of deism and theism that once upon a time the world did not exist and then God created it. "Pantheism" literally means "all God"; this is an appropriate name for the position of pantheism because according to pantheism God is all there is, and all there is is God. God and Reality (including the physical universe) are one and the same thing. Therefore, since God is the world, and since nothing can bring itself into existence, God did not bring the world into existence. Because God has always existed, and the world is identical with God, therefore the world (the universe) has always existed. (At the risk of inflicting mental whiplash on you I must add that according to some versions of pantheism the world is an illusion, so it has never existed, except as an illusion. For now we will focus on those versions of pantheism according to which the world is real and is identical with God.)

According to this kind of pantheism, God cannot exist apart from a world, since God is the world – nor can the world exist apart from God, since the world is God. The world at any given moment is the form that God is taking at that moment. To be sure, God can exist apart from the specific individuals which constitute the world at any given moment, but when those individuals perish, they will and must be succeeded by others – just as a lump of clay can exist apart from the shape in which it is currently molded but cannot exist apart from shape altogether; in losing one shape, it acquires another. Hence, particular individuals come and go, but God/the World continues forever.

The word *"atheist"* is used most commonly in our culture to refer to a person who believes that there are no supernatural intelligent beings (not even one), so the atheist rejects polytheism, deism, theism, and pantheism. However, there is a potentially confusing overlap between pantheism and atheism. Atheists, like pantheists, believe that the world is all there is to reality, but pantheism is not just atheism by another name. Pantheists think of the world as a unified, intelligent, purposive being that has goals and is pursuing them intentionally. By contrast, atheists reject that analogy and believe that (1) the physical world is all there is to reality, (2) intelligence and purpose are rare, emergent properties of a few things in Nature – not properties which characterize Nature in general, and (3) the basic processes of Nature proceed without purpose or intention. The laws of gravity and electromagnetism, for example, have no purpose or intention. They grind along blindly and, together with the other blind forces of Nature, sometimes accidentally produce living beings.

Further, pantheists feel that Nature as a whole is sacred and worthy of reverence. Contemporary atheists do not typically have those kinds of feelings toward the world – though of course they may feel awe in the presence of its enormity, love its beauty, and feel astonishment in the presence of its complexities and surprises.

An *agnostic* is someone who says, "I am not confident as to whether there is a God or not – maybe there is and maybe there isn't." Note that agnosticism is not a metaphysical position; it does not make a claim as to whether God exists or not. It is an epistemological position; it consists of the belief that one's evidence is inadequate for drawing a conclusion about the existence of God.

Hard and Soft Beliefs

Now let's distinguish two types of agnosticism: hard and soft. The hard agnostic says, "I am not confident as to whether there is a God or not, and I believe that humans will never have decisive arguments for or against the existence of God." There are different reasons why one might become a hard agnostic; usually those reasons have to do with the limits of the human mind and the difficulty of the question of the existence of God. The soft agnostic, by contrast, is not convinced that it is *impossible* for humans to ever develop a good argument that God does or does not exist. Too many times in history people have thought something was impossible which later was achieved. The soft agnostic is someone who says, "Right now I am not confident whether there is a God or not, but someday something might convince me one way or the other."

If we apply the hard/soft distinction to atheism and theism, we discover additional subtle but important differences in the attitudes that people can take (and do take) toward their religious beliefs. A soft theist is someone who says, "At present it seems to me that the evidence available favors the position that there probably is a God." A hard theist says, "It seems to me that the evidence available proves conclusively that there is a God." The difference here turns on the distinction between proof and probability.

That which is probable is that which the evidence indicates to be more likely to be true than false, but probability always falls short of certainty. Something may have an exceedingly high probability, such as that I will live for another 60 seconds, but the fact that that possibility is only a probability means that it is not certain and that, therefore, I may not live for another 60 seconds! Many an otherwise healthy person has been felled in mid-sentence by an unsuspected brain aneurysm, an unexpected heart attack, a criminal attack, or an accident 'out of nowhere' (for example, the falling airplane part that plunges into a private home). Similarly, that which is exceedingly improbable may nonetheless occur; for example, I may swim 100 meters on my 100th birthday. (By contrast, if something is necessary or impossible, then it is not probable; it is certain.)

The soft theist, then, believes that there probably is a God but acknowledges that new evidence or a better analysis of the evidence than ze currently has might persuade mer that there is not a God. (Just for this chapter, to give you a feel for how they work, I will use "ze," "zer," and "mer," in reference to humans, as well as God.) The hard theist, by contrast, believes that the evidence at which ze has looked constitutes a proof that there is a God. To be sure, a mature hard theist knows that people can make mistakes in their reasoning, and so is willing to reconsider what ze takes to be a proof, but until the hard theist becomes convinced that that argument is not a proof, ze will continue to believe that it is conclusive and therefore shows that there definitely is a God (and not merely *probably* is).

The hard/soft distinction applies with equal illumination to atheists. A soft atheist is someone who believes that the evidence that ze has examined favors the position that there probably is no God. A hard atheist believes that the evidence ze has examined shows that there definitely is no God.

The preceding distinctions can help us understand better what our own current position is, what other people's positions are, what the range of alternatives is that is available to us, and what questions we need to ask when we are not sure what we or someone else believes.

Religious Theism and Philosophical Theism

The most influential world view in western history for the last 1500 years has been classical theism. At the center of theism is the concept of God. That concept has taken different forms in the religions of different peoples. Hence, it is important to keep in mind that although the abstract concept of God in philosophical theism is at the heart of religions such as Judaism, Christianity, and Islam, these religions add to the abstract, philosophical concept of God whatever more they believe has been *revealed* to them by God but cannot be discovered by reason.

Orthodox Jews, for example, believe that one day God will send a messiah to overthrow evil and institute God's kingdom; Orthodox Christians believe that God is three persons in one Godhead; Muslims believe that the Koran is God's greatest and final revelation to humankind. These beliefs are held by these different groups on the basis of what they consider to be special revelations from God; they are not beliefs that were arrived at by means of the procedures of reason – nor could they have been.

Because we are exploring philosophy, we will focus on how the concept of God has developed within western philosophy, rather than on how it has developed within communities based on belief in divine revelation. Keep in mind, however, that some of the most important philosophical work on the concept of God has been done by people of deep religious faith. When they work philosophically on the concept of God, however, they do not appeal to revelation. Now let's begin our philosophical analysis.

How can we tell what is God like?

First we should note that in recent philosophy the concept of God has been of interest largely in relation to the question as to whether God exists. Clearly, before we can attempt to answer the question, "Is there a God?," we must have a reasonably clear understanding of what is meant by "God"; without such an understanding we could not get started with the project of looking for God, or for evidence relevant to the question of the existence of God. If, for example, I asked you to tell me whether there is a tachistoscope within a 100 yard radius from where you are now, before you could begin to try to find the answer, you would probably have to ask me what such a gizmo is. Without such knowledge, you could stare a tachistoscope right in the face and yet not know that you were looking at one!

Similarly, we need to determine what is meant by "God" before we can look for and evaluate evidence pertaining to the existence of God. But how can we go about finding out what God, or anything, for that matter, is like? Here are three basic ways of establishing what a thing is like: observation, induction, and definition. The first way, "observation," consists of finding out what a thing is like by examining the thing itself. If, for example, I take you to see a tachistoscope, tell you that it is a tachistoscope, and show you how it works, then you will develop a conception of what a tachistoscope is and be able henceforth to answer the question, "Is there a tachistoscope in this room?"

Note that in using observation to find out what a thing is like, first you encounter the thing itself; from that encounter, you construct your idea of what the thing is like. Can we use that approach to discover the nature of God? It seems not. The word "God" is not used by theists to refer to a part of the physical world or to the physical world as a whole, so we cannot find out what God is like by observation, by "taking a look," so to say.

Fortunately, there is a second way of approaching the question of the nature of a thing, namely, by constructing a definition of it; then we can look to see whether there is adequate evidence to think that such a thing exists. For example, we can designate the term "chiliagon" to stand for a polygon of exactly 1000 sides, and then we can go looking to see whether a chiliagon exists. Or, for another example, if some of the first British who settled Australia decided, in playful inebriation, to construct the idea of the weirdest creature they could think of, agreed to call it a "weirdwon," and said that it would have the body of an otter, the tail of a beaver, a bill like a duck, a venomous stinger like a stingray, would lay eggs, and yet be a mammal, they would indeed have conjured up a very unlikely creature! Yet, if they had struck out to see whether they could find a weirdwon in Australia, then, given enough time, they would have discovered that such a thing does exist – today we call it a duckbill platypus! If on one of their less imaginative days these wags had conjured up only a unicorn, it seems that their search of Australia, and, indeed, of the world, would have led to the conclusion that there is no such thing – even though on the basis of general background evidence, such as the existence of horses and rhinoceroses, the existence of a unicorn seems much more likely than that of a platypus!

Now what about God? As we have seen, if we begin with the assumption that God is a supernatural being, then we cannot use the observation approach to tell what God is like, since we can observe only physical beings; to go at this point from the other end, if we use "God" to mean something that we can observe, then God is going to have to be a physical entity – but that does not fit with what theists in particular or people in general mean by "God," so let's try the second approach.

The *definitional* approach to the nature of God says that we must begin by spelling out what we mean by "God" – just as we did with "chiliagon," "weirdwon," and "unicorn" – so that then we can think about what kinds of evidence would be relevant to deciding whether there is or is not a God. Perhaps the two most widely used definitions of God in western philosophy are those formulated by St. Anselm and René Descartes. Anselm recommended that we use "God" to mean a being than which none greater can be conceived. This means, by definition, that nothing can be superior to or even equal to God in any respect. Descartes agreed with this and added that such a being can be thought of also as the supremely perfect being. This means, by definition, that God possesses every quality that it is good to possess and possesses each of those qualities in a perfect way to a perfect extent. The concept of God is, then, when defined in this way, the concept of a being that is perfect in all ways in which a single being can be perfect.

Note that God is perfect as a being, and not merely as a certain kind of being. A specific tomato can be a perfect tomato, that is, perfect of its kind, but it is not perfect as a being because it lacks good things such as knowledge and indestructibility. Therefore, the notion of God is the notion of a being who is perfect in power, in knowledge, in goodness, and in all additional ways that would be necessary in order for a being to be supremely perfect.

These necessary attributes of God – perfect knowledge, perfect power, etc. – can be inferred also from Anselm's conception of God as a being than which no greater being can be conceived. This concept implies that whenever we are thinking of a being that is less than omnipotent or omniscient, etc., we are not thinking of God – though of course we might mistakenly think we are. If we believe we are thinking of a being with more power than God, then we simply are not thinking of God in the first place. Why? Because, by definition, we cannot think of a being that has more power than God. We can think of a being that is more powerful than another being only when that other being is less than omnipotent. Therefore, when we think of a being that is less than omnipotent, we are not thinking of God – just as when we think of a polygon that has fewer than 1,000 sides, we are not thinking of a chiliagon. The same reasoning pertains to God's knowledge, goodness, duration, etc.[3]

If you ask, "But how do we *know* that God is perfectly good and powerful?," it may be that you have not yet understood the second approach adequately. According to that approach you *know* that God is perfectly good and powerful because that is the way we defined "God" (just as we know that a chiliagon has 1000 sides because that is the way we defined the word "chiliagon"). In many cases, however, that question ("How do we *know* that God is like that?") suggests that a person is following a third approach to understanding the nature of God. This third approach, *the inductive approach*, is kin to the first approach because it is based on observation, rather than definition, but it is based on indirect observation plus inference, rather than on direct observation and description. In the absence of an opportunity to observe directly what a thing or person is like, we can gather evidence connected to that person or thing and then infer from the evidence what that thing or person is like. When we do that we are reasoning from effects to their cause. For example, let's say you have a garden that has been damaged. From the kind of damage done to your garden you might infer that it was caused by an animal rather than a human, and by a deer rather than a rabbit. From letters by a pen pal whom you have never met or seen pictures of, you might infer that you are corresponding with someone who is male rather than female, of average height, musically gifted, prone to exaggeration, etc.

Note that the inductive approach proceeds on the assumption that the object of inquiry exists; the aim of the inquiry is to figure out more exactly *what* the object of inquiry is like, not whether it exists. *Something* must have caused the damage to the garden; *someone* must be writing those pen pal letters to you. Similarly, the inductive approach applied to the question of the nature of God usually proceeds on the assumption that God exists; it is just trying to get clear on what God is like. This approach is often taken when people use "God" to mean "the creator of the world." Those who take this approach think it is obvious that *something* had to create the world. They call that something "God." The interesting question to them is not *whether* there is a God; it is *what* God is like (in addition to being the creator). Hence, they look at the way the world is and from it try to infer more fully what the creator is like – just as you looked at the damage to your garden and from it inferred more fully what the damager was like.

This inductive approach to discovering the nature of God by reasoning from the way the world is to the way its creator is can lead to interesting results. For example, we might conclude because of all the evil in the world that God, the creator, is very smart and very powerful but not very good (or ze would not have allowed human history to be so horrible). Or we might conclude that God is perfectly good but is not all powerful (or ze would not have allowed human history to be so horrible).

The results of the inductive approach to establishing the nature of God can be even more radical. To see this, let's return to your garden. We might have to conclude from our examination of your garden that nothing damaged it – that what happened was just a natural process. For example, I once reported to my wife that something had seriously damaged the Lamb's Ear in her herb garden. She asked what the damage looked like. I told her it looked like it had been scalded – like a deer had peed on it or a human had thrown something caustic on it. She explained that nothing had damaged the Lamb's Ear. That's just what Lamb's Ear does after awhile. She later showed this phenomenon to me in someone else's garden. To an amateur it sure *looked* like something had scalded the Lamb's Ear, but I had to give up my belief. There was no damager, and therefore no damager whose identity and nature I could discover by clever detective work. Similarly, we may start out to find the nature of God by examining God's creation but then conclude from our examination of the world that it probably had no creator and just runs on its own – just as I had to conclude that nothing damaged the Lamb's Ear. (It is also possible, of course, that someone who does not believe in God might examine the universe and conclude that there probably is a God.)

Four arguments for belief in the existence of God

Now that we have examined various ways of establishing what we mean by "God," let's take the definitional approach (which is the approach that is most common in philosophy) and turn to the question of the *existence* of a supremely perfect being who is creator of the universe. Keep in mind as we go along that most arguments for the existence of God are based on *the principle of sufficient reason*. According to this principle there is an explanation satisfactory to reason for everything that exists and everything that occurs (although, of course, we do not always know what that explanation is). The general strategy of most arguments for the existence of God is to show that the best explanation for the existence or occurrence of certain things is that God exists.

The first argument we will consider is called "the cosmological argument," though it is really a family of arguments rather than a single argument. The most famous set of these arguments was produced by Thomas Aquinas. What distinguishes cosmological arguments from other arguments for the existence of God is that they appeal to the mere existence of the world or to the existence of some simple thing in the world. Let's look at an example of each, beginning with the existence of the world as evidence for the existence of God.

It is conceivable that there could have not been a world, that is, that absolutely nothing might have existed. Why, then, is there something rather than nothing? Why is there a world rather than none? Nothing that we know of in the world exists necessarily – which means that everything in the world could have not existed. Hence, the existence of the world is not self-explanatory. To explain its existence we must go beyond the world to a force capable of bringing the world into existence, and that is what God is – a being who has the knowledge and power to create a universe. Hence, the very existence of the world is evidence for the existence of God. In brief, unless we can come up with a better explanation of the existence of the world, then in order to have a satisfactory explanation of the existence of the world we must assume the existence of God.

A critic might reply, "That doesn't help because now we have to ask who created God, and then we have to ask who created the creator of God, and so on *ad infinitum*, so we should just stop with the world." A theist could reply that the critic has not understood the concept of God adequately. A supremely perfect being is one which would exist by its very nature and not depend on anything else for its existence; anything which depends on something else for its existence is not a supremely perfect being. Consequently, when we properly understand God, we understand that the explanation of God's existence is internal to God's own nature, whereas the explanation of the existence of the world is not. Hence, we do not have to go beyond God to have an explanation of God's existence whereas we do have to go beyond the world in order to have an explanation of its existence.

A second cosmological argument states that the existence of motion in the world is evidence that God exists. Why? Because things in the world are not self-moving. One and all they depend on other things to get them going. Billiard balls on a level table sit motionless unless they are struck or the table is moved; seeds sit dormant unless they are stimulated by moisture and light; animals die unless energized by food; etc. But, since the whole world consists of nothing but things that depend on other things to get them moving, and nothing in the universe is self-moving, we must ask, "Why are things in motion? Why aren't all things suspended motionless in space?"

The most plausible explanation, according to the theist, is that there is a First Mover which got things moving in the first place and is an Unmoved Mover, so that its injection of motion into the universe does not have to be explained in terms of something else beyond it (otherwise we would not yet have an explanation as to how motion got started in the first place). The concept of God *is* the concept of a being that is capable of injecting motion into the universe yet does not have to have motion injected into it. Therefore, the existence of motion in the universe is evidence for the existence of God. In contemporary terms, the existence and decree of God explain the Big Bang that launched the universe into existence and onto its course.

The second type of argument at which we will look is called "the teleological argument." It is also called "the argument from design." As with the cosmological argument, there is a family of teleological arguments. What distinguishes them from other arguments for belief in the existence of God is that they focus on the orderliness of the parts and processes of the world and on what seems to be the purposiveness of many things in Nature. Let's start with an argument from the orderliness of the world.

Plato, in Book Ten of his *Laws*, notes that orderliness in human affairs is generally a sign of intelligence at work. The things to which we intentionally apply ourselves usually take on an orderly form, structure, or arrangement. By contrast, the things we neglect fall into disarray. For example, the amorphous shape of a large block of wood gradually takes on a distinct and meaningful shape under the hands of an able sculptress. But what is happening on the floor as the sculptress works? A disorderly mess of wood chips and sawdust is piling up, and tools are strewn about randomly. The difference between what is happening to the block of wood and what is happening on the floor results from the fact that the sculptress is focusing her attention on the block of wood and ignoring what is happening on the floor. What happens by accident, what happens unintentionally (such as the debris falling on the floor around the sculpture) is usually disorderly.

Now think about the world. What is it like? It is a highly orderly affair. The relations of the sun and the planets to one another are highly regular; our seasons change in an orderly fashion; plants germinate and develop in an orderly fashion. Indeed, because the things that make up the universe are arranged and change in such orderly ways, science is possible – that is, it is possible to discover dependable regularities in the universe. Such widespread and pervasive orderliness makes it much more plausible that the universe is a result of intelligent, intentional design than of mindless, unintentional processes. Hence, the orderliness of Nature is good evidence for the existence of an extraordinarily intelligent and powerful being who designed it. Thus, it is good evidence for the existence of God.

William Paley was sympathetic to Plato's point and added an observation from his own era, the mid eighteenth century. As a result of the rapid progress of science and technology in the seventeenth century, machines in the eighteenth century took a quantum leap forward in complexity. Consider, for example, a wind-up watch. It is typically constituted of several different kinds of stuff, such as brass, steel, and glass. Moreover, these different kinds of stuff are formed into many different shapes and sizes for many different purposes. Furthermore, these many different parts – such as gears, a stem, a spring, numbers, hands, a case, and a cover – are coordinated into precise relations with one another so that the hands will turn at a constant speed and thereby enable a person to use the movement of the hands to measure the passage of time.

Now, says Paley, note that if we were out for a walk in the woods and by accident discovered a watch, or some other comparably complex device, such as a camera, we would not for a moment believe that that device had been created by accident. We might believe that it had been lost or left behind by accident, but we would not believe it had been *created* accidentally by the mindless forces of nature rubbing various materials against one another over time – not even for thousands of years! Why would we refuse to believe that such a device had been created by accident? Because of the way that its many different, precisely connected parts enable us to achieve some end, such as keeping track of time (in the case of a watch) or making a photograph (in the case of a camera). Such coordinated complexity just does not seem to happen by accident. Accident destroys complexity; it does not create it.

Paley reasoned by analogy that if we are not willing to consider it plausible that a watch or a camera would come about by accident, then neither should we consider it plausible that the human eye or a fruit tree or any number of other complex things in Nature are the results of accident. Such things are far more complex than anything humans have yet created, and their parts are intricately interrelated for obvious purposes – seeing the physical world in the case of the eye, and producing fruit in the case of the tree. Through telescopes and microscopes we keep encountering ever more vast and intricate "machines" in Nature.

Hence, because of the immensity and orderliness of the universe, and because of the coordinated complexity and goal-oriented nature of so many things in the universe, it is more plausible to think that it was created by an extremely intelligent and powerful agent, namely, God, than that it all just happened by accident. To be sure, the teleological argument does not prove for certain that there is a God, but at the very least, according to its supporters, it shows that it is quite plausible that the world was created by an intelligent being. Some supporters go further and say that these analogies show that it is more *probable* that the world was created by an intelligent being than that it was not.

A third type of argument for the existence of God is the argument from religious experience. Some people have been convinced of the existence of God by having a personal experience which seemed to be an experience of God. It is important to realize that such an experience does not usually serve as evidence from which the experiencer infers that God exists. A religious experience usually has such a compelling quality that during the experience the experiencer can no more seriously doubt that the experience is being caused by God than I can now doubt seriously that I am typing on a computer or than you can doubt seriously that you are reading a book. Hence, the experiencer usually believes zer experience to be *veridical*, that is, revealing of the truth. But is it rational to believe in the existence of God because of such an experience?

According to *the principle of credulity* it is rational for a person to believe that things are the way they seem to be as long as ze knows of no good reason why ze should doubt the veridicality of the experience (such as that ze was delirious because of drugs or fever). The justification for the principle of credulity is this: If we were to suspend judgment about or reject the veridicality of the experiences that seem true to us until we could prove that they were veridical, we would never be justified in believing anything about the world, and so we would perish in uncertainty. If something *seemed* like it would slake our thirst, we would have to suspend judgment, and therefore action, until it was proven that it would slake our thirst, and everyone else should do the same, so we would all die of thirst. If something *seemed* like a dangerous animal about to pounce on us (and was), we would have to suspend judgment until something demonstrated that the animal was or wasn't dangerous, and so we would be injured or killed.

From experience, though, we know that the way things seem to be often turns out to be the way they are, so by trusting our experiences we form a reliable understanding of the world. From those instances in which what seems to be the case is not so, we learn how to distinguish what is real from what appears to be real but is not. If we did not assume that things are generally the way they seem to be, we could never form a background of normal experiences against which to identify which experiences are abnormal, such as hallucinations and perceptual illusions. Consequently, we are justified in believing that things are the way they seem to be – unless there is good reason for thinking otherwise. Hence, the person who has a religious experience, and has no adequate reason for rejecting or doubting it, is justified in accepting it as veridical, as revealing the truth.

To appreciate the extent to which religious experiences have caused belief in God, it is important to realize that there are two rather different types of religious experience: interruptive and suffusive. Some people are convinced of the existence of God because of an interruptive, intense, and transient experience. Such an experience is *interruptive* in that it is not expected or under one's control. It breaks into one's life in a startling way. Such an experience is *intense* in the sense that it overwhelms one's attention, somewhat as a powerful noise or brilliant light seizes our attention. Such an experience is *transient* in that it normally does not last very long; usually minutes; in rare cases, perhaps hours or a few days.

Consider, for example, Simone Weil's description of her first religious experience:

> In a moment of intense physical suffering, when I was forcing myself to feel love, but without desiring to give a name to that love, I felt, without being in anyway prepared for it (for I had never read the mystical writers) a presence more personal, more certain, more real than that of a human being, though inaccessible to the senses and the imagination.[4]

This was not an experience that caused Weil to begin believing in God; she already did. Her experience did, however, confirm and enrich her belief in God, and it is an excellent example of how powerful and real such an experience can be. It was unexpected, overwhelming, and brief. During that experience her sense of the reality of God was more intense than her everyday sense of the reality of people, so she could henceforth no more doubt the reality of God than she could doubt the reality of other people.[5]

When many of us think of religious experience, we think of the kind that Weil reported, but there is another kind. It is often overlooked because it does not involve the emotional pyrotechnics of the interruptive religious experience. This second type of religious experience does not suddenly and unexpectedly break into one's life; rather, it is a gentle, suffusive, enduring way of seeing and feeling the world. It suffuses one's life in the sense that virtually all of one's life is experienced as being lived in the presence of God or in God's world. Some people who feel this way say they cannot remember ever beginning to feel this way; they just always have.

This way of seeing and feeling the world is gentle in the sense that it is like hearing a continuous background whisper rather than being startled by an unexpected shout. To be sure, people who have this kind of experience might not always be aware of it (just as one might not be aware of a soft breeze or a distant object unless one focuses on it), but they find that it is almost always there when they search for it, and they often feel it without searching for it.

Note that this way of experiencing the world, like the interruptive way, is not an argument for the existence of God. Rather, one simply feels that God is quietly present, or one simply experiences the world as God's creation, or both. This kind of conviction that there is a God is sometimes called "a basic belief" because it is not based on or inferred from any other belief; it is just the way things seem to some individuals. If we put that conviction together with the principle of credulity (the principle that ordinarily we are justified in thinking that things are the way they seem to be), then it seems that the person who has a suffusive religious experience is justified in thinking that there is a God – unless, of course, that person has a good reason for doubting the veridicality of that experience.

In brief, according to the principle of credulity it is rational for people who have had what seemed to them to be an experience of God to believe that God exists – unless they know something which undermines the credibility of that experience.[6]

A fourth argument for belief in the existence of God is *the argument from authority*. This is not an argument that has received a lot of discussion by professional philosophers, but I have met it so often in discussions with people who are not professional philosophers that it seems worth mentioning. There are two prongs to this argument. The first prong begins by noting that millions and millions of normal people over thousands of years of time, right up to the present day, have believed in God – many on the basis of interruptive or suffusive religious experience. This prong concludes that it is unlikely that so many people over so many centuries, right up to the present, would believe in God if there were no God. This is a "where there is so much smoke there is probably fire" kind of argument.

Some people reject the first prong of the argument from authority by retorting that "the masses are asses"; that is, most people are ignorant, emotional, and superstitious, so their opinions should not be respected. That is when the second prong of the argument pops up. It points out that in addition to the masses of ordinary people who believe in God, many of the geniuses of humankind – indeed, probably most of them – have believed in God. In areas of life other than religion it would be irrational to reject such widespread agreement among ordinary and outstanding people without extremely good reasons. Hence, such widespread confidence in the reality of God shows at the very least that it is rational to believe in God and that it may be irrational not to believe.

There are additional arguments for the existence of God that we do not have time to explore, and the arguments that we have studied could be presented at much greater length. Nonetheless, the arguments that we have examined, and the extent to which we have examined them, should give you a sense of why some people believe in the existence of God, and how a person might go about developing an argument to show that God definitely or probably exists, or that it is plausible that God exists, or that it is rational to think that God exists.

Notes

1 The word "ze" is not a mistake. Other than the male pronouns "he," "his," and "him," there are no singular pronouns in the English language that are used to refer generally, without regard to gender, to men, women, androgenous persons (such as we find in science fiction and perhaps will find in fact on other planets), and persons without gender, such as God. For good reasons, this practice of using male pronouns to refer generally to all persons is less and less common. I, too, have tried to avoid that practice by using various devices to try to be fair to persons in all of these categories. Unfortunately, some of these devices make for awkward or ungrammatical language – though awkward and even superficially incorrect language is better than unfairly offensive language.

Still, it seems clear that the sooner we come up with a set of non-gendered personal pronouns that are widely adopted, the better off we will be. We need to go beyond (1) the confusion and incorrectness of using "he" (or "she") generically for both sexes, e.g., "Everyone who pays his taxes by check should write his social security number on his check," (2) the awkwardness of written and spoken locutions like "he/she" and "he or she"; for example, "Everyone who pays by check should put his or her social security number on his/her check," (3) the incorrectness of using a plural pronoun to refer to an individual, as in, "When an individual is in distress, we should help them," and (4) the incorrectness of referring to non-gendered persons, such as the biblical God, as "he" or "she" or "it", e.g., "If we are faithful to God, he will bless us." Hence, we need gender-neutral personal pronouns that refer indifferently to persons who are female, male, both, or – as in the case of God, some angels, and perhaps extra-terrestrial beings – none of the above.

I propose "ze" for the nominative case, "zer" for the possessive case, and "mer" for the accusative case. "Mer" is a blend of hi*m* and h*er*; "ze" is a blend of "*s*he and h*e*" ("*s*e"), and "zer" is a blend of "hi*s* and h*er*" ("*s*er"). However, "z" has been substituted for "s" so as to avoid such homophones as "see" and "sea" in the case of "se", and "sir" in the case of "ser". Making the appropriate substitutions, the illustrative sentences above would read: "Everyone should pay zer taxes by check"; "When an individual is in distress, we should help mer"; "If we are faithful to God, ze will bless us."

"Ze", "zer", and "mer" may seem awkward now, but if used regularly, in a decade they will seem quite natural. Meanwhile they will enrich the categories of our language and will improve our ability to communicate smoothly, precisely, and grammatically.

I would like to have used these new pronouns throughout *Thinking Philosophically* where appropriate, but that would have been too distracting and annoying to some readers. Still, these non-gendered personal pronouns seem particularly appropriate in philosophical discussions of God, so I hope you will indulge my use of them in that context.

2 "Nature" is used frequently in philosophical discussions but with two very different meanings. To distinguish those meanings I will capitalize "nature" when it means the physical universe. I will leave it uncapitalized when it means the essence of something,

as in, "It is the nature of gold to be heavy."
3 The word most commonly used to describe the perfect *power* of God is "omnipotence." It literally means that God is all-powerful: "omni" (all) "potent" (powerful). The word most commonly used to describe the perfect *knowledge* of God is "omniscience." It literally means that God is all-knowing: "omni" (all) "scient" (knowing).
4 Simone Weil, *Waiting for God* (NY: Capricorn Books, 1951), p. 24.
5 Three other excellent examples of interruptive religious experience, all in the Bible, are Moses' experience of God and the burning bush (Exodus 3), young Samuel being called by God (I Samuel 3), and Saul on his way to Damascus (Acts 9).
6 If there is a God and ze reveals merself to humans in religious experiences, then religious experience, along with deduction and induction, can be a source of knowledge of what God is like. Indeed, religious experience could be thought of as a form of *observation* whereby we learn something of what God is like, since in an authentic religious experience we would be directly aware of God, just as in a sensory experience we are directly aware of a physical object. There are, however, significant differences between sensory observation as a source of knowledge and revelatory observation as a source of knowledge. For those reasons I did not think it appropriate in this context to include observation as a source of knowledge of what God is like. God is not an entity or being which we can in any straightforward way publically observe when and how we wish so as to test and confirm or disconfirm our beliefs, based on perception, as to what God is like. When an experience of God begins, when it ends, and what it reveals is entirely up to God, according to classical theism. Hence religious experience is not a source of knowledge which is public, testable, and accessible at will in the way in which knowledge gained by sensory observation or induction or deduction is. That is not to say that religious experience is not a source of knowledge about God. It is only to say that if it is a source of knowledge about God, it falls outside the methods of knowledge acquisition that humans control and which, therefore, seem appropriate for inclusion in a philosophical survey of sources of knowledge at to what God is like. For an extensive, sophisticated discussion of these issues, see William Alston's *Perceiving God* (Ithaca, NY: Cornell University Press, 1991).

Reading Further

John Hick, *Philosophy of Religion*, 4th edition or later.
William Rowe, *Philosophy of Religion*, 2nd edition or later.
William Wainwright, *Philosophy of Religion*, 2nd edition or later.

Chapter 18: Metaphysical Materialism

- Criticisms of arguments for belief in the existence of God
- Four arguments against belief in the existence of God
    - Seeing is believing; we cannot see God
    - Science is adequate; we do not need God
        - Ockham's Razor
    - The argument from evil
        - Natural evils and moral evils
    - The irrationality of religious belief
        - Evidentialism
- Materialism as a worldview
    - The popular sense and the metaphysical sense
    - Arguments for materialism
    - Implications of materialism
    - Historical origins of metaphysical materialism
    - Two types of modern materialism: Naturalism and Marxism

As you know very well by now, the presentation of a position should be followed by evaluation of it. The worldview known as "materialism," and sometimes as "naturalism" or "physicalism," is atheistic and therefore critical of the theistic way of understanding reality, so let's make a transition to metaphysical materialism by listening to some of its reasons for rejecting theism. That will set the stage for materialism's presentation of its own position.

Criticisms of arguments for belief in the existence of God

First, the materialist is not impressed by the argument from religious experience to belief in the existence of God. Regarding suffusive religious experience, many materialists would say that that kind of experience is foreign to them. Their suffusive experience is that there is not a God, and to some of them it has never seemed otherwise. Moreover, regarding the principle of credulity – which states that if something seems to be the case, then that in itself is a good reason for assuming that it is the case – atheists have as much of a right to that principle as theists have, so suffusive atheists have as much of a right to conclude that there is no God as suffusive theists have to conclude that there is. More specifically, suffusive atheists might say they have a gentle, suffusive, enduring experience that there is no God. Hence, appeals to religious experience have no significance for them.

This does not mean that there is just a stand-off or a tie game between theists and materialists, however. Materialists have much more to say. For example, intensive religious experiences commonly manifest characteristics that individuals obviously have learned from their culture. What goes on in the religious experience of an American Indian and of a Hindu Indian are quite different, but the ways in which those experiences are different are what one who knows both cultures would expect. Those experiences, then, can be explained in terms of things in this world – namely, acculturation – so we need not accept a supernatural explanation of religious experience.

As to the various forms of the teleological or design argument for the existence of God, they all fail to recognize that given enough time (and the universe has existed for at least 15 billion years, and perhaps forever), it was highly probable that the accidental but continual mixing of the basic physical particles of the universe would eventually produce the combination of particles which is the current state of the universe. After all, if you roll ten dice long enough, you will eventually hit all the possible combinations. Furthermore, some combinations of particles are more durable than others, so we should expect that, over time, more and more durable combinations would come about by accident and remain on the scene for considerable periods – and that is exactly what we observe when we look at the history of the universe. Order is a more efficient, enduring way for things to be related to one another than is disorder, so it is natural that even by accident the universe should become more and more orderly.

As to the cosmological argument and its various forms, they all assume that the existence of the universe needs an explanation and that such an explanation can succeed only if it assumes the existence of a creative force that exists beyond the universe. Those assumptions, however, are not obviously true. First, it is not obvious that the principle of sufficient reason is true. Some things, such as the speed of light and the movement of an electron from one ring of an atom to another, may have no explanation. They may just be "brute facts" – facts with absolutely no explanation, and the existence of the universe may be another brute fact.

Second, even if the universe is not a brute fact, we cannot rule out the possibility that it is the nature of the universe to exist and that, therefore, its existence does not need an explanation from beyond itself. When theists are asked, "If God created the universe, then who created God?," they usually say, "No one. It is the nature of God to exist." Well, materialists can say the same thing about the universe: "No one created it. It is the nature of the universe to exist." To be sure, we do not *know* that that is true, but neither do we know that it is false, so reason does not compel us to accept God as the only possible explanation of the existence of the universe.

As to the first prong of the argument from authority, the materialist provides psychological and sociological explanations as to why we needn't take the religiousness of the masses seriously (we will look at those explanations soon). As to the second prong, the materialist points out that public expression of atheism was persecuted until very recently in western history, and that since such persecution began to subside, there has been a surge of brilliant thinkers who have argued vigorously for atheism and agnosticism – for example, Feuerbach, Marx, Nietzsche, Freud, Russell, and Sartre. Hence, because of the extensive religious indoctrination and persecution in western history, we do not know what percentage of the intelligentsia would have been atheists if they had been free from childhood to form their own opinions.

Four arguments against belief in the existence of God

A good criticism of an argument *for* the existence of God shows only that that particular argument is not a good argument for believing in the existence of God. It does not show that God does not exist (just as a good criticism of a solution to a mathematical problem shows only that that solution does not work; it does not show that there is no solution to the problem). Consequently, let's now turn from criticisms of arguments *for* believing in the existence of God to four arguments *against* believing in the existence of God.

First consider *the argument from the imperceivability of God*. We know that lakes and trees and dogs and clouds and stars exist. We see them and in some cases can touch them. God, however, we can neither see with our eyes nor feel with our fingers nor hear with our ears, so why should we believe ze exists? This difficulty for theists was framed in a poignant way in the early years of travel in outer space. The first American astronauts to go into outer space reported back to Earth that on looking out of their spacecraft's windows, they saw God everywhere in the splendor of space. Not much later, the first Soviet cosmonauts went into outer space. They looked out of their windows and reported back to earth that they saw God nowhere.

Presumably the Americans were using the word "see" in a figurative sense; in seeing the beauty and magnificence of the heavens, they felt that they were seeing the very handiwork of God. In terms of our earlier terminology, they were having a suffusive religious experience. The Soviets, by contrast, seemed to be using "see" in both a figurative and a literal way. They did not see God anywhere in the literal sense in which they saw thousands of stars, and they did not "see" the heavens as the handiwork of God.

To be sure, theists themselves say that God is by nature imperceptible, that is, unable to be apprehended by our senses or by physical instruments, such as telescopes and oscilloscopes. God, the Bible says, is pure spirit. But that itself, according to materialists, undermines the rationality of believing in God. First, there is a serious question as to whether we can even make sense of the idea of a non-physical being. How are we to understand it? Some materialists say that the concept of such a thing is empty of meaning. (Recall the physical monist's argument against the intelligibility of a non-physical soul.)

Second, we know how to go about demonstrating, or at least showing the probability of, the existence of physical things – whether we are speaking of unicorns, or Sasquatch, or the Loch Ness Monster, or black holes, or quarks, or a suspected tenth planet in our solar system. But God, according to theists, is not a physical entity and so is not part of our world – which is the only world about which we can have knowledge. Therefore, even if we can make sense of the idea of God, we cannot be justified in believing that God exists.

The second argument, which is related closely to the first argument, I will call *"the adequacy of science argument."* It points out that one of the reasons for the emergence of the concept of God was the natural and powerful desire of people to believe they understood why things happened in the world. We, today, can sympathize with that desire; it is our desire, too. However, we now know we do not need the concept of God or belief in God in order to explain anything in the universe. Hence, we should abandon belief in God; otherwise we are like a person who carries around a big, awkward tool that he thought he needed, but which, it turns out, he didn't need and which doesn't work anyway!

This point was made exquisitely by Pierre Simon de Laplace when he was summoned to appear before the French military leader and ruler, Napoleon Bonaparte. Napoleon had read Laplace's essay on astronomy, *The System of the World* (1798). He noticed that in Laplace's explanation of the movements of the heavenly bodies, no mention of God was made. Napoleon found that puzzling and summoned Laplace to explain his omission. Laplace's succinct, eloquent, courageous reply was, "We have no need of that hypothesis."

Laplace's point was that the relations and movements of the heavenly bodies can be explained perfectly well by means of physics and mathematics, with no reference to God. And so it has seemed to materialists before and since: the idea of God is not needed to explain anything about the world. Science can explain it all, including the creation and development of the universe, and so-called 'miracles,' 'revelation,' and 'answered prayers.' There is no longer a need for "the God of the gaps," that is, for using the idea of God to explain things that we do not yet understand. It is understandable that pre-scientific people engaged in religious explanations of things, but since the advent of modern science, we have no excuse for giving or accepting such explanations.

Ironically, one of the main weapons used by modern materialists against theism was forged by William of Ockham, a devout fourteenth-century Christian who was also a brilliant logician. Ockham pointed out that one of the traits of a good explanation is that it is economical; that is, it is not more complicated than it must be in order to explain satisfactorily whatever it is trying to explain. In his own words, translated from Latin, "What can be done with fewer [assumptions] is done in vain with more," and "Plurality is not to be assumed without necessity." This point, commonly called "*Ockham's Razor*" because it is used to cut away unnecessary beliefs, has been popularized in the following statement: "Entities should not be multiplied beyond necessity."[1]

A clear implication of Ockham's principle is that if the world can be explained just as well without God's existence as with it, then it should be explained without reference to God. Why make an explanation more complex than it needs to be? From this it follows that if we can explain in terms of natural laws alone such phenomena as lightning and thunder, and such extraordinary events as the ancient Israelite escape across the Red Sea, then we should. Ancient people believed that in seeing lightning and hearing thunder, they were witnessing battles between the gods and hearing their angry roars – now we know better.

We also know now that the southern end of the Red Sea is quite shallow and that strong southern winds occasionally push the waters back from the sea floor, as when an ocean tide goes out. Surely Moses, who was reared in that region, knew this and realized that if he timed things just right, the Israelites could flee on foot across the Red Sea while the water was pushed back by the winds, whereas the Egyptians, pursuing them hours later in heavy, horse drawn chariots with narrow metal wheels, would first bog down in the soft bottom of the sea floor and then be covered by the water as the winds died down later in the day. Hence, rather than praising God for the miracle of the Israelites' escape, we should admire Moses' brilliance as a military strategist.

A third argument against the existence of God is called "*the argument from evil.*" It goes like this: God is by definition all-good, all-knowing, and all-powerful. Because God is all-good, ze does not want anything evil to exist or occur in the world. Because God is all-knowing, God knows how to prevent anything evil from occurring in the world. Because God is all-powerful, God can prevent any occurrence of evil in the world. Therefore, if there is a God, then there is no evil in the world – since God knows how to prevent all evil, has the power to do what ze knows will prevent all evil, and wants to prevent all evil. However, there is evil in the world; therefore, there is no God – at least not in the classical sense of a supremely perfect being.

Some atheists think that the argument from evil is a proof that God does not exist (they are hard atheists). Other atheists would not go that far, but do believe that the enormous amount of human and animal suffering throughout history, plus all of the evil in the world today, make it unlikely that there is a God (these are soft atheists).

Theists sometimes defend God by pointing out that most human suffering is caused by human immorality, not by God. Critics have two replies to that point. First, in addition to numerous *moral evils* (evils caused by humans who choose freely to act immorally), there are many *natural evils* (that is, causes of suffering and deprivation for which humans are not responsible). Typical examples of natural evils are earthquakes, tornados, hurricanes, floods, droughts, blizzards, diseases, birth defects, and harmful animals, plants, and natural substances. These things afflict children as well as adults, the innocent as well as the guilty, and the just as well as the unjust; it seems highly improbable that a supremely perfect being would cause or even allow such things, so it is highly improbable that there is such a being.

Second, because God is omniscient and omnipotent, God could have made humans so that they were good by nature and therefore unable to act immorally or even to want to do so. Hence, it seems that if God were good and omnipotent, then God would have made humans naturally good and moral, but ze did not. Therefore, either God is not good or not omnipotent or not omniscient or does not exist. But by definition God is good, omniscient, and omnipotent. Therefore it seems that God does not exist.

A fourth argument against belief in the existence of God is *the argument from irrationality*. According to this argument, belief in God is based on emotion, not on evidence or justification. Most people believe in God because of fear: fear of the future, fear of death, fear of loneliness, fear of responsibility, etc. People fear their precarious situation in this morally indifferent world that smites us with diseases, earthquakes, droughts, blizzards, and a thousand other cruelties. It is understandable, then, that people unwittingly conjure up the illusion of a powerful being who can protect them from such assaults, or at least give them peace of mind and strength to go on after tragedy has occurred.

People also do not want to accept the imperfections of this life and the finality of death, so they make up the idea of a being who will carry them beyond death and this imperfect world to a perfect world. Some people feel profoundly small and lonely in the cold, dark infinitude of space, so they imagine for themselves an ever present companion. Sometimes people do not want to accept responsibility for their actions so they say, "Don't blame me; I was only doing the will of God." Sometimes people do not want to accept responsibility for solving their personal problems and the problems of the world, so they say, "Humans are weak and cannot solve such problems, so let us pray to God to solve them," or they say, "We humans are incorrigibly sinful, so if we try to fix our lives or the world, we will only make things worse, so let us pray to God to fix them."

Notice that none of these attitudes involves evidence for the existence of God. According to *evidentialism*, however, it is wrong to believe anything except on the basis of adequate evidence. As David Hume put it: "Belief should be apportioned to evidence." As W. K. Clifford stated: "It is wrong always, everywhere, and for anyone, to believe anything on insufficient evidence." To be sure, the argument from the irrationality of religious belief does not show that there is no God, but it does show that belief in God is unjustified at best and disreputable at worst because it is based on emotion, not reason. Therefore we should not believe that God exists.

Materialism as a worldview

What, then, should we believe? The worldview at the heart of the preceding criticisms of theism is *metaphysical materialism* – also called "physicalism" (all things are physical) and "naturalism" (Nature is the whole of reality). According to this worldview, reality consists of matter and space, nothing more and nothing less. There are no souls, no angels, no God, and no world or dimension other than the one that we apprehend by our senses. It is up to science to tell us what forms matter takes, such as frogs, crystals, DNA molecules, black holes, etc., and what the laws of nature are that govern the behavior of all things.

The popular sense and the metaphysical sense

A person who thinks this way is called "a materialist." Keep in mind, however, that this meaning of "materialist" has little, if anything, to do with the popular usage according to which a materialist is a person who greatly values material possessions, such as expensive jewelry, clothes, cars, etc. A materialist in the metaphysical sense may or may not care about owning expensive material possessions. Some metaphysical materialists are content with a simple life and devote themselves to noble causes such as fighting world hunger and promoting world peace.

Arguments for materialism

Two arguments that materialists often set forth in support of materialism are the following. The first is the "seeing is believing" argument. It goes like this: "I believe in the existence of matter and material objects because I see, feel, and hear them all around me. Indeed, I, myself, am a physical entity. If you show God to me, then I will believe in God; until then it will be unreasonable for me to believe in the existence of God but reasonable for me to believe in the existence of matter and the material world."

If the theist replies, "But without God we have no explanation for the existence of the world and the mysteries of life," the materialist responds, "There is no reason to think that all questions have an answer. Those questions that can be answered can be answered by science; those questions that cannot be answered by science simply cannot be answered. Some things, such as the existence of the world and the speed of light, are *brute facts* – facts for which there is no explanation. That's just the way things are."[2] And that is the second argument.

The preceding position, you may recall, is the heart of the "adequacy of science" argument that we examined earlier, and it complements nicely the "seeing is believing" argument. Sensation and science, say materialists, are the only reliable, decisive paths to knowledge about reality. Mystical and religious claims that refer to the supernatural are at best the delusions of well-meaning people who are wasting precious time on wishful thinking and fantasies; at worst such claims are frauds perpetrated by wicked people who for their own benefit are trying to manipulate the fears and hopes of other people.

Implications of materialism

Notice that with a bit of reflection we can infer from these basic beliefs of materialists what their beliefs are on some other topics. With regard to philosophy of religion, materialists are atheists; with regard to philosophical anthropology, they are physical monists. Because they are physical monists, they generally believe that metaphysical behaviorism is true, that libertarianism is false, and that there is no life after death.

As to epistemology, because materialists believe that there is nothing more to reality than Nature, they also believe that our only sources for knowledge of the world are ordinary sensation and science – and science is merely ordinary sensation *fortified* by systematic methods of inquiry and the power of scientific instruments, and *supplemented* by the imaginative construction of physical forces and entities that cannot be observed (such as gravity and quarks) but which explain those things which can be observed.

As to axiology, materialists generally believe there is no meaning to life in general, and there is no meaning to an individual's life except for the meaning which that individual *gives* to his or her particular life. For example, a medical researcher might decide, "The meaning of my life will be to try to find a cure for AIDS." Furthermore, the only true values are the physical, social, intellectual, and aesthetic values of *this* world. Because there is no God and no dimension of reality other than the one that is revealed to us by sensation and science, either spiritual values are empty and delusive, or they are natural values that some people mistakenly think are supernatural in origin or significance.

Historical origins of metaphysical materialism

There are two more important points to make about materialism, one having to do with the past and one with the present. The point about the past is that the roots of metaphysical materialism in the west can be traced to ancient Greece. Metaphysical materialism did not emerge as a consequence of modern science. The emergence of modern science gave enormous credibility and momentum to materialism, but materialism as a way of understanding reality was alive and well in ancient Greece, where it was encouraged by the ideas of Thales, Heraclitus, and Empedocles, and was developed explicitly by Leucippus, Democritus, and Epicurus.

As we saw in chapter 2, the last three thinkers concluded that although physical objects can be cut into smaller and smaller pieces, it is unreasonable to think that they can be cut into ever smaller and smaller pieces. Eventually, they reasoned, we must come to the smallest parts of physical objects; those smallest parts of the world are what they called "atoms." (Physicists today think that "quarks" or "strings" may be the ultimate building blocks of the universe.) Recall that "atom" in Greek literally means "not cut" or "not cuttable." Atoms, then, are not cuttable because they are not composed of parts; they are the smallest pieces of the world – so small as to be imperceptible. By contrast, the ordinary physical objects that we perceive are cuttable because they consist of a multitude of atoms which are connected but separable.

These "uncuttables" were thought of as alike except for a few variations, such as speed and size. Some moved more swiftly than others; some were larger than others; but all had tiny hooks which caused them to cluster together or break apart as they collided with one another. (In Chapter 2 I suggested it might help to think of atoms as covered with the contemporary substance "velcro," which would cause atoms to cling to one another, but not so tightly that they could not be knocked apart by sufficient force.) History, according to the ancient atomists, is the process of the coupling and uncoupling of these ever-moving, uncreated, indestructible atoms.

Contemporary materialists, of course, have a much more sophisticated understanding of the atomic realm than the ancient Greeks did. Indeed, materialist *philosophers* leave it to scientists to tell us what matter is like. The contemporary metaphysician's contribution is to point out, in kinship with ancient materialists, that the history, structure, and processes of the universe can be explained adequately in terms of the eternal existence of uncreated, indestructible matter/energy and the laws which characterize its behavior. As a result of the ceaseless motion of matter/energy, accidents occur and novelty results. Over billions of years, zillions of combinations have occurred accidentally. Some of those combinations were simple; others were complex; some lasted only a nanosecond; others have lasted millions of years. Notice, then, that by accident we get some combinations of material particles that are simple and short-lived (such as a bubble of water in the froth of an ocean wave); some that are simple and long-lived (such as a granite boulder); some that are complex and short-lived (such as a mayfly, which in a single day is born, reproduces, and dies); and some that are complex and long-lived – such as trees, turtles, whales, and human beings.

One might think that the more complex a being is, the less likely it is to survive, since the more complex a thing is, the more joints there are through which the world can drive a destructive wedge – whether it be a mechanical, chemical, or electrical "wedge." From this conviction one might infer that it is highly improbable that beings so complex as humans would have arisen and survived for so long just by accident – which suggests that more than accident has been at work in the emergence and survival of humans.

However, to think that way would be, as we saw earlier, to overlook the fact that there are different kinds of complexity. To be sure, some kinds of complexity increase the vulnerability of a thing to destruction. Consider, for example, "a house of cards." The more complex a structure of playing cards is, the more likely it is to fall apart because of gravity, a tiny tremor of the surface it is resting on, or a slight movement of the air around it. Some kinds of complexity, however, make a thing more able to survive because they give it more ways to respond to the environment than simpler things have. Consider, for example, the lungfish, which can travel and live out of water for brief periods, whereas its simpler relatives cannot. If the food supply in a body of water dwindles, the lungfish can seek food on land or in another body of water nearby, whereas its simpler relatives cannot. If the water dries up, the lungfish can survive while its simpler relatives die. There is, then, a kind of complexity that can make a creature more durable, and that is the kind of complexity which humans have, especially because of their large complex brains, opposable thumbs, and upright posture – though there are, of course, limits to the environmental adaptability of any creature.

Two types of modern materialism: Naturalism and Marxism

Having examined ancient Greek materialism, now let's look at two schools of contemporary materialism. Most materialists today divide into two schools of thought regarding the dynamics of human history. The first of these schools is most often called "naturalism"; the second is called "Marxism" or "dialectical materialism." Naturalists and Marxists agree that there is nothing more to reality than the space/time world revealed by sensation and science. What they disagree about are philosophy of history and social philosophy, or, more specifically, what the driving forces of history are and how social progress comes about.

Naturalism is linked with the liberal democratic philosophy that most humans are deeply moved by the ideals of justice, charity, and excellence, and it urges that by education we should inspire and prepare the young to live rational, moral lives, according to these ideals. Social progress toward these ideals should be pursued by identifying and attacking problems one by one, in a straightforward, linear fashion. By means of a multitude of small but definite steps of progress, social problems will be diminished, social goods will be increased, and society will move closer and closer to perfection. A Chinese aphorism states the effectiveness of this incremental approach in the following way: "A journey of 10,000 miles begins with but one step." And, of course, such a journey is completed by taking step after step after step.

However, the naturalist urges, whether we will complete or make substantial progress on our "journey of 10,000 miles" toward a perfect society is up to us as individuals. It depends on whether we, as individuals, dedicate ourselves to the goal of improving society, whether we act intelligently, courageously, energetically, and persistently on behalf of that goal, whether we enlist others in the cause, whether we encourage those who are tired or distracted to not give up the journey, and whether we appeal to the consciences of those whose backgrounds or vested interests put them in opposition to the goal of a just and generous society.

The preceding statement sets forth the social philosophy that prevailed in much of the modern west until the middle of the nineteenth century, when Karl

Marx set forth a radically different understanding of the causes of social problems and how they must be overcome. According to Marx, the piecemeal problem-solving approach of liberalism can improve a society, but only within narrow limits. Why? Because the most significant social problems are products of *the economic structure* of the society in which they exist. Specifically, they are products of the fact that, in one way or another, the society is divided into two economic classes: masters and slaves, or lords and serfs, or bourgeoisie and proletariat (owners and workers). The latter class in each pair is ruled and exploited by the former class. In each case, the exploiting class minimizes the power and welfare of the subordinate class in order to maximize its own power, wealth, and security. Moreover, it creates and uses a police force, an army, and a bureaucracy to keep the underclass peaceful, orderly, and obedient.

In such a society, Marx argued, the most important social change that is called for – justice for the masses of people – cannot be brought about by the liberal program of improving the system step by step. Why not? Because it is the system that is the problem! That was true under slavery and feudalism, and it is true under capitalism – which allows a few people to gain enormous wealth and leisure by controlling and exploiting the economic situation of the masses.

What is called for is the replacement of unjust, exploitive economic systems. Moreover, it is inevitable that they will be replaced. Social revolutions come about due to historical circumstances, just as surely as seeds are germinated and brought to fruition by the influence of water, sun, and soil. Consequently, every oppressive economic and political system is doomed because it bears within itself the seeds of discontent; eventually the members of its oppressed classes will become sufficiently angry, equipped, and organized that they can and will overthrow those who have been exploiting them. Rousing speeches for social revolt can speed the coming revolution, but nothing can prevent it; its eventual occurrence is as natural and inevitable as the eruption of a volcano. To be sure, such an overthrow need not be violent, but more often than not it is violent because the members of the class that is in power will not peacefully give up their power and wealth, and the sense of superiority and self-importance that go with them.

There is, then, according to dialectical materialism, little value in appealing to the consciences of those in power. Nor, it adds, is there much to be gained by urging the masses to revolt. The very circumstances of their lives will eventually enable them to revolt successfully and make them unhappy enough to do so. To be sure, able and inspiring political leadership can move things along more quickly, and such leadership is critical when the time for revolution comes, but social revolution is generated primarily by social circumstances, not by political charisma and not by appeals to economic parasites to give up their prey.

In brief, every social system except a perfectly just one bears within itself oppressive features that will eventually germinate the seeds of revolution, leading to an overthrow of that system, freeing the oppressed class to establish another kind of social system which overcomes some of the problems of the preceding system. This dialectical pattern of social progress by revolution will continue until a society is established in which there are no classes, no one is oppressed, everyone receives what he or she needs for a good life, and everyone contributes freely to the society according to his or her ability. Such a society is what a communist society is.

Hence, Marxists argue, it will only be with the establishment of communism in a society that the struggle of its social classes with one another will come to an end — because only then will there be no exploited class, no seething masses resentful of the structure of their society and eager to overthrow it. Moreover, only with the world-wide establishment of Marxism will there be world peace — because only then will there cease to be exploiter nations and nations that are exploited. It was with this understanding of the dynamics of human history that Nikita Khrushchev, once premier of the USSR, said to the capitalist nations of the world: "History will bury you."

In comparing the merits and demerits of liberalism and Marxist communism, it would be a mistake to dismiss Marxism because of the recent collapse of the USSR and its European allies. The economically caused alienation and anger that gave rise to Marxism in the first place still exist, and masses of people still find the social vision proposed by Marx to be profoundly attractive, just, and compassionate. Soviet Marxism was only the first attempt to actualize the Marxist ideal. The attempt was bungled, as many first experiments are, but much was learned. If massive economic discrimination and exploitation continue within nations and between them, then we should not be surprised to see in the twenty-first century a resurgence of Marxism under leadership that has learned important lessons from the Soviet experiment. Perhaps as a result of the clashes of liberalism and dialectical materialism in the twentieth century we will also see new forms of liberalism that are more sensitive to, realistic about, and effective at overcoming the injustices of certain economic systems, and perhaps we will see new forms of capitalism which treat workers with dignity and the desperate with charity.

Notes

1 See Ernest A. Moody, "William of Ockham," *The Encyclopedia of Philosophy*, Paul Edwards, ed. (NY: Macmillan Pub. Co., 1967), vol. VIII.
2 This aspect of materialism is sometimes called "scientism." One version of scientism says there is nothing more to reality than the kinds of things that can be discovered and

studied by science. Hence, only scientific knowledge is genuine knowledge of reality. For a critique of this point of view see *Scientism* by Tom Sorell (London: Routledge, 1991).

Reading Further

Karl Marx, *The Communist Manifesto*.
Abraham Maslow, *Religions, Values, and Peak Experiences*.
Bertrand Russell, *Religion and Science*.

Chapter 19: Metaphysical Idealism

- Popular idealism and metaphysical idealism
- A general justification of metaphysical idealism
- Subjective Idealism (theistic): Berkeley and Hartshorne
    Perception and the will of God
    Shared perceptions and the will of God
- Objective Idealism (pantheistic): Hegel
    Ockham's Razor applied to Berkeley's pluralistic world
    The Absolute Mind, finite individuals, and history
        The synchronic and the diachronic perspectives
    Externalized ideas and the commonality of perceptions
    Criticisms
- Phenomenalism (atheistic): Hume, Ayer, Buddha
    Experience: yes. Minds: no.
    Reality as a sequence of experiences
    Rejection of the principle of sufficient reason
- Solipsism and the problem of other minds
- Metaphysical Nihilism
    Criticism: Descartes
        Reductio ad absurdum
- Why consider "crazy" positions?
- *Criticisms* of solipsism
    Lack of control
    Argumentum ad ignorantiam
    Moral implications of solipsism: Sartre
    The value of the social dimension of life
- Desert landscapes and Tropical forests
- Affirming, improving, or replacing a worldview

Popular idealism and metaphysical idealism

To say that a person is an idealist is, ordinarily, to say that he or she is committed to very high standards of behavior or to goals that are noble or ambitious (and perhaps unrealistically lofty). As we discovered with the term "materialist," however, the philosophical meaning of "idealist" is quite different from its ordinary meanings. To be sure, someone could be an idealist in one of the ordinary senses *and* in the philosophical sense, but as we saw with "materialist," it is not necessary that someone who is an idealist in the one sense also be an idealist in the other sense. Therefore, when someone says that he or she is an idealist, or is said by others to be an idealist, we need to make sure that we understand the sense in which that is meant so that we will know what to think about that person and how to respond to him or her.

To add to the difficulties of communicating about idealism, there is the further fact that there are various forms of metaphysical idealism, as there are of metaphysical materialism. How many forms of idealism one thinks there are depends on how one defines the essence of idealism. Perhaps the broadest definition that has been given is that idealism is a worldview according to which mind or spirit is the ultimate factor that determines what happens in reality. In this sense, theism (because of the divine mind) and Platonism (because of the eternal "ideas" and divine beings in which Plato believed) have been said to be forms of idealism. We will focus on a narrower definition of idealism which excludes classical theism and Platonism but includes a fascinating set of positions that begins in modern western philosophy with the Irish thinker George Berkeley.

According to this narrower, more common philosophical definition, *idealism* is a worldview which holds that the whole of reality consists of nothing but minds and their experiences and actions. The most controversial implication of this position is that matter does not exist – at least not in the way in which most people think it does. The last point should remind you of psychic monism. Metaphysical idealism is usually the worldview of psychic monists. Psychic monism is usually the mind/body view of metaphysical idealists. Because the idealist understanding of reality is so foreign to most westerners, let's begin our study of idealism by examining general reasons for adopting it. Then we will examine four versions of idealism, paying special attention to how each position *develops dialectically* from the one that precedes it (that is, develops as a response to the problems of the preceding position and is proposed as a solution to them).

A general justification of metaphysical idealism

William of Ockham, as you may recall, taught that when we explain anything, we should not make our explanation more complicated than it absolutely must be to get the job done. This principle of simplicity or economy in explanation is widely respected and known as "Ockham's Razor." Here are some examples of how it has been applied to worldviews. Platonism explains everything by means of *three* eternal things that exist independently of one another: Matter, Essences, and God. Classical theists responded that with an adequate conception of God, we do not need to think of eternal Essences as existing outside the mind of God, so they lopped off those Forms, moved them into the mind of God, and explained the existence and nature of the physical world entirely in terms of *two* principles: God and the matter from which God created the physical world.

Materialists say they can explain reality entirely in terms of matter, so they use Ockham's Razor to lop off the concept of God from the theistic explanation of reality, leaving us with only *one* thing, matter. By contrast, idealists agree with materialists that something needs to be lopped from the classical theist's understanding of reality, but they think it is matter, not God, which should be eliminated. Let's review some of their reasons for thinking this.

The typical concept of matter, as we saw in the mind/body chapter, is the concept of something which is supposed to be the cause of our sensory experiences. It is the idea that there is some mind-independent stuff "out there" that impacts our senses and causes us to have experiences. Yet as we saw in chapter 16, whereas we know for certain that our experiences exist, we can never know for certain that matter exists. Hence, to explain mind by matter, as physical monists do, is to explain the certain by the uncertain, and that is a flawed form of explanation. The less certain should be explained in terms of the more certain, not vice versa. Minds and their volitions and experiences are all we ever know directly and all that we can make sense of, according to idealism. Hence, the nature and existence of matter – which is neither a mind nor anything that a mind is directly aware of – must be a permanent mystery. From this the idealist concludes that the most intelligible and plausible understanding of reality is that it consists entirely of minds and what they experience and will.

This point of view was captured nicely in the following statement by American idealist Charles Hartshorne:

> That experiences do occur cannot be denied; hence, the only open question is, does anything else occur? One may safely defy critics to prove the affirmative. Nonhuman experiences occur, no doubt, but that things constituted by no sort of experience,

however different from ours, occur, this no science, no philosophy, can possibly establish. And an intelligible world-picture results from so modulating the idea of experience as such that it coincides with that of reality. At no lesser price can such a picture be had.[1]

But if nothing occurs except experiences and things that are parts or aspects of experiences, how should we think of physical objects – such as the computer screen at which I am looking as I type these words, or the sheet of paper at which you are looking as you read these words? Does idealism mean they are illusions? Not at all.

According to idealism a physical or material object is a cluster of *properties* such as color, shape, size, weight, and texture, but there is no need or reason to think that those properties are backed up by or caused by or reside in some non-mental stuff called "matter." To recall a point from the section on psychic monism, what has gone wrong is that people have taken a perfectly good adjective, "material," and *reified* it; they have assumed that the noun version of "material," namely, "matter," must refer to something that exists in its own right.

According to idealism, that assumption is a mistake. Just because the noun "evenness" can be constructed from the mathematical adjective "even" (as in "Four is an even number") does not mean that evenness is a thing (like a lemon) or a stuff (like clay) out of which even numbers are made. When we think carefully about the concepts of evenness and matter, we see that even though there are such things as even numbers and material objects (that is, objects with physical properties), there are no good reasons to think that evenness and matter exist as things or as stuff.

Matter, then, in the sense of a stuff that exists apart from perception and is neither a mind nor anything which exists in a mind, is at best, as physician/philosopher John Locke put it, "something we know not what" – if it exists at all. The mysteries, confusions, and contradictions that result from believing in matter as materialists and mind/body dualists do, can be overcome only by conceding that belief in matter causes more problems than it solves – and so we should use Ockham's Razor to cut it off and throw it away – which leaves us with mind or spirit as the sole constituent of reality.

Critics will want to know, if matter does not exist, what causes our experiences of physical objects – such as seeing trees, hearing birds, and smelling smoke? Also, how can idealism account for the commonality of perceptions, that is, the fact that different people experience the same object at the same time – as when thousands of fans in a stadium watch Mark McGuire knock a baseball out of the park or see Mia Hamm kick a soccer ball into the opponent's net?

Subjective Idealism (theistic): Berkeley and Hartshorne

One set of answers to these questions was given by the eighteenth-century Irish philosopher George Berkeley (after whom Berkeley, California, is named). Berkeley, a theist, said that it is God who causes us to have experiences of physical objects. When you have the experience of seeing an apple, it is because God wills that you see an apple. God does not need to create matter in order to cause you to see an apple. God can just directly will that you have the experience of seeing an apple.

The will of God accounts for the *involuntary* nature of seeing an apple on a tree. We see the apple whether we like it or not because God is causing us to have the experience of seeing the apple. Idealism does not imply that because matter does not exist, apples are just something we make up in our minds. Of course we can choose to *imagine* an apple; that is a voluntary action – something *we* do. When we *perceive* an apple, however, that is involuntary – it is something that happens to us, and it happens to us because God makes it happen. Indeed, it is God who causes us to see, hear, smell, etc., the whole realm of Nature.

Now you might be thinking, "Yikes! How did God get in here? Isn't belief in God even more problematic than belief in matter?" Berkeley says, "Not at all." As we have seen, we do not know and cannot know what matter is like, but each of us knows what a mind is like because each of us is one. Therefore we can understand what God, the supreme mind, is like. Of course we cannot understand *fully* what God is like – God is infinite and we are finite. But we know what it means to have knowledge, to have a will, to be just, to be merciful, and to have power, so we can make sense of the idea of a being who has unlimited power, knowledge, and goodness. Hence, we can make sense of the idea of God, and we can understand that God, the supremely perfect being, can directly cause us to have any experience that ze wants us to have. By contrast, we cannot understand what matter, which is totally non-mental, would be like, or how matter (whatever it would be if it does exist) could cause minds to have experiences.

God can, then, directly cause us to have all the experiences we have without creating and using matter to do so. It would be unreasonable to think that God would take the extra, unnecessary step of creating a stuff (matter) to use to cause us to have physical experiences when God can just will that we have those experiences. Hence, we should take out Ockham's Razor and lop off matter from our worldview. We don't need it, and God certainly doesn't.

It seems, then, that subjective idealism can explain perceptual experiences without postulating the existence of matter, but how can it account for the commonality of our perceptions in the absence of matter? Again the answer is "God." God sometimes causes two or more of us to have the same perception at the same time. It is as though God were simultaneously, but separately, running a private showing of the same movie for each of us. There can be two copies of one film; you can watch one copy in one theater while I simultaneously watch the other copy in another theater halfway around the world. Immediately after the film ends, we can get in touch by telephone and discuss that movie just as though we had seen it while sitting next to one another.

Berkeley was saying that all so-called common perceptions are like the two theater example. Each mind is a separate theater. Every time two or more people think they are seeing one and the same thing at the same time (such as two opposing basketball players looking at a basketball that a referee is about to toss up), they are mistaken. What is happening is that God is causing each of them to have such similar experiences that they can communicate about them just as though they really were in the same situation. That is certainly conceivable, and it should be no problem for God to do.

Objective Idealism (pantheistic): Hegel

In order to critique subjective idealism, let's hand Ockham's Razor to the objective idealist. The objective idealist who has had the most profound and enduring influence was G. W. F. Hegel, a late eighteenth, early nineteenth-century German philosopher. Hegel agreed with Berkeley that there is no such thing as matter in the materialist's sense, and that spirit is the essence and whole of reality. However, Hegel objected to the idea that God is separate from the world and that, therefore, reality consists of many things (God and the minds that God creates). Hegel thought it more plausible that reality consists of one thing, not many, so he used Ockham's Razor to eliminate the separation or distance between God and what God creates.

The most plausible way to think about reality, according to objective idealism, is to think of reality as a single, absolute, all-inclusive mind. This absolute mind, which Hegel often referred to as "The Absolute Spirit" – or simply "The Absolute" – includes everything that ever is or occurs. Nothing – no space or time or thing or event or relation or dimension – ever exists or occurs outside of The Absolute. The Absolute Spirit is like a single gigantic organism that not merely occupies all of reality but is all of reality.

Although the Absolute Mind is absolute (all-inclusive and unsurpassably great), it is not static. It is constantly changing and progressing. The Absolute contains within itself all the possibilities that there are, and they are infinite in number. A history of reality would be a narrative of how the Absolute Spirit has been unfolding, developing, and actualizing its infinite potentialities from forever to forever. The self-actualization of the Absolute on higher and higher levels is a process that had no beginning and will have no end.

This position is clearly pantheistic. It holds that there is a supreme being which is spiritual in nature and identical with all of reality. Hegel shied from using the word "God" to refer to the Absolute Spirit because although there are similarities between his conception of the Absolute Spirit and the classical conception of God, there are also profound differences: the God of theism is different from the world and does not progress; the Absolute Spirit progresses and is identical with all that exists.

Now let's ask, "If the Absolute is everything, then what is the relation of the Absolute to a particular finite thing, such as a planet – or you or me?" Finite individuals, such as you and I and Jupiter, are not separate, self-existing individuals anymore than are the heart and the kidneys. We are not complete in ourselves; we are parts of something larger. We depend on the Absolute for our emergence, nature, and duration. However, we are not separate from or even different from the Absolute – we are forms that the Absolute is taking at this time. We are related to the Absolute somewhat as simultaneously existing cells are related to a single organism of which each cell is a part. While each cell exists, it is a real part of the organism and makes a unique contribution to it. At any given time there is nothing more to the organism than the cells which make it up. Similarly, at any given time there is nothing more to the Absolute than the universe as it is at that time.

At every moment of time, however, there are two aspects to the Absolute. First there is the *synchronic* (simultaneous) aspect, which consists of the individuals which exist with one another at a specific moment of time and which constitute the actuality (as distinguished from the potentiality) of the Absolute at that moment. This aspect can be thought of as a photograph or a time-slice of the Absolute. Second is the *diachronic* (historical) aspect of the Absolute. This aspect is generated by the "nisus" or dynamic drive of the Absolute to actualize new potentialities by going beyond each moment to a new moment – a new moment that always emerges from, builds on, and transcends the preceding moment. By means of such incessant self-creativity, the Absolute progresses beyond each moment of time to a new moment, forever and ever. The diachronic (historical) aspect of the Absolute is analogous to a movie rather than a photograph. The synchronic aspect of the Absolute is like a still-frame taken from a movie; it is but a snapshot of an ongoing process. The essence of the Absolute, and therefore of the universe, is eternal, wholistic, progressive self-transcendence. That is why history is evolutionary.[2]

Analogies can be illuminating, but the analogues (the things being compared in an analogy) are always different as well as similar, so there are always points at which an analogy "fails" or "breaks down." Those points can be as illuminating as the similarities, so let's look at a disanalogy between the Absolute and an organism.

The cells that constitute an organism continually emerge, make their contribution, and then die or are replaced bit by bit, whereas the organism itself continues to exist. We have compared the cells to you and me, who come and go, and the organism to the Absolute, which continues. Eventually, however, the organism perishes, too, and that is where the analogy between the Absolute and an organism breaks down because although every finite organism eventually perishes, the Absolute, which is reality itself, will never perish. The Absolute is, to speak metaphorically, the one organism that never came into existence, will never pass out of existence, and is that of which all other things are partial and passing parts. The Absolute Spirit is uncreated, unending, ever unfolding creativity.

You and I and all other finite things come into existence as forms of the Absolute. We make a unique contribution to the advancement of the Absolute, and then we perish, opening the way for new individuals and further advances. The Absolute is the only individual that can and does endure everlastingly. Because it does, then although our existence be ever so brief or humble, we shall have made an everlasting difference to the Absolute, and by the Absolute we shall be remembered forever.[3]

An implication of Hegelian pantheism is that everything that exists or occurs is a part of the Absolute and plays a role in the Absolute's progressive self-development. The universe at any given moment is the form or "shape" that the Absolute Spirit is taking at that moment. The history of the universe is the history of the Absolute evolving and expressing itself from moment to moment, from forever to forever. This means that everything, including you, is a form that God is taking; this means that history is meaningful and purposeful, and the universe is sacred because its essence is living spirit, not mindless matter.

Earlier we saw that subjective idealism explains mutual perceptions by supposing that God is different from us and sometimes causes the same experience at the same time in several of us. Objective idealism cannot explain the commonality of experience that way because it denies that God (the Absolute) is different from us. So how does objective idealism explain common perceptions? It says that materialists are correct that there are things that exist outside us or externally to us, like clouds and stones. Materialists are mistaken, however, in thinking that clouds or stones or anything else is made of matter – in the sense of a stuff that is not mental in nature and continues to exist even though no one is aware of it or thinking about it. Subjective idealists agree with objective idealists on the latter point, but the objective idealists say that subjective idealists are mistaken in thinking that objects do not exist externally to us. Objective idealists say we should think of physical objects as *externalized ideas*. A mental exercise should help clarify this concept. Close your eyes and imagine an apple as vividly as you can (really! please do this before reading on).

Welcome back! Now consider the fact that *you* were directly aware of the apple that you imagined, but no one else was. If you tell me that you did what I asked, I don't know that you did; I just have to accept your word for it. If you did not do what I asked, yet you tell me that you did, I can't truthfully say, "I knew you were trying to fool me; I could see that you were really imagining an orange instead of an apple"; nor, if you *did* what I asked, could I say, "You don't have to tell me that you did what I asked; I could see that you were imagining an apple."

A normal human is able to create objects in his or her own imagination, but those objects remain private to each of us; we cannot, merely by willing it, *externalize* the objects in our imaginations for other people to perceive. I cannot cause you to see the apple that I am imagining, nor can you cause me to see the apple that you are imagining. The Absolute Spirit, however, does not suffer from our limitations. It *can* project its ideas into the public realm for all to see. However, when the Absolute's ideas are externalized (projected into the public realm), they continue to be mental entities; they do not become hunks of matter. Indeed, since the Absolute Spirit is all that exists, the apple is one form that the Absolute is taking – and so are the perceivers of the apple!

In brief, sometimes the Absolute diversifies itself simultaneously into various forms in order to experience something from various points of view. What seem to be separate entities (you, me, and the apple) are really only diverse forms that one thing – the Absolute – is taking. The apple really is external to you and me – though of course it is not external to the Absolute. It is an externalized idea (somewhat like a hologram), not a hunk of non-mental stuff (matter). Indeed, we are all simply forms or aspects of the one thing that exists. This understanding of reality is similar to the scientific theory according to which the universe is a single field of energy, so that although we can distinguish things in the universe (you, me, and the apple), and many things in the universe seem entirely unconnected, in fact the universe is a unified field of energy and is the only real individual. What we see as multiple, distinct individuals are simply areas of the energy field that are dense enough for us to perceive. The appearance of things being separate and unconnected is an illusion resulting from the limitations of our senses.

Criticisms

How might the subjective idealist object to objective idealism? First the subjective idealist might object to the substitution of the Absolute Spirit for God. The objective idealist's claim that we humans are forms that the Absolute Spirit is taking means that the Absolute gets diarrhea, headaches, stomach flu, and can't drive straight when it's drunk. Why would the Absolute inflict such things upon itself? And if it does or can't help doing so, is it worthy of being called "The Absolute"? It is certainly not worthy of being called "God." That name, says the theist, should be reserved for the highest conceivable being – a being which is truly absolute.

Second, the subjective idealist objects that the objective idealist's conception of the world does not seem simpler than that of the subjective idealist. For example, the notion of externalized ideas does not provide a simpler explanation of common perceptions than does the notion of God causing similar experiences in different minds at the same time. Indeed, the subjective idealist claims, it would be simpler for God to simply will that different people have similar experiences than it would be for God to create matter or externalized ideas in order to cause different people to have the same experience. Moreover, the notion of an eternally perfect God who does not evolve or need to evolve is simpler than the notion of an Absolute which needs to evolve because it is not perfect. Finally, the notion of a God who can simply will things into existence seems less complicated than the notion of an Absolute which must *become* whatever it wills into existence.

Third, it is puzzling as to *how* an idea could exist outside our minds. If the mind is not a physical entity, then it does not have an inside for anything to be outside of. So what would it mean for a physical entity, even in the form of an externalized idea, to exist "outside" of a mind? If externalized ideas exist outside our minds, it would seem that we must use our senses in order to perceive them; yet according to objective idealism, our senses must be externalized ideas! But how could one externalized idea, for example, an apple, impact other externalized ideas, namely, our senses, and then how could those externalized ideas, namely, our senses, have an impact on our minds? On analysis the objective idealist's explanation of the commonality of perception turns out to be more complicated and problematic than does the subjective idealist's account. Physical objects, even the ones in our waking experiences, says the subjective idealist, exist *subjectively*, that is, within our minds, not *objectively*, that is, outside our minds – and so Berkeley's position is called "subjective" idealism in contrast to Hegel's "objective" idealism.

Phenomenalism

A fourth and more radical criticism of both objective and subjective idealism is aimed at the notions of God and the Absolute. Thinkers such as David Hume,

A. J. Ayer, and Buddha (Siddhartha Gautama) have used Ockham's Razor (or something like it) to lop off the concept of God from idealism. They agree with the basic idealist convictions that the materialist conception of matter is obscure at best and empty at worst, and that we could not have direct awareness of matter even if it existed. The phenomenalists add that we could not have direct awareness of God or the Absolute, either. So even if the idea of God or the Absolute or Brahman (in the Hindu tradition to which Buddha was objecting) is more intelligible than the idea of matter, we would be just as unjustified in postulating the existence of God or Brahman or the Absolute as we would be in postulating the existence of matter. All that we can know directly are finite minds and their experiences, so they are all that we are justified in believing in. Anything more is sheer speculation. Consequently, we should dispose of belief in God and matter and think of reality as consisting of nothing but finite minds and their actions and experiences.

You may be wondering, "But if there is no God or matter, if all that exists are minds like my own, where do our experiences of natural objects (clouds, trees, animals, etc.) come from?" The phenomenalist's response is that there is no explanation of these experiences. They just happen. That's just the way things are. The question "What causes my experiences?" presupposes that there must be an answer. The phenomenalist's position is that there is not always an answer of the kind we want or expect – that is, an explanation in terms of some cause or reason. (The phenomenalist's bumper sticker says: EXPERIENCE HAPPENS!)

To some questions there are answers; to others there are not. For example, why am I writing this book? Because I believe that philosophy is important for everyone to explore, and because I believe that my many years of working with students have taught me some useful ways to clarify the concepts of philosophy and show their relevance to life. Note that in what I just said I gave *reasons* for my writing of this book. But why does the filament in a light bulb glow when the switch connected to it is closed? Because the closed switch allows electrons to flow through the filament, causing it to heat up considerably. Note that the light bulb explanation speaks of *causes*, not reasons. These are *two very different kinds of explanation*: the causal and the intentional.

Note what happens when we pursue the light bulb illustration further: Why do the electrons flow through the wire rather than staying in place or backing up when the switch is closed, and why does the filament glow rather than turning darker or staying the same color when electrons course through it? The phenomenalist says that *ultimately* there is no answer to these questions. That's just the way things are. There is no intelligible necessity to the laws of nature. Science can only discover and tell us *what* the laws of nature are; it cannot tell us *why* the laws of nature are as they are, or why there is a universe rather than none.

Science must start with the existence of the universe and its laws; it cannot get beyond or behind them. In brief, the phenomenalist rejects *the principle of sufficient reason*, according to which there is an explanation satisfactory to reason for everything that exists and occurs. Some facts, such as the existence and basic nature of the universe, says the phenomenalist, are *brute facts* – facts that have absolutely no explanation.

Before we leave phenomenalism, we need to note that phenomenalists typically cut away not only the notion of an Infinite Mind but also the notion of finite minds. They do not deny the existence of minds in every sense, but they do deny that the mind is a thing or substance or entity. They argue that it is a mistake to think that experiences must exist *in* something or must be modifications *of* something. Many people think that shape, for example, must exist in something or be a modification of something, such as a lump of clay or a piece of wood, each of which they take to be a hunk of matter. But why think that? Objects in dreams have shape, but the shape isn't a property of a hunk of matter. More generally, we know from dreams, imagination, and hallucinations that physical properties such as shape, color, texture, size, etc., do not have to existence in a thing or a substance. They can just exist as pure objects of awareness. Similarly, experiences do not have to exist "in" a mental entity or substance. They can just happen.

According to phenomenalists, then, there is no such thing as a mind in the usual sense of "a thing"; a mind is nothing more than a connected sequence of experiences (or as David Hume put it, "a bundle of impressions"). All that we *know* reality to consist of are moments of experience succeeding one another. It is these strands of flowing experience of which you, I, and other people consist. There is no way to know whether there is anything more to us than our experiences, and there is no reason to think that there is. Nor is there good reason to think that there is more to reality than the sum total of experiences that have occurred, are occurring, and will occur (recall the quotation from Charles Hartshorne). Therefore, it is illuminating to think of reality as a sequence of diverse strands of experience: I am one strand of experiences; you are another; and so on. This is sometimes called an *event* understanding of reality as distinguished from a *substance* understanding – that is, reality consists of events of experience rather than substances.

Solipsism and the problem of other minds

It may seem that Ockham's Razor has now cut as deeply as it possibly could. There is, however, at least one more possible cut. A phenomenalist believes there are many minds or streams of experience in addition to his or her own. The *solipsist* responds like this: "I can no more know for certain that there are minds or experiences other than my own than I can know for certain that God or matter exists. All I can know directly of what seems to be another person are physical properties, such as the color of their eyes, hair, and skin, the shape and height of their body, the sounds and movements they make, etc. I am never directly aware of the mind or the experiences of anyone else. Hence, I am not justified in believing in God, matter, or other minds."

What, then, does the solipsist think reality consists of? "Reality," the solipsist says, "consists of nothing but me – and so far as I can ever know, I am only a mind which has experiences." Note: the solipsist is not merely saying, "*My* reality consists of nothing more than me and my experiences" – as though there are other people, and their reality consists of their experiences. What the solipsist is denying is that he or she has sufficient grounds for believing that there are other people or that there is anything other than his or her mind and its experiences. ("Solipsism" comes from the same Latin word from which "sole," meaning "only," as in "sole survivor," comes. The solipsist is saying, "I am the sole thing that exists.")

To the solipsist, other people and things, and his or her own body, are nothing more than figments of his or her mind. Until a person comes to realize the truth of solipsism, she is like someone who is dreaming and believes that the people and things in her dream have an existence of their own apart from her mind, but when she wakes up, she realizes that those people and things were entirely dependent upon her mind for their existence. The solipsist realizes that the people and things in her waking experiences are just as totally mind dependent as are the people and things in her dream experiences, so she realizes that all that she experiences, whether in waking or dreaming, exists only in her mind. It is, of course, possible to believe more than this – such as that there is a God or matter or other minds – but only on the basis of blind faith, and that, says the solipsist, would not be rational.

Metaphysical Nihilism

Can we go beyond solipsism? Is there anything left that Ockham's ruthless razor might cut away? Having denied the existence of everything but himself, the only thing left to which an uncertain solipsist might apply Ockham's Razor is his own existence. That is, the solipsist might turn Ockham's Razor against himself and conclude that even he does not exist. By this move the solipsist would move from solipsism to metaphysical nihilism, according to which *nothing exists* – not even oneself.

Criticism

A powerful way to critique a claim is to ask whether it *implies* anything absurd, that is, anything which seems obviously false.[4] If it does, then that is a good reason for rejecting that claim. To critique a claim in this way is called, in Latin, "reductio ad absurdum." Literally that means "to reduce a claim to absurdity" by showing that it has an implication which is obviously false. Let's see if we can derive an absurdity from believing that metaphysical nihilism is true. If I am a metaphysical nihilist, then I believe that nothing exists. An implication of my belief that nothing exists is that I do not exist. But if I did not exist, I could not be thinking that metaphysical nihilism is true – since only something which exists can think. Hence, because it would be absurd (obviously false) for me to think that I do not exist, therefore it is absurd for me to think that metaphysical nihilism is true, since if it were true, I could not be thinking that it is true. In this way an obviously false implication of metaphysical nihilism reveals to me that metaphysical nihilism itself is false.

With solipsism, then, we reached metaphysical hard rock, against which Ockham's Razor can cut no further. Why? Because one can deny one's existence only if one exists. If one did not exist, one would not be able to deny that one exists. In René Descartes' famous formulation: "I think; therefore I exist" (in Latin: "*Cogito ergo sum*"). That is, whenever I think I exist, I cannot be wrong, as I have to exist in order to think that I exist. And if I think I do *not* exist, I cannot be right, as I have to exist in order to think that I do not exist. Hence, whenever anyone thinks that metaphysical nihilism is true, necessarily he or she is wrong, and whenever anyone thinks that metaphysical nihilism is false, necessarily he or she is right. Metaphysics does, then, lead to some definite truths, so let's return to solipsism, the simplest, least complex, of metaphysical positions that might be true, and see what objections might be raised against it.

Why consider "crazy" positions?

Solipsism is so foreign to the way that most people think and feel about reality that they have a hard time taking it seriously. I once posed the following question to a class of students at the end of an "Introduction to Philosophy" course: "Of the things we covered, on which do you wish we had spent less time?" I had spent only one fifty minute class on solipsism, yet one student answered, "Things I viewed as ridiculous, such as solipsism. How can anyone in their right mind truly believe that?!!" I can sympathize with that response as a knee-jerk reaction, but there are several good reasons for giving serious consideration to what seems ridiculous.

First, recall that it is important to "run out the permutations" on a problem, examining each and every possibility with as much objectivity as we can muster – no matter how absurd some of the possibilities might seem. After all, human history is studded with ideas that were ignored or ridiculed because they seemed obviously false – yet later proved to be true. Then there are the sad stories of people who have been persecuted or executed because they put forth ideas that seemed ridiculous or outrageous at the time but which later were discovered to be true. Conversely, human history is littered with ideas which for a long time seemed obviously true to many people (and sometimes to *all* people, including the experts) yet later proved to be false.

Clearly, then, when we do not explore all possible solutions to a problem, and when we are not tolerant of people who see things differently than we do, we are putting our search for truth in jeopardy. When because of a crisis or deadline there is not enough time to explore all possibilities yet action must be taken, then, of course, we should proceed in terms of the possibility that seems most likely to be true. But when time does permit a systematic pursuit of the truth, every possibility should receive fair consideration because, so far as we know, any of them might be the truth.

A second lesson of history is that much value can be gained from fair-minded, vigorous consideration of possibilities that prove to be wrong. Thoughtful consideration of possibilities that prove to be wrong often yields important insights and results along the way. This lesson has been confirmed over and over in science, as well as in philosophy. I have heard that Thomas Edison, the great inventor, once showed a visitor around his laboratory. Edison showed his visitor experiment after experiment which had failed to produce the results that he had hoped for. The visitor finally said, "That must be very discouraging." Edison replied, "Not at all. Now I know a hundred things that do not work!" And, of course, in the process of discovering that something does not work, we sometimes discover something interesting and valuable that we were not looking for at all.

Third, the effort to understand and refute what initially seems ridiculous has shown many times that there is more merit to a 'ridiculous' idea than we initially thought, and that it is not as easy to refute it intellectually as it is to reject it emotionally. Hence, sincere efforts to correctly comprehend and fairly evaluate positions that initially strike us as ridiculous, absurd, ludicrous, silly, etc., help instill into us appropriate intellectual discipline and humility. (Note: I said *appropriate* intellectual humility. We should not be philosophical bullies or dogmatists, but neither should we be philosophical wimps or jellyfish!)

Criticisms of solipsism

Keeping in mind the preceding justification for taking solipsism seriously, now let's look at some reasons for rejecting it. The first reason for rejecting solipsism that occurs to many people is its seeming inability to account for involuntary experiences. This objection says: "If I am all that exists, why can I not control my experiences and make them occur as I wish? Doesn't the fact that I do not have control over most of what I experience mean that there is something that exists independently of me and imposes itself on me? The fact that I often cannot make my experiences what I want them to be gives at least some plausibility to the idea that there is more to reality than just me."

Perhaps the solipsist's strongest reply is to claim that he or she may be a much more complex and resourceful being than he or she is as yet aware of. After all, it is generally accepted that one's dreams are created by one's mind, yet what happens in one's dreams is generally not under one's voluntary control. Similarly, the contents of one's waking life may be generated entirely by one's mind yet not be entirely under one's voluntary control – at least not yet. The solipsist might reason that as she experiences more and more, she will gain greater and greater control of her experiences. As an adult, I have much more control over what I experience than I did as a baby. Perhaps I will continue to gain more and more control.

Still, the fact that one's experiences are not entirely under one's control should, at the very least, constrain a rational person to accept as a reasonable hypothesis the possibility that those experiences (such as seeing rain when one does not want to and feeling pain when one does not want to) are being caused by something other than one's own mind.

A second criticism of solipsism is that it involves a fallacy known as *the argument from ignorance* (in Latin the *argumentum ad ignorantiam*). An argument from ignorance consists of claiming that because someone else's position has not been proven to be true, it is false, and that, therefore, its opposite (one's own position) must be true (because of the principle of excluded middle: a position must be either true or not true). We have encountered this kind of reasoning quite a few times: the idealist says, "No one has been able to prove that matter exists, so we should cease believing that it does"; the phenomenalist says, " We cannot prove that God exists, so we should cease believing in God"; the solipsist says, "The existence of minds other than my own cannot be proved, so I should cease believing that there are minds other than my own." But notice that even if the solipsist is correct that none of these things (God, matter, and other minds) can be proved to exist, that does not prove that they do not exist!

Furthermore, let's suppose for the sake of argument that the solipsist is correct that there is no way at present to prove that other minds exist or even to show that they probably exist. Does that mean that our only alternatives are (1) to suspend our judgment on this issue until decisive evidence is found or (2) to choose for or against solipsism on blind faith? There is (3) a third option to consider, namely, to choose for or against solipsism on the basis of values. We can ask: "Are there axiological or ethical reasons because of which we could be justified in rejecting solipsism?"

Here are two reasons for rejecting solipsism that have to do with values rather than with evidence about the truth or falsity of solipsism. The first of these reasons (which constitutes our *third* criticism of solipsism) builds on the criticism that solipsism commits the fallacy of *argumentum ad ignorantiam*. This third criticism states that if it is *possible* that there are other minds, that is, other creatures who are capable of experiencing pain and pleasure, humiliation and fulfillment, etc., then it would be *immoral* for anyone to live as a solipsist should be willing to live (a solipsist should be willing to treat other people and animals as phantasms that have no real feelings, and to whom, therefore, one can do whatever one pleases, since phantasms feel no more pain, anxiety, or grief than do the people or animals in our dreams and hallucinations).

If the solipsist cannot prove that other minds do *not* exist, then it may be that the people and animals who appear in his or her experiences are real individuals who *do* have feelings, desires, aspirations, responsibilities, and emotional needs that are just as real and important to them as the solipsist's feelings, desires, etc., are important to him or her.[5] Hence, until solipsism is proven to be true, it should be rejected for moral reasons.

Along this line, Jean-Paul Sartre pointed out that many people who are not metaphysical solipsists *live* as though they are. Sartre called these people "ethical solipsists." An ethical solipsist is someone who treats other people as though they have no feelings, desires, plans, or needs – or, in a weaker version, the ethical solipsist treats other people as though their feelings, desires, plans, and needs are not as important to them as the ethical solipsist's feelings, desires, plans, and needs are important to him or her.

If, however, there are about as many other people (with minds of their own) as there seem to be, and if their feelings, plans, etc., are just as important to them as mine are to me, then it would be wrong for me to treat other people as though they have only dull mental and emotional lives, or none at all. Consequently, in the absence of very strong reasons to think that the emotions and mental lives of other people are less intense or important to them than mine are to me, it would be wrong for me to proceed as though that were the case – that is, to live as an ethical solipsist. In moral matters, the benefit of the doubt should be given to the other person. (This argument can also be extended to our treatment of animals.)

A fourth reason for rejecting solipsism is based not on the possibility that others may have feelings, but on one's own feelings. This reason is axiological in nature. To be a solipsist is, in a sense, to regard emotional reality as one-dimensional (solitary) rather than two-dimensional (social). According to solipsism, there are no genuine personal relationships; there is only the appearance of them. There is no personal depth to the people I meet, come to know, to respect, and to love, nor is there any affection or pleasure in the cat that rubs against me and purrs when I stroke it, or in the dog that boisterously welcomes me home. If solipsism is true, then I may love these people and animals, but they do not love me. Phantasms have no thoughts or feelings. Hence, solipsism entails a lonely, emotionally one-dimensional view of life.

To be sure, maybe metaphysical solipsism is true, but until it has been proven to be true, why should I opt for it? Why should I live according to a point of view that impoverishes my life rather than enriches it? Perhaps it is time to close Ockham's Razor and put it away. It has not proven that other minds do not exist, so unless we are forced by reason to conclude that other minds do not exist, why not act on our natural faith that they do? Why not proceed on faith and hope that affectionate behavior from friends is motivated by feelings of affection on their part, and that praise and approval from people whom we respect are expressions of their considered judgment, not figments of our self-flattering imaginations, and that people who appear to be suffering from hunger, illness, or oppression really are suffering, need our help, and shouldn't be brushed aside as figments of our minds? Meanwhile we can strive to see to what extent reason can justify what we feel and hope to be the case, and to what extent reason requires modifications in what we feel and hope, but we need not abandon those hopes and feelings just because the evidence in support of them is mixed or inconclusive.

Desert landscapes and Tropical forests

Ockham's Razor is a powerful, important idea because it forces us to examine our theories for unnecessary complexity and to defend the complexities that we insist on. But if reality really is rich with other minds, matter, God, and perhaps more, then there is no reason why the epistemologically humble cannot at least through faith and hope live richly with some or all of those things while continuing to distinguish carefully between what we know and what we do not know, what we have sufficient evidence for believing and what we do not have sufficient evidence for believing.

To be sure, some people prefer desert landscapes, so they prefer the simplest worldview they can get by with. Other people prefer tropical forests, so they prefer the most complex worldview that seems plausible. We started in the tropics, with complex theories such as dualistic interactionism and theism, and we ended in the desert, with solipsism. Perhaps now it is time to do a 180° turn and begin working back from solipsism to more and more complex positions to see just how complex a theory you can find defensible. There isn't time for me to accompany you on your return journey, but now you are well prepared for that journey, and I hope you will have other philosophical companionship along the way. Remember: philosophy at its best is not merely thinking about philosophical issues – it is rational dialogue about those issues. Therein lies the wisdom and power of Socrates' statement that the greatest good for a human being is to *discuss* some philosophical topic every day.

Affirming, improving, or replacing a worldview

The first worldview that we acquire is not acquired by rational deliberation on our part. It is acquired by osmosis from our family, religion, community, and culture, or some combination thereof. Once we reach the age of reason and realize how we acquired our initial worldview, we should want our worldview to be something that *we* believe is rational. Consequently, we should want to evaluate the worldview we have inherited, and then, according to our best judgment, we should (1) keep it as it is, (2) keep it but modify it, or (3) reject it and replace it. But what are the criteria by which we should evaluate our inherited worldview? They are the six criteria we examined long ago in our study of skepticism.

The worldview or philosophical position we ally ourselves with should be: (1) coherent (not self-contradictory and not implying anything absurd), (2) comprehensive (accounts for all relevant data), (3) simple (not more complex than it needs to be), (4) resilient (responds satisfactorily to criticisms and alternatives), (5) fruitful (leads to new, valuable insights), and (6) useful (yields expected, valuable consequences when acted on). In so far as a worldview or philosophical position is self-contradictory or implies what is absurd, or cannot account for relevant facts, or posits things that are not necessary, or does not stand up well to criticisms or compare well to alternatives, or has no interesting implications, or cannot be acted on (lived with) satisfactorily, it needs to be improved or replaced.

Our motivation for doing all this, for seeking the truth about reality, is twofold. On the one hand we want the truth in order to satisfy the natural desire of our intellect. On the other hand we want the truth in order to be able to live as well as we can – or, as Whitehead put it, to live, live well, and live better. Fortunately, the truth and the practical value of a worldview are not unrelated – just as the correctness and the practical value of a map are not unrelated. A worldview that is false is not likely to help us live, live well, and live better anymore than a map that is mistaken is likely to guide us to our destination. When a worldview is not true, it will almost certainly sooner or later send us into a head-on crash with reality; it will tell us there is satisfaction where there is not; it will tell us there is not satisfaction where there is. By contrast, a true worldview will guide us well, just as an accurate map will. Consequently, if a worldview guides us well in the living of life, that is some evidence in favor of its truth.

If, then, on the basis of evidence we are not confident which worldview is true, we are justified in choosing among plausible worldviews on the basis of the six criteria discussed above plus our axiological preferences and moral convictions (which, of course, should also have survived close philosophical scrutiny). Living according to our choice will generate new evidence as to whether our preferred worldview is true or not, and that new evidence should become part of our ongoing deliberations.

Notes

1 Charles Hartshorne, "The Development of Process Philosophy," in *Process Theology: Basic Writings*, ed. Ewert H. Cousins (NY: Newman Press, 1971), p. 64.
2 This point raises an issue that pertains to materialism, as well as to idealism. Materialists divide into atomists and organicists. The atomists think of reality as consisting of individual particles that have no overall organization or direction. The particles of the universe are not moving toward any common end, and there are no significant causal connections between areas of the universe that are greatly removed from one another. The universe is somewhat like a pool table the size of a football field that is continually tilting back and forth in random directions with a hundred sticky, regulation size pool balls on it. In such a situation, accidents happen; sometimes they produce quite astonishing results – such as a perfect pyramid of balls. By contrast, materialistic organicists think that the universe is more like a living plant in which all of the parts are significantly connected to one another and tend toward a common end. Another good analogy for the organicist's conception of the connectedness of reality is a spider web, in which all of the parts are delicately connected to and make a difference in one another. Objective idealists have an organic conception of reality, though they think that mind, not matter, is the connecting and driving force at the heart of the process.
3 Charles Hartshorne, an American idealist who is a theist rather than a pantheist, and who does not believe in life after death, says we should care very deeply about how we live our lives because they influence the quality of life not only of other creatures but also of God. Because God is exquisitely sensitive to everything, never forgets anything, and is the only individual who lives forever, the joy and sorrow that our lives cause God will be remembered and felt by God eternally. Hence, the highest significance of

our lives consists of the opportunity we have to contribute to the quality of the eternal life of God.
4 In philosophy, to say that a statement is absurd is not meant as an insult; it simply means that to the speaker that statement seems obviously false. However, because "absurd" is used so frequently outside of philosophy as a term of insult, it is probably best to say that a statement or implication seems "obviously false" rather than "absurd."
5 Josiah Royce, a nineteenth-century American idealist, claimed that the realization that other people's feelings and aspirations are as real to them as ours are to us is our most basic moral insight. See *The Religious Aspect of Philosophy*.

Reading Further

George Berkeley, *Three Dialogues Between Hylas and Philonous*.
G. W. F. Hegel, *Reason in History*, trans. Robert Hartman, or *Introduction to the Philosophy of History*, trans. Leo Rauch, chs. 1–4.

Part V: Parting Remarks

- Socrates' advice
- The difficulties and inconclusiveness of philosophy
- The personal importance and intimacy of philosophy
- Benefits of philosophy
- The spiral of philosophical growth

Socrates – one of the wisest and most courageous people in history and therefore someone worth listening to – said that the best thing for a human to do everyday is to discuss virtue and other such topics. One of my hopes is that you will take his advice and discuss everyday at least one such topic as virtue or goodness or beauty or justice or truth or love or morality or friendship or knowledge or God or (fill in the blank!). The conviction of Socrates was, and perhaps the deepest conviction of philosophers is, that far more good comes from our *discussing* such topics than from ignoring them or thinking about them only in solitude.

If that conviction about the importance of discussing, as well as thinking about, philosophical topics is true, then it is very important for each of us to express to other people our philosophical beliefs, thoughts, and feelings, and to ask about and listen sensitively to their philosophical beliefs, thoughts, and feelings. It is equally important for us to be mutually respectful of one another and to be cooperative rather than combative in our common effort to discover the best philosophical positions that are available and to create even better ones. Being cooperative does not, however, exclude being candid and vigorous in debate. Indeed, finding the truth often requires candor and vigor – just as being well-matched in competition brings out the best in athletes.

Among the benefits of philosophy is that it helps save us from dogmatism, superstition, and naiveté by showing us what knowledge is and exposing us to ideas that we would not have thought of or appreciated adequately on our own. It helps save us from indifference, timidity, and arrogance by showing us why truth is important, why a cooperative community of truth seekers is important, and why we should seek to be a participating member of such a community. By struggling with the issues of philosophy, especially in conversation with others, we sharpen, strengthen, and expand our minds just as surely as we improve our bodies by frequent, challenging exercise.

Nonetheless, it is not unusual at the end of my introductory course in philosophy for a student or two to tell me that he or she is disappointed with philosophy because it did not give them answers. In a light-hearted way I usually reply, "That's not true at all. Philosophy gave you *plenty* of answers. It just didn't tell you which one was true!" What those students mean, of course, is that philosophy did not give them "THE" answer. Understandably, they found that discouraging. After all, we engage in philosophy to find the truth about some very important questions, yet all too often we wind up with numerous answers to the same question, insufficient evidence as to which answer is true, plus more questions! Furthermore, philosophical thinking and dialogue can be confusing, frustrating, intimidating, annoying, controversial, and sometimes deeply disturbing.

Still, I hope that now there are two things you will never say sincerely. The first thing is, "I took a philosophy course, but I'm just not interested in philosophy." The second thing is, "I like philosophy; I see its importance, but it's just too difficult for me." As to the first comment, I don't see how we can be fully functioning humans and not be interested in the questions and theories of philosophy – such as what truth is, what knowledge is, whether there is a God, what is morally right and wrong, whether our behavior is determined, whether life has a meaning, whether we survive death, and so on. Moreover, we have and cannot help but have beliefs about various philosophical topics, so how can we not care to understand clearly what our present beliefs are and whether there are respectable alternatives to those beliefs – alternatives that we ourselves might prefer if only we were aware of them?

As to complaints about the difficulty of philosophy, I am very sympathetic. I struggle with the difficulties of philosophy on a daily basis and have done so for decades. In a sense, philosophy is too difficult for everyone – even the best and brightest of thinkers, as well as for us ordinary mortals. But difficult as it is, philosophy is personally and intimately relevant to each of us. We might want to hire an accountant to figure our taxes, or a mechanic to fix our car, but surely we don't want anyone else to do our philosophizing for us – to tell us what to believe about philosophical issues. Why not? Because it is in and through our philosophizing about these issues that we discover who we are and decide who we want to be – and surely that is something that we each want to do and ought to do for ourselves.

Nonetheless, to some people the fact that philosophy does not tell us what is true and how we ought to live is depressing and disturbing. There is, however, a positive side to this fact about philosophy. Sometimes we want an opportunity to choose for ourselves. So, in a way, the inability of philosophy thus far to prove in many cases what is true is nice because it means that we as individuals get to choose the alternative that *we* prefer, rather than having an alternative forced on us. In arithmetic, science, and history, we usually do not have a choice. The cube root of 8 is 2; heat is a function of molecular activity; the Berlin Wall was torn down in 1989.

With regard to philosophical questions, the evidence available usually does not dictate an answer. The answer, we say, is "underdetermined" by the evidence. Consequently, we are free – within the limits of rational integrity – to choose the philosophical position we will live by. Hence, the richness and openness of philosophy is in a way a positive thing. Philosophy introduces us to numerous rational alternatives regarding various issues, and it leaves us free to choose among those alternatives according to our own judgments, insights, intuitions, and values – insisting only that we choose in a rational way.

Furthermore, even if philosophy does not give us knowledge of the truth in the way that mathematics or science does, it does give us alternatives, and knowledge of those alternatives is valuable. Apart from the work of philosophers over the centuries we would not be aware of so many philosophical alternatives, and ignorance is not bliss – at least not usually. More often than not, ignorance leads us to make choices that hurt or impoverish ourselves or others or both, and it leaves us unaware of valuable alternatives we would have preferred if we had been aware of them. How many times have you said to yourself, "I wish I had known about that. If I had, that's what I would have chosen!"?

Hence, philosophy helps liberate us from ignorance about alternative ways of understanding reality and living our lives. Philosophy also engages us in rational evaluation of those alternatives. The more alternative ways we have of thinking about life and reality, and the more adequately we evaluate them, the more likely it is that we will find a satisfying philosophy of life. Hence, the ways in which philosophy enriches our options and sharpens our minds makes it worth the confusion, frustration, anxiety, and struggle that it can involve.

In the beginning of this book I noted that you would not have to learn philosophy from the beginning – that it was something you were already involved in, even though you may not have been aware of it. I also pointed out that it doesn't matter where you begin in philosophy. If you start where you are most interested and just keep going, you will eventually move through all the major areas of philosophy. That is what we now have done. In a sense we have gone full circle through philosophy, and you are ready to begin again. But the circle has spiraled into a wider loop than you were in before – a loop that includes new possibilities, insights, and methods that you were unaware of before. Now you are on a higher level and can look back with a better understanding of where you were and where you now can go.

The poet T. S. Eliot wrote in the last stanza of his poem *Four Quartets*: "We shall not cease from exploration / And the end of all our exploring / Will be to arrive where we started / And know the place for the first time." The place is reality. We begin by thinking about reality and we end by thinking about reality. There is nothing else to think about. If we proceed by means of rational dialogue, then at the end of each loop of the spiral we will find ourselves thinking about reality in a more comprehensive, penetrating, satisfying way. I hope your philosophical spiral continues to move and grow. It will if you make it so.

Reading Further

On nearly any topic in philosophy consult *The Routledge Encyclopedia of Philosophy* (1998), 10 volumes, edited by Craig and Floridi, or *The Encyclopedia of Philosophy* (1967), 8 volumes, edited by Paul Edwards. Also available are excellent dictionaries and mini-encyclopedias of philosophy. For example, *The Blackwell Companion to Philosophy*, *The Cambridge Dictionary of Philosophy*, *The Oxford Companion to Philosophy*, and *The Oxford Dictionary of Philosophy*.

Index

Absolute Mind, the   315–18
absolutism, ethical
    nature of   164, 171–2, 174
    non-theocentric   181–93
    theocentric   175–81
absurdity   323, 329 n.4
absurdum see reductio ad
action, nature of an   160–1
Adler, Alfred   197
adversarial, philosophy is not   88
agnosticism (hard and soft)   282, 283
Alcibiades   197
alternatives
    importance of considering   46–7, 323–4, 333
    package theory of   141–2
    to philosophy *see* philosophy, alternatives to
ambiguity   30–3
ambivalence   33–4
Andronicus of Rhodes   38
Anselm, St.   285
anthropology, philosophical   17, 209, 276 n.1
Aquinas, St. Thomas   131, 288
argument
    nature of in general   113–15
    deductive
        nature of   115–16
        sound or unsound   118–19
        valid or invalid   116–18
    inductive
        nature of   119–20
        warranted or unwarranted   120–4
argument from
    authority   293, 298
    imperceptibility   298–9
argumentum ad ignorantiam see fallacies
Aristotle   3, 24, 26, 95, 130–1, 132–3, 138, 143, 154 n.2, 198
art, philosophy of   18, 19, 23 n.1
assertions and assertiveness, importance of   47, 73–4
atheism (hard and soft)   281, 282, 283
atoms   13–14, 304–5, 329 n.2
attitude, a healthy philosophical   69–74
Augustine, St.   56, 212, 234 n.4
authority, argument from   293, 298
axiology   6, 17, 19, 22–3, 137–8, 326–7

behaviorism, linguistic   258–9, 261
belief
    nature of a   106
    hard and soft   282
    irrationality of religious belief   301–2
    justified   78–81, 106–8, 328–9
    truth and   95–6, 106–7
Bellow, Saul   87
Bentham, Jeremy   145–50, 156 n.26, 183–6, 200 n.8

Index

Berkeley, George 265, 314
Bible references 18, 67 n.15, 85, 156 n.20, 176–7, 295 n.5
Blanshard, Brand 210, 222
Boethius 138
Bradley, F. H 26–7
Brunner, Emil 175
brute facts 297, 303, 320–1

calculus, hedonic *see* hedonic 145–9
category mistake *see* fallacies
causation
    nature of 263–4
    correlation and 121
    determinism and 216–17, 219–24, 226–8, 231–2
    mind/body 245–6, 248–9, 251–4, 255, 263–4, 270, 274
cause *see* first cause argument for God
causes versus reasons 228–9, 232
character, importance of 194, 198
Clifford, W. K. 302
cogito ergo sum 260, 323
coherence 78
commonality of perceptions 271
compatibilism/soft determinism 223–6, 235 n.15
comprehension, importance of 48
comprehensiveness 78
concepts
    nature of 207–8
    analysis of 56
conclusion of argument 115–16
conditions *see* necessary *and* sufficient
conscience 161–2, 171–5
consequences
    importance of 69–70, 141–2, 145–9, 156 n.18
    utilitarianism and 184
consistency 78, 99–100
continuity, personal *see* personal identity
contradiction 78, 99–100, 224, 276 n.8
correlation, accidental and causal 121

cosmological argument for God 288–9, 297–8
cost 141–2, 148–9, 156 n.18
Cotta 60
counterexamples, nature of 139–40
courage, philosophical 70–1
crazy positions 323–4
credulity, principle of 230, 291, 293, 296–7
criteria of a justified belief 78–81, 107
criticism, nature of a 208–9
curiosity 33

debate
    nature of 113
    how related to an argument 115
decoy effect 136
deductive argument *see* argument, deductive
deism 280
Democritus 13–14
deontological ethics 181–3
Descartes, René 39, 46, 48–9, 85, 191, 285
    mind/body position 240, 260, 276 n.4, 323
desire theory of the good 139–40
determinism
    fatalism and 217–19
    hard determinism *see* incompatibilism
    naturalistic 219–23, 235 n.10
    physical monism and 264–5
    science and 220–1, 232, 235 n.10, 320–1
    soft determinism *see* compatibilism
    theistic 217–19, 234 n.8
    universal (complete versus partial) 216–17
Dewey, John 86, 97, 133, 138, 148–9, 151, 188, 195, 198, 199
diachronic perspective 316
dialectic 87, 88, 311
dialectical materialism *see* materialism

dialogue, rational 47–51, 79, 84–6, 331–2
Diotima, priestess at Delphi 143
divine command ethics 175–8
dogmatism 81–3
dual aspect theory *see* monism, neutral
dualism
   mind/body 239–42, 247–8, 250–4
   God/world 279–80

Eddy, Mary Baker 191
education, moral 193–9
Eliot, T. S. 58, 334
Empedocles 13, 80
empirical methods 37–8, 41
enjoyment of the good 138–9
enjoyment, the good of 138–9
environmental history 226–9
Epictetus 27–8
epiphenomenalism 250–5
epistemology 6, 18, 19, 22–3
ethics
   nature of 6–7, 17, 19, 22–3
   absolutistic *see* absolutism, ethical
   altruistic *see* utilitarianism
   anthropocentric 181–2
   divine command 175–8
   eudaemonistic 186–92
   metaethics 163
   normative 163
   perfect being 178–81
   rationalistic *see* deontological
   relativism *see* relativism, ethical
   theocentric 175
eudaemonia 131, 155 n.13
euthanasia of the will 136–7
evaluation, the nature of 44, 49–50, 64
evidence, the nature of 124 n.3
evidentialism 302
evil
   evidence against God 300–1
   moral and natural 301
excluded middle *see* laws of thought
existentialism 16, 17, 59

experience, interaction theory of 141
explanation
   causal versus intentional 320–1
   justification versus 171–3
externalized ideas 318, 319
extrospection 244–5

faith 105–6
fallacies
   nature of a fallacy 116
   argumentum ad ignorantiam 48, 140, 325
   category mistake 259
   hasty generalization 136
   post hoc ergo propter hoc 121
   reification 256–7, 313
fallibility 71, 82
fatalism 217–19
first cause argument for God 288–9
freedom, objective 210–12, 233 n.4
freedom, subjective
   according to compatibilism 224–5
   according to libertarianism 211–12

generalization, hasty *see* fallacies
genetic endowment 226–9
given, interpretation of the 32
God
   arguments against 296–302
   arguments for 288–93
   definitional approach to nature of 284–6, 287
   inductive approach to nature of 286–7
   observational approach to nature of 284, 295 n.6
golden rule 183
good 143–4
   enjoyment of the good 138–9
   the good of enjoyment 138–9
   three conceptions of the good 139–141
gratitude
   morality and 195–6
   rational dialogue and 72–3

guilt *see* conscience

happiness   130–7, 157
  Aquinas   131
  Aristotle   130–1, 132–3
  Kant   134–5, 155 n.8
  Pascal   131
  Schopenhauer   135–7
hard determinism *see* incompatibilism
Hartshorne, Charles   312–13, 329 n.3
hasty generalization *see* fallacies
hedonic calculus   145–9
hedonic treadmill   144, 156 n.25
hedonism   155 n.13, 200 n.8
hedonistic paradox   135–7, 155 n.13
Hegel, G. W. F.   27, 77, 88, 315–8
Heidegger, Martin   20, 27
history and philosophy   40–1
Hobbes, Thomas   32, 103, 210
hope, the nature of   104–5
Hume, David   195, 245, 260, 263–4, 275, 302, 319
Husserl, Edmund   59

idealism
  absolute *see* idealism, objective
  arguments for   312–13
  metaphysical   9, 10 n.1, 311
  objective   315–19, 329 n.2
  phenomenalistic *see* phenomenalism
  solipsistic *see* solipsism, metaphysical
  subjective   314–15
ideas, externalized   318, 319
identity
  law of *see* laws of thought
  metaphysical   241–2, 246–7, 257–8, 261–3
  personal   246–7, 262–3
identity thesis, the   257–8, 261–2
ignorance, consequences of   142–3, 333
ignorantiam, argumentum ad *see* fallacies
imperatives, categorical and hypothetical   182–3

imperceptibility, argument from   298–9
incompatibilism/hard determinism   226–9, 235 n.15
inductive argument *see* argument, inductive
instrumental value   150–4
intelligence as method and instrument   142–3, 198–9
interaction theory of experience   141
interactionism, dualistic   243–4
interconnectedness of philosophy   8–9, 16, 18–20
interpretation of the given   32
intrinsic value   150–4
introspection   244–5
irrationality of religious belief   301–2

James, William   61, 97
Jaspers, Karl   86
Johnson, Samuel   269
justification versus explanation   171–3
justification, nature of   47–8, 208
justified belief, criteria of a   78–81, 107

Kant, Immanuel   7–8, 24, 39, 134–5, 155 nn.8, 11, 181–3, 200 n.6, 206, 234 n.6
Kierkegaard, Søren   177
knowledge
  cannot be false   106–7
  causal theory of   109–10
  justification theory of   108–9
Kushner, Rabbi Harold   233

language, philosophy of   18, 100
LaPlace, Pierre Simon de   299–300
law
  morality and   215–16
  philosophy and   88
  philosophy of   17
laws of thought
  excluded middle   100
  identity   99

non-contradiction 99–100
Leibniz, G. W. 247–8, 250
libertarianism 212–16
  hard and soft 213–14, 234 n.4
linguistic behaviorism 258–9, 261
logic, nature of 19, 22–3, 63, 112–13
love and morality 196–7

Malebranche, Nicholas 247–8, 250
Marx, Karl 306–7
Marxism *see* materialism, dialectical
material/physical objects, nature of 265–70, 276 n.6
materialism
  arguments for 303
  atomic versus organic 329 n.2
  dialectical 307–8
  implications of 303–4
  metaphysical 9
  naturalism as a form of 306
  origins of 304–5
mathematics and philosophy 39–40
matter, the nature and knowability of 265–70, 312–13
Maugham, W. Somerset 218–19
metaethics 163
metaphilosophy 4, 163
metaphysics 7–9, 17, 19, 22–3, 38
methods for doing philosophy
  phenomenology 58–9, 67 n.9
  rational dialogue 47–51
  running out the permutations 45–7
  the Socratic method 43–5
mid-wife, Socrates as 43–4, 54
Mill, John Stuart 149–50, 184–6, 188
Mind, Absolute 315–18
mind, philosophy of 16, 17
mind according to
  dualisms and psychic monism 239–40
  epiphenomenalism 250–1
  neutral monism 271–3
  phenomenalism 275, 321
  physical monism 255–9

mind/body relations according to
  idealism *see* physical/material objects
  interactionism 243–4
  occasionalism 248
  parallelism 248–9
  physical monism 257–8
monism
  the nature of 13, 255
  neutral 271–5
  physical 255–9
  psychic 265–71
moral education 193–9
moral realism 189
morality distinct from ethics 182
morality distinct from prudence 181–2
motives for doing philosophy
  cathartic 25–6
  mystical 26–7
  noetic 24
  sport 28–9
  wisdom 27–8

naturalism 306–7
natural philosophy 39, 57
Nature 294 n.2
necessary and sufficient conditions 116–17, 118, 217
neglect *see* philosophy, alternatives to
neutral monism 271–5
Niebuhr, Reinhold 99
Nietzsche, Friedrich 66, 133
nihilism
  ethical 162–3
  metaphysical 322–3
normative ethics 163
non-contradiction *see* laws of thought
noumenon 273, 275

object *see* material objects
occasionalism 247–8
Ockham's Razor 300, 312–13, 315, 320–3, 327
omnipotent 295 n.3

omniscient 295 n.3
openness, philosophical 71–2
other minds, problem of 322

package theory of alternatives 141–2
Paley, William 290
pantheism 281, 316
paradox, hedonistic *see* hedonistic paradox
parallelism 248–50
Pascal, Blaise 29, 39, 76, 131
Peck, M. Scott 70–1
Peirce, Charles 71
perception *see* commonality of perceptions *and* principle of credulity
perfect being ethics *see* ethics, theocentric
permutations, running out the 7–8, 45–7, 324
personal identity and continuity 246–7, 262–3
perspectives, diachronic and synchronic 316
phenomenalism 275, 319–21
phenomenology 58–9, 168
phenomenon 273, 275
philosophy
  alternatives to
    dogmatism 81–4
    neglect 76–7
    skepticism 77–81, 83–4
    solitude 84–6
  ambiguity as cause of 30–4
  areas of 16–18
  benefits of 86–7, 331–2
  clue words 22–3
  curiosity as cause of 33
  definition of 5, 20–1, 35
  historical beginnings 11–14
  inescapable 75–7
  interconnectedness of 8–9, 16, 18–20
  kinds of
    analytic 56
    critical 64–6
    descriptive 57–9
    expository 54–5
    prescriptive 62–4
    speculative 60–1
    synthetic 56–7
  law and *see* law
  literal meaning of 6, 15
  methods for doing *see* methods
  motives for doing *see* motives
  not adversarial 88
  other fields of study and 35–42
  problems of 6–7, 16
physical/material objects 265–70, 276 n.6
physical monism 255–9
Plato 3, 26, 40, 289
pleasure theory of the good 140
pluralism 13
polytheism 278–9
Popper, Karl 54, 64
position, nature of a 208
premise 115
principle of
  credulity 230, 291, 293, 296–7
  sufficient reason 287, 297, 321
probability 120, 124 n.2, 148–9, 282
propositions 94
prudence 181, 184, 194–5
psychic monism 265–71

rational dialogue 47–51, 79, 84–6, 331–2
rationalistic ethics *see* deontological ethics
Rawls, John 192–3
razor *see* Ockham's razor
real objects *see* material objects
realism, moral 189
reason 60
reasoning to the best explanation 41
reason and philosophy 35–42
reasons versus causes 228–9, 232
reductio ad absurdum 323
rehabilitation not retribution 226
reification, fallacy of, *see* fallacies

relativism, ethical
  nature of   164
  individual   164–6
  social   166–70
religion
  philosophy and   35–6
  philosophy of   16, 17
religious experience
  evidence for God   291–3, 296–7
  interruptive and suffusive   291–3
  as perception of God   295 n.6
responsibility, legal, moral, and physical
  according to libertarianism   214–16
  according to soft determinism   225–6
revelation as a source of knowledge
    102–3
Rousseau, Jean-Jacques   210–11, 211
Royce, Josiah   330 n.5
Russell, Bertrand   95, 188, 198
Ryle, Gilbert   259

Santayana, George   132, 142
Sartre, Jean-Paul   20, 78, 155 n.6, 212–13
Savant, Marilyn vos   98
Schaff, Adam   55
Schopenhauer, Arthur   135–7, 155 nn.14, 15
science
  God and   37, 299–300
  philosophy and   37–9
  philosophy of   16, 17, 96, 264
  process and product   235 n.18
scientism   308 n.2
self-contradiction *see* contradiction
silver rule   183
simplicity, theoretical   80
skepticism *see* philosophy, alternatives to
Skinner, B. F.   220, 226–8, 232
Socrates   3, 25, 63, 65, 85–6, 157, 179–80, 191, 331
  as mid-wife   43–4, 54
  mind/body position   240, 242
Socratic method   43–5

soft belief *see* belief, soft
soft determinism *see* compatibilism
solipsism
  ethical   326
  metaphysical   321–2, 324–7
solitude *see* philosophy, alternatives to
sophistry   65
soundness and unsoundness *see* argument, deductive
Spinoza, Benedict   26, 196, 221–2, 222–3, 223, 244, 271
Starkweather, Charles   215
Steinbeck, John   162
subjective idealism *see* idealism
sufficient and necessary conditions
    116–17, 118, 217
sufficient reason, principle of   287, 297, 321
Swinburne, Richard   200 n.5, 235 n.17
synchronic perspective   316

teleological argument for God   289–90, 297
teleological suspension of disbelief   177–8
teleological worldview   154 n.2
Thales   12–13
theism
  hard and soft   283
  philosophical   9, 284–93
  religious   280–1, 283–4
transcendental argument   260
treadmill, hedonic   144, 156 n.25
truth
  assertions, beliefs, propositions and
    94–6
  coherence theory of   97–8
  correspondence theory of   95–6
  importance of   69–71, 98–9
  pragmatic theory of   96–7
  reality and   95–6
  sources of   100–3

universal determinism   216–17

Upanishads 144
utilitarianism 183–6

vagueness 33–4
validity and invalidity *see* argument, deductive
values, instrumental and intrinsic 150–4
value theory
    nature and objectives of 137–8, 160
    basic concepts in 153–4
veil of ignorance 192–3
veridicality 291, 293
virtue 85, 194, 200 n.13
voluntary actions as free actions 225

warranted and unwarranted *see* argument, inductive
Watts, Alan 27, 157
Weil, Simone 292
Whitehead, A. N. 137–8
Wilson, John 112–13
Wittgenstein, Ludwig 25–6
words 207–8
worldviews
    nature of 279
    affirming, improving, replacing 328–9
worth theory of the good 140–1

ze, zer, mer 294 n.1